W9-AJD-124
T 84130

Donated in
Memory of

ByJoe Cookout
(Robert Miller)

by

James & Judith Feerst.

SCIENCE

100 SCIENTISTS WHO CHANGED THE WORLD

JON BALCHIN

Enchanted Lion Books
New York

CONTENTS

First American Edition published in 2003 by
Enchanted Lion Books, 115 West 18th Street, New York, NY 10011

26/27 Bickels Yard, 151–153 Bermondsey Street, London SE1 3HA

Edited by Paul Whittle
Cover and book design by Alex Ingr

Balchin, Jon.
Science: 100 scientists who changed the world/ Jon Balchin.--1st Amer. ed
p. cm.
ISBN 1-59270-017-9 (alk. paper)
1. Science--History. 2. Scientists--Biography. I Title.

Q125.B377 2003
509.2'2--dc21
[B]
Printed in China

Acknowledgements
The publishers would like to thank Glen Carlstrom and Virginia Ingr for their invaluable contributions to this book.

The author would like to thank the following people: Anne Fennell, Paul Whittle, Matthew Smith and everyone at Arcturus Publishing, KTB, my parents, Jon, Iain, Faye, Grace, Alice, Alan, Irene, my extended family across the world, and last but never least Merryn.

▸ The Eighteenth Century

▸ The Nineteenth Century

▸ The Twentieth Century

Foreword

To be alive today is to be confronted by the products of science. Science has given us television, the internal combustion engine, the aeroplane, and the computer, to name but a few. Yet consumer products such as these are but one aspect of the benefits science can bring to mankind. Too often, for example, the field of medicine is overlooked in favour of more 'glamorous' fields, such as astrophysics or rocketry.

As recently as the last century, death from disease was an everyday occurrence. Both small-pox and polio killed millions until Edward Jenner made the simple yet life-changing discovery that milkmaids infected with cowpox were immune from smallpox, and Jonas Salk developed the polio vaccine. That both diseases continue to be killers in the modern world is due not to science, but to a tragic reluctance on the part of richer countries to share its benefits with their poorer counterparts.

Science has also produced less beneficial developments; the tank, machine gun, and atomic bomb, but science does achieve results, however morally questionable some of those results may be, and it is this which sets it apart from superstition, witchcraft and religion.

Important though the products of science may be, what is perhaps more significant is the scientific method itself, proceeding as it does from empirical observation to theory, to modification of theory in the light of further evidence.

We may still pray for rain, but we understand the physical causes of the weather, and to an extent can predict it; we no longer ascribe it to the actions of some unknowable deity and sacrifice our first-born in the hope of a favourable outcome.

This method contrasts with the previous means of discovering truth 'by authority', which claimed beliefs as true not on the basis of what was claimed, rather on the basis of who was making the claim.

Rejecting the notion of truth by authority, the scientists in this book observed the world around them, proposed theories to explain it, and modified these theories to account for further observations.

The road out from the darkness of superstition into the light of reason has not always been an easy one. When Vesalius dared to contradict the authority of Galen, he was abused as a liar and madman; the Montgolfier brothers' claims met only scepticism. Galileo and Copernicus both narrowly avoided following Giordano Bruno to the stake for proposing the heliocentric theory of the solar system, in opposition to accepted Church dogma. Yet they perservered, and in so doing, lit a beacon for the rest of humanity to follow.

The men and women who make up this book have blazed, in Bertrand Russell's poetic phrase, 'with all the noonday brightness of human genius'. How far the beacon they have lit will guide us, and how far science will yet progress, however, we shall leave to the next generation of scientists who will change the world.

100 SCIENTISTS WHO CHANGED THE WORLD

ANAXIMANDER

c. 611–547BC

A NOTE ON DATES Beyond the fact that Anaximander was born in the Greek city of Miletus, on the coast of Asian Turkey, probably around 611 BC, we know very little of his life. This is mainly because he wrote very little down, a task he preferred to leave to his pupils. What we do know has come down to us secondhand through later Greek scientist-philosophers, who naturally took an interest in their illustrious predecessors.

Imagine a world everyone knew to be flat, supported in the vastness of space by pillars. It was widely accepted that this world sat at the centre of a tent-like universe, with stars, equidistant from the earth, stuck around the edges. Now imagine being informed that, contrary to popular opinion, the world had 'depth', it was completely unsupported, and the stars, moon and sun were not only at different distances away, but also cycled around this three dimensional earth. It would be a revolutionary concept, completely altering existing preconceptions of the universe, and is exactly the giant leap in scientific understanding with which Anaximander is credited.

▸ A THEORY OF THE INFINITE

Often known as the founder of modern astronomy, Anaximander is the beginning point of the current Western concept of the universe. A Greek, he was born and died in Miletus, now in modern Turkey, although he is also thought to have travelled widely as he formulated his views

Anaximander effectively discovered the idea of space: that is, a universe with depth

of the cosmos. Anaximander was a pupil of Thales of Miletus, himself credited with original work in physics, philosophy, geometry and astronomy. As with Thales, very little detail is known about Anaximander's life and only one passage of his original texts remains. The rest of what we know comes from descriptions by later Greeks, in particular **Aristotle** and Theophrastos. If anything, they remembered Anaximander to be a philosopher rather than a scientist, expressing a bold theory on the 'infinite' or 'boundless'. This idea was his 'first principle' of all things, with no origin and no end but 'from which came all the heavens and the worlds in them' (from Theophrastos's description of Anaximander's work). Yet it is his ideas in astronomy which had the long-term impact, presenting theories which changed the world.

A TOPOGRAPHICAL UNIVERSE

Arguably Anaximander's most important leap was to conceptualise the earth as suspended completely unsupported at the centre of the universe. It had been assumed by other Greek thinkers that the earth was a flat disc held in place by water, pillars, or another physical structure. Although Anaximander obviously had no notion of gravity, he supported his argument by supposing that the earth, being at the centre of the universe, at 'equal distances from the extremes, has no inclination whatsoever to move up rather than down or sideways; and since it is impossible to move in opposite directions at the same time, it necessarily stays where it is' (Aristotle explaining Anaximander's theory). Moreover, because the earth was suspended freely, it allowed Anaximander to propose the idea that the sun, moon and stars orbited in full circles around the earth. This explained, for example, why the sun would disappear in the west and rise in the east. When you add to this the idea that the earth had depth (although Anaximander envisaged it being cylindrical in shape and still with a flat disc on the top which was the only 'surface') a revolutionary view of the universe emerges.

THE VOID BETWEEN THE STARS

Anaximander effectively discovered the idea of 'space' or a universe with depth. Rather than view the earth caged in a planetarium style 'celestial vault', he argued the 'celestial bodies' (the sun, moon and stars) were different distances away from the earth, with space or air between them. He attempted to ascribe distances for these bodies from the earth as they rotated around it, although he wrongly proposed the stars were closest, then the moon, with the sun furthest away. Anaximander may have drawn a map of his version of the universe. Although wrong in its detail, this would have been a steep change in its graphical representation.

FURTHER ACHIEVEMENTS

Anaximander was not just an astronomer. He is thought to have introduced the sundial to Greece from Babylon, using it to determine the solstices and equinoxes. In geography, he is thought to have drawn the first map of the known world, a hugely important breakthrough in itself. Meanwhile, in biology, he may have pre-empted Darwin's theories of evolution, albeit unwittingly, through his belief that mankind grew out of the original animal inhabitants of the earth. Anaximander believed these to be primitive kinds of fish taking their form from rising water heated by the sun.

PYTHAGORAS

c. 581–497BC

CHRONOLOGY

• **c. 525 BC** Pythagoras is taken prisoner by the Babylonians • **c. 518 BC** establishes his own academy at Croton, now Crotone, in southern Italy, where he is regarded by many as a cult leader • **c. 500 BC** as Croton becomes increasingly politically unstable, Pythagoras makes a final move to Metapontum

Very little is known for certain about the life of this Greek mathematician and philosopher. One obstacle is that many of the mathematical discoveries credited to Pythagoras may have actually been discovered by his disciples, the Pythagoreans, members of the semi-religious, philosophical school he founded. Moreover, because of the reverence with which the originator of the brotherhood was treated by his followers and biographers, it is sometimes difficult to discern legend from fact.

EXPERIMENTAL MATHEMATICS

It is, however, fairly clear that Pythagoras himself did undertake practical experiments concerning the relationship between mathematics and music. It is thought he either attached different weights to a series of strings, or alternatively experimented with different string lengths, examining the mathematical relationship between the resultant notes when plucked and the weights or lengths applied. What he discovered was that simple, whole number relations, for example a string of one length and another of twice that length, produced harmonious tones. These obser-

'All physical things the stars and the universe, are mathematically related'

vations ultimately led to the determination of musical scales as we know them today. Not only was this a momentous musical discovery, but was probably the first time a physical law had been mathematically expressed. As a result it began the science of mathematical physics.

▸ THE WORLD AS A SPHERE

This idea of a harmonious relationship between physical entities also enabled Pythagoras to conceptualise the world as a sphere, even if he had limited scientific basis at that time with which to back up his belief. For Pythagoras and his followers the idea of a 'perfect' mathematical interrelation between a globe moving in circles and the stars behaving similarly in a spherical universe (just as musical tones harmoniously danced around and depended on each other) seemed much more pleasing than **Anaximander's** cylindrical earth, or one composed of a flat disc. The view was so powerful that it inspired later Greek scholars, including **Aristotle**, to seek and ultimately find physical and mathematical evidence to reinforce the theory of the world as an orb.

▸ PYTHAGORAS AND HIS SCHOOL

Pythagoras founded his school at Croton in Italy, one of its objectives being to further explore the relationship between the physical world and mathematics. Indeed, of the five key beliefs that the Pythagoreans held, one was dominant: the idea that 'all is number'. In other words, reality is at its fundamental level mathematical and that all physical things, like musical scales or the spherical earth and its companions the stars and the universe, are mathematically related. The experiments of the Pythagoreans led to numerous discoveries such as 'the sum of a triangle's angles is the equal to two right angles (180°)'. 'The sum of the interior angles in a polygon of n sides is equal to 2n-4 right angles' was another. Yet arguably their most important arithmetical discovery was that of irrational numbers. This came from the realisation that the square root of two could not be expressed as a perfect fraction. This was a major blow to the Pythagorean idea of perfection and according to some accounts attempts were even made to try to conceal the discovery.

▸ PYTHAGORAS' THEOREM

Pythagoras' famous Theorem was probably known to the Babylonians but Pythagoras may well have been the first to mathematically prove it. 'The square of the hypotenuse on a right-angled triangle is equal to the sum of the squares on the other two sides' can otherwise be expressed as $a^2+b^2=c^2$, where a and b are the shorter sides of the triangle and c is the hypoteneuse.

FURTHER ACHIEVEMENTS

It is perhaps ironic that Pythagoras is remembered today for his Theorem, the principles of which had previously been known for over a thousand years, and yet his more original discoveries are obscure. As the discoverer of the musical scale, in effect creating a rule book for the musical harmonies that we take for granted, it is arguable that this has had a much more profound impact on the history of the world than a simple, largely borrowed, mathematical formula. Equally, some 2,000 years before Christopher Columbus was credited with the idea, Pythagoras proposed that the world was a sphere.

HIPPOCRATES OF COS

c. 460–377BC

A NOTE ON DATES Beyond the fact that Hippocrates was born on Cos, probably around the middle of the fifth century BC, such dates as we have, as for Anaximander, are generally so vague as to be scarcely worth mentioning.

Much of what is attributed to Hippocrates is contained within *The Hippocratic Collection*, a series of sixty to seventy medical texts written in the late fifth and early fourth century BC. It is widely acknowledged, though, that Hippocrates himself could not have written many of these works, which is why precise details about his life and achievements remain unclear. Written over the period of a century and varying hugely in style and argument, it is thought they came from the medical school library of Cos, possibly put together in the first instance by the author to whom they later became attributed.

Given the sobriquet of the 'Great Physician' by Aristotle, Hippocrates is today more commonly referred to as the 'Father of Medicine'. Without question, Hippocrates of Cos, in spite of the limited factual details actually known about his life, helped lay the foundation stones of the science of medicine and greatly influenced its later development, even up to the present day.

▸ A COMMON SENSE APPROACH

For Hippocrates, disease and its treatment were entirely of this earth. He cast aside superstition and focused on the natural, in particular observ-

The answers he prescribed are still good medicine two thousand years later

ing, recording and analysing the symptoms and passages of diseases. The prognosis of an illness was central to Hippocrates's approach to medicine, partly with a view to being able to avoid in future the circumstances which were perceived to have initiated the problems in the first place. The development of far-fetched cures or drugs, however, was not so important. What came from nature should, in Hippocrates's mind, be cured by nature; therefore rest, healthy diet, exercise, hygiene and air were prescribed for the treatment and prevention of illness. 'Walking is a man's best medicine,' Hippocrates wrote.

▸ THE THEORY OF HUMOURS

He regarded the body as a single entity, or whole, and the key to remaining healthy lay in preserving the natural balance within this entity. The four 'humours' he believed influenced this equilibrium were blood, phlegm and yellow and black bile. When present in equal quantities, a healthy body would result. If one element became too dominant, however, then illness or disease would take over. The way to treat the problem would be by trying to undertake activities or eat foods which would stimulate the other humours, while at the same time attempting to restrain the dominant one, in order to restore the balance and, consequently, health.

Although this approach may still seem a little unscientific by today's standards of medicine, the fact that Hippocrates was prescribing such a natural, 'earthly' solution at all was a major advancement. Moreover, the concept and treatment of humours endured for the next two thousand years, certainly as far as the seventeenth century and in some aspects as far as the nineteenth. In addition, the answers he prescribed for healthy living such as diet and exercise are still 'good medicine' two thousand years later. Language introduced by Hippocrates also endures: an excess of black bile in Greek was 'melancholic', while someone with a too dominant phlegm humour became 'phlegmatic'.

▸ THE OATH OF HIPPOCRATES

Ironically, Hippocrates may not have even written his own most enduring legacy. The Hippocratic Oath, probably penned by one of his followers, is a short passage constituting a code of conduct to which all physicians were henceforth obliged to pledge themselves. It outlines, amongst other things, the ethical responsibilities of the doctor to his patients and a commitment to patient confidentiality. It was an attempt to set physicians in the Hippocratic tradition apart from the spiritual and superstitious healers of their day. Such has been its durability that even students graduating from medical school today can still vow the Oath.

The Legacy of Hippocrates

Before the time of Hippocrates there had been virtually no science at all in medicine. Disease was believed to be the punishment of the gods, Divine intervention came not from the natural, but the supernatural. The only 'treatment', therefore, also came from the supernatural: through magic, witchcraft, superstition or religious ritual.

Hippocrates confronted this notion head on, with a conviction remarkable given the age in which he lived. His approach brought the rational to the previously irrational and with it medicine strode into the age of reason. 'There are in fact two things,' said Hippocartes of Cos, 'science and opinion; the former begets knowledge, the latter ignorance.'

DEMOCRITUS OF ABDERA

c. 460–370BC

NOTE ON DATES Like many of his contemporaries, Democritus left behind no written record of his work; the details of his approach have only been passed down to us through the writings of later Greeks, most notably Aristotle who opposed it, and Epicurus who endorsed it. The only vaguely reliable date is that of his birth, around 460 BC, although some authorities give this as 490 BC.

John **Dalton** is widely remembered today as the founder of atomic theory for his work in the nineteenth century which proposed that elements were made of tiny, indivisible particles. Yet the concept of the 'atom', and a systematic argument for how it formed the physical world had been in existence for over two thousand years before him, expounded by Democritus of Abdera, in Thrace.

ATOMIC THEORY

The word 'atom' comes from the Greek *atomon*, meaning 'indivisible'. Dalton acknowledged this two millennia later by using the same word in his thesis. But even Democritus was not the first. His teacher, Leucippus, as well as Anaxagoras, had all considered this notion of the indivisible particle.

Democritus, however, was the first to propose an all-encompassing argument for the primacy of the atom in the make up of the universe. Although not based on scientific evidence, as was Dalton's, instead being simply a reasoned hypothesis, many aspects of Democritus's theories are still resonant.

Democritus presents a systematic argument for the primacy of the atom

▸ ATOMS, BEING AND THE VOID

For Democritus, there were only two things: space and atoms. Space consisted of the 'Void', an infinitely sized vacuum, with infinite numbers of atoms making up 'Being', or the physical world. The atoms and space had both always existed and always would exist because nothing could come from nothing. The atoms, which were the constituent parts of everything on earth, as well as of the planets and the stars, had always and would always be the same: solid, impenetrable, invisible blocks which never changed. They simply combined with other atoms in the Void to form different things from rocks to plants to animals. When these things died or fell apart the structure disintegrated and the atoms were free to form new things by combining again in a different shape with other atoms.

Democritus reasoned that the method by which the atoms could combine was through their different shapes. While all atoms were the same in substance, liquids were thought to have smooth, round edges so they could fall over each other, while those which made up solids had toothed, rough edges which could hook on to each other. As with physical form, Democritus argued that other perceived differences in things, such as their taste, could also be explained by the edges of the atoms: sweet tastes were caused by large round atoms, sharp ones by jagged, heavy atoms. Likewise, the colour of things was explained by the 'position' of the atoms within a compound which would result in darker or lighter shades or shadows being cast.

Democritus's thesis is all the more remarkable because it completely rejects the notion of the spiritual or religious. The soul, for example, was explainable through a fast-moving group of atoms brought together by encasement in the body. These atoms reacted to disturbances by other atoms inside and outside that body. The motion produced sensations which interacted with the mind (itself just a collection of atoms) to produce thoughts, feelings and so on. Once the body was dead, however, Democritus argued that the soul ceased to exist because the object which held the fast-moving atoms together had disintegrated. Thus released, they could separate and interact with other atoms to form new things. This left no place for abstract notions of the supernatural or an afterlife.

▸ DETERMINISM

Even the concept of freedom of choice could not exist under Democritus's model. All human actions were determined by atoms striking the human body, not as part of any grand design or plan, but simply because motion and collision with other atoms within the Void is what they did and always would do, thereby leaving no free will to the human at all.

A Mathematical Legacy

Although many elements of Democritus's thesis have subsequently been tested and often discredited by modern science, his remains one of the earliest attempts to explain the universe with a few simple physical and mathematical laws. This represents an important change in thinking towards the subject and is a notion that has pre-occupied scientists ever since.

Democritus is also credited with discovering the mathematical law that a cone's volume is a third that of a cylinder sharing the same sized base and height, as well as a similar respective relationship between a pyramid and a prism.

PLATO

C. 427–347BC

CHRONOLOGY • **427** BC Plato born in or around Athens • **399** BC On the execution of Socrates, Plato leaves Athens in disgust • **387** BC On his return to Athens, Plato founds his Academy, a bastion of intellectual achievement until its closure on the orders of the emperor Justinian in 529 AD • **389** BC Plato visits Sicily for the first time

To understand how Plato came to the conclusions which have exercised such a profound impact on Western thinking, it is necessary to understand his own influences. Born in or around Athens at a time when the city-state was flourishing as one of the most dominant and culturally enlightened places on earth, he was strongly affected by the arguments of another great philosopher, Socrates, who also lived there. Socrates' approach was to constantly strive for clearer definitions of words and people's perception of those words in order to get nearer to 'the truth' that lay behind their often irrational and ill-thought-out use of them. This introduced to Plato the notion of 'reality' being distorted by human perceptions, which would become important in his approach to science and, in particular, metaphysics.

▸ SOCRATES' INFLUENCE

Socrates' was executed in 399 BC for allegedly 'corrupting' the youth of Athens with his 'rebellious' ideas. Reacting to this, Plato fled the city-state and began a tour of many countries which would last more than a decade. On his

'Let no one ignorant of geometry enter here'

INSCRIPTION ABOVE THE ENTRANCE TO PLATO'S ACADEMY

travels he encountered a group of people who would become another major influence, the Pythagoreans. Begun by their founder **Pythagoras**, the school of disciples in Croton continued to promote their 'all is number' approach to everything.

▸ THE THEORY OF FORMS

The combination of these two major forces on Plato plus, of course, his own work, brought him to his Theory of Forms, his main legacy to scientific thought. This consisted of an argument that nature, as seen through human eyes, was merely a flawed version of true 'reality' or 'forms'; in an instructive metaphor, he compares humanity with cave dwellers, who live facing the back wall of the cave. What they perceive as reality, is merely the shadows thrown out by the sun. There is, therefore, little to be learnt from direct observation of them. For Plato, there had always existed an eternal, underlying mathematical form and order to the universe, and what humans saw were merely imperfect glimpses of it, usually corrupted by their own irrational perceptions and prejudices about the way things 'are'. Consequently, for Plato, like the Pythagoreans, the only valid approach to science was a rational, mathematical one which sought to establish universal truths irrespective of the human condition. This validation of the numerical method strongly impacted on modern science; disciples following in its tradition 'made' discoveries by mathematical prediction. For example, arithmetic calculations would suggest that future discoveries would have particular properties, in the case of unknown elements in Dmitry **Mendeleev's** first periodic table for instance, and subsequent investigative work by scientists would prove the mathematics to be true. It is an approach still used by scientists today.

▸ THE ACADEMY

Plato also helped to influence scientific thought in a much more physical sense by founding an Academy on his return to Athens in 387 BC. Some commentators claim this institute to be the first European university, and certainly its founding principles as a school for the systematic search for scientific and philosophical knowledge were consistent with such an establishment. Plato's influence was pervasive; it is said there was an inscription over the entrance to the institute which read, 'Let no one enter here who is ignorant of geometry.' Over the subsequent centuries, the Athenian Academy became recognised as the leading authority in mathematics, astronomy, science and philosophy, amongst other subjects. It survived for nearly a thousand years until the Roman emperor Justinian shut it down in 529 AD, around the time the Dark Ages began.

THE LEGACY OF PLATO

Plato is best remembered today as one of the greatest philosophers of the Western tradition. He might not, therefore, be an obvious candidate for inclusion in a book on famous scientists. But in exactly the same way that the influence of Plato's work stretched into many other academic areas such as education, literature, political thought, epistemology and aesthetics, so it is the case with his science.

Although Plato's scientific and philosophical legacy has undergone significant revival and reinterpretation over the course of history, his logical approach to science remains influential, standing testament to his far-reaching ideas.

ARISTOTLE

C. 384–322BC

CHRONOLOGY • **367 BC** Aristotle enters Plato's Academy in Athens • **347 BC** On Plato's death, he leaves the Academy for Lesbos • **342 BC** Becomes tutor to the young Alexander the Great at Macedon • **335 BC** Returns to Athens and founds his own school, the Lyceum • **323BC** Accused of impiety: to 'prevent the city sinning twice against philosophy', Aristotle returns to Chalcis, where he dies the following year

Aristotle's work in physics and cosmology dominated Western thought until the time of **Galileo** and **Newton**, when much of it was subsequently proved to be wrong. He began with the accepted Greek notion that everything was made up of one of four elements: earth, water, air or fire.

▸ THE FOUR ELEMENTS

He also accepted the notion of the earth at the centre of the universe, with the moon, planets, sun and stars all orbiting around it in perfect circles. He believed that the four elements always sought to return to their 'natural place'. This was why a rock, for example, would drop to the earth as soon as any obstacles preventing it from doing so were removed – because 'earth' elements, being denser and heavier, would naturally seek to move downwards towards the centre of the planet. Water elements would float around the surface, air would rise above that and fire would seek to rise above them all, explaining the leaping, upward direction of flames. By the same method,

Aristotle's scientific proposals have at times been accorded an almost god-like authority

Aristotle could explain why a rock would travel through the air first before heading downwards when thrown, rather than straight towards the earth, as one would expect. This was because the air, seeking to close the gap made by the invasion of the rock, would propel it along until it lost its horizontal speed and it tumbled to the ground.

▸ THE FIFTH ELEMENT

Aristotle encountered a problem, however. This notion of everything tending towards its 'natural place' was not consistent with his view of the rest of the cosmos which was rotating in perfect, uniform order, with none of the disturbances or jostling for position associated with earthly elements (otherwise the planets and stars would tumble towards earth at the centre of the universe). To explain this, he added a fifth element to the traditional four, that of 'aether' which naturally had a circular motion. Everything beyond the moon was regulated by aether, explaining both its perfect movement and stability, while everything below it was subject to the laws of the four other elements. Although this explanation may seem far-fetched to a modern audience, it was widely accepted for the next two thousand years. In so doing it made a lasting impact on the development of scientific thought, if only in slowing down its progress due to the unchallenging acceptance with which Aristotle's laws were accepted for so long.

In other physical areas Aristotle was more accurate in his assessment. For example, he reinforced the view initially espoused by **Pythagoras** that the earth was spherical. He noticed every time there was a lunar eclipse, an arc-shaped shadow consistent with a globe was cast upon the moon. In addition, he noted correctly that when travelling north or south along the earth, stars 'moved' on the horizon until some gradually disappeared from view. He concluded that this would also be consistent with the idea of a spherical planet.

▸ TOWARDS BIOLOGY

Some of Aristotle's biology was faulty, such as the notion of the heart, not the brain, as the seat of the mind. However, consistent with his empirical approach he undertook detailed dissections to dispel certain myths, for example, that an embryo is formed at the moment of fertilisation, and that the sex of an animal is determined by its position in the womb.

Aristotle was also one of the first to attempt a methodical classification of animals, using means of reproduction, differentiating between those animals which gave birth to live young, and those which laid eggs, a system which was the forerunner of modern taxonomy.

THE LEGACY OF ARISTOTLE

In contrast to his teacher and mentor Plato, Aristotle believed there was much to be learnt from observing nature. He applied this approach to vast areas of existing knowledge to validate, reject, or add to what was already known in subject areas like physics, philosophy, astronomy and biology. Although a pupil at Plato's Academy for nearly twenty years, the two great thinkers were diametrically opposed on a number of subjects, but Aristotle's theses had just as profound an effect on Western thinking as his master's.

In the area of scientific thought, in particular, Aristotle had an even more fundamental influence, to the point that over the ensuing centuries his proposals were attributed an almost god-like, unchallengeable authority, not always with beneficial results.

Euclid

c. 330–260BC

Chronology Although we possess extensive knowledge about the thoughts of many of the ancients, as we have seen, it is often the case that their lives and times are more obscure; this is certainly true in the case of Euclid. Although a name familiar to every schoolchild, almost nothing is known about his life, when and where he studied, or even when and where he was born and died: a true international man of mystery!

It is said that King Ptolemy I Soter of Egypt asked Euclid if it was possible to master geometry by a more direct route than reading his thirteen volume definitive work on the topic. Euclid famously replied, 'There is no royal road to geometry, your Majesty.' Yet what Euclid had provided was one of the most majestic routes to the subject. It would go on to be revered for over two thousand years.

▸ THE ELEMENTS

Euclid's legacy is well known, and yet, as we have seen, much of the life of the Greek mathematician remains a mystery. He probably studied under **Plato** at Athens and certainly spent most of his time in Alexandria where he founded a mathematics academy. Whether all the works credited to him, including *Data*, *On Divisions*, the *Optics* and *Phaenomena*, were actually compiled solely by Euclid, or were produced with assistance from students at his school, remains unclear, but the impact of the texts is known to be great. In particular, *The Elements*, Euclid's masterwork on geometry, had a phenomenal influence on

After the Bible, The Elements *has probably been more studied than any other book in history*

Western academic thinking. This is best illustrated by the suggestion that after the *Bible*, *The Elements* has probably been more studied, translated, and reprinted than any other book in history.

The reason for this is twofold: not just what Euclid said, but also the way that he said it. Indeed, the latter is arguably the more enduring of the two because it profoundly influenced the presentation of almost every future mathematical, scientific, theological and philosophical text, amongst others. The reason is because Euclid took a systematic approach to his writing, laying out a set of axioms (truths) at the beginning and constructing each proof of theorem which followed on the basis of the proven truths which had gone before. This logical, 'building block' method set the accepted academic precedent for proving knowledge and continues as a standard model today.

▸ A GEOMETRIC SYNTHESIS

The compilation of knowledge that Euclid brought to the thirteen volumes of *The Elements* was so comprehensive and persuasive that it remained virtually unchanged and unchallenged as a teaching manual for over two millennia. Certainly, many of the theories outlined were not his; he was simply seeking to assimilate all geometric (and much other mathematical) knowledge into a single text. For example, the ideas of previous Greek mathematicians such as Eudoxus, Theaetetus and **Pythagoras** were all evident, though much of the systematic proof of theories, as well as other original contributions, was Euclid's. The first six of the thirteen volumes were concerned with plane geometry, for example laying out the basic principles of triangles, squares, rectangles and circles and any issues around these, as well as outlining other mathematical cornerstones, including Eudoxus's theory of proportion. The next four books looked at number theory, including the celebrated proof that there is an infinite number of prime numbers. The final three works focused on solid geometry.

▸ NON-EUCLIDEAN SPACE

Ironically, it is with some of the text's initial axioms that later mathematicians have found fault. The last axiom in particular has proved to be controversial. This 'parallel' axiom states that if a point lies outside a straight line, then only one straight line can be drawn through the point which never meets the other line in that plane (i.e. the parallel line). This was examined in the nineteenth century by the Romanian mathematician Janos Bolyai. Taking on his father's life work, he attemted to prove Euclid's parallel postulate, only to discover that, in fact, it was unprovable. This began a new school of thought and later, given further weight by Albert **Einstein's** belief that the geometry for space was also non-Euclidean, it was subsequently proved to be true.

THE LEGACY OF EUCLID

Although the discoveries of the last two hundred years have shown time and space to be other than Euclidean under certain circumstances, this should not be seen to undermine his achievements. To have constructed The Elements *in the manner he did, to have had an effect of such magnitude on the development of Western thought, and to have been accepted as the only authority on geometry for so long, (and for most practical purposes still attain such a status) is a profound legacy that few have equalled.*

ARCHIMEDES

C. 287–212BC

CHRONOLOGY • **213 BC** Archimedes' war machines ensure Roman attack on Syracuse is unsuccessful. Siege opens • **212 BC** The Romans capture Syracuse; Archimedes is killed by a Roman soldier during the sack of the city • **75 BC** Archimedes' tomb is discovered and restored by the Roman statesman Cicero

'Give me a place to stand on, and I will move the earth,' Archimedes is reputed to have declared to the people of Syracuse. The practicalities of an earth-bound life may have denied him that particular pedestal but arranging for his patron King Heiron to move a ship by pushing a small lever was considered only a slightly less miraculous feat. With such audacious displays, along with his brilliance as an inventor, mechanical scientist and mathematician, it is no wonder Archimedes was so popular and highly regarded among his contemporaries.

THE MATHEMATICIAN

It was not only his peers, however, who benefited from Archimedes' work. Many of his achievements are still with us today. First and foremost, Archimedes was an outstanding pure mathematician, 'usually considered to be one of the greatest mathematicians of all time,' according to the *Oxford Dictionary of Scientists*. He was, for example, the first to deduce that the volume of a sphere was $4\pi r^3 \times 3$, where r is the radius. Other work in the same area, as outlined in his treatise *On the Sphere and Cylinder*, led him to deduce that a sphere's surface area can be worked out by

> *'Give me a place to stand, and a long enough lever, and I will move the earth'*

multiplying that of its greatest circle by four; or, similarly, a sphere's volume is two-thirds that of its circumscribing cylinder. He calculated pi to be approximately 22/7, a figure that was widely used for the next 1500 years.

THE ARCHIMEDES PRINCIPLE

Archimedes also discovered the principle that an object immersed in a liquid is buoyed or thrust upwards by a force equal to the weight of the fluid it displaces. The volume of the displaced liquid is the same as the volume of the immersed object. Legend has it that he discovered this when set a challenge by King Heiron to find out whether one of his crowns was made of pure gold or was a fake. While contemplating the problem Archimedes took a bath and noticed that the more he immersed his body in the water, the more the water overflowed from the tub. He realised that if he immersed the crown in a container of water and measured the water that overflowed he would know the volume of the crown. By obtaining a volume of pure gold equivalent to the volume of water displaced by the crown and then weighing both the crown and the gold, he could answer the King's question. On making this realisation, Archimedes is said to have leapt from his tub and run naked along the street shouting 'Eureka!', 'I have found it!'

LEVERS AND PULLEYS

Indeed, it was the practical consequences of Archimedes' work which mattered more to his contemporaries and for which he became famous.

One such practical demonstration allowed King Heiron to move a ship with a single small lever – which in turn was connected to a series of other levers. Archimedes knew the experiment would work because he had already prepared a general theory of levers. Mathematically, he understood the relationship between the lever length, fulcrum position, the weight to be lifted and the force required to move the weight. This meant he could successfully predict outcomes for any number of levers and objects to be lifted.

Likewise he came to understand and explain the principles behind the compound pulley, windlass, wedge and screw, as well as finding ways to determine the centre of gravity in objects.

ARCHIMEDES GOES TO WAR

Perhaps the most important inventions to his peers, however, were the devices created during the Roman siege of Syracuse in the second Punic War. The Romans eventually seized Syracuse, due to neglect of the defences, and Archimedes was killed by a Roman soldier while hard at work on mathematical diagrams. His last words are reputed to have been, 'Fellow, do not disturb my circles!'

FURTHER ACHIEVEMENTS

Inventions

- Archimedes' Screw: *a device used to pump water out of ships, and also to irrigate fields.*
- Archimedes' Claw: *a huge war machine designed to sink ships by grasping the prow and tipping them over, used in the defense of Syracuse.*
- Compound pulley systems: *enabled the lifting of enormous weights at a minimal expenditure of energy.*
- The method of exhaustion: *an integral-like limiting process used to compute the area and volume of two-dimensional lamina and three-dimensional solids.*

Discoveries

- *Archimedes was responsible for the science of hydrostatics, the study of the displacement of bodies in water (see Archimedes' Principle). He also discovered the principles of static mechanics and pycnometry (the measurement of the volume or density of an object).*
- *Known as the 'father of integral calculus', Archimedes' reckonings were later used by, among others, Kepler, Fermat, Leibniz and Newton.*

HIPPARCHUS

C. 170–125BC

CHRONOLOGY The most significant date pertaining to Hipparchus is 134 BC, when he observed a new star in the constellation of Scorpio. Most of the detail of Hipparchus's life that has come down to us is taken from Ptolemy's record of his achievements. The vast majority of Hipparchus's original work has been lost. He was born in Nicaea, Bithynia, now in modern Turkey, where he undertook some of his astronomical observations, along with sustained periods in Rhodes and to a lesser extent in Alexandria.

Hipparchus spent long periods taking measurements of the earth's position in relation to the stars. The results enabled him to make a number of important findings and calculations.

▸ THE 'PRECESSION OF THE EQUINOXES'

He discovered what is now known as the 'precession of the equinoxes' by comparing his own observations with those noted by Timocharis of Alexandria a century and a half previously together with earlier recordings from Babylonia. What Hipparchus realised was that even taking into account any observational errors made by his predecessors, the points at which the equinox (the two occasions during the year when day and night are of equal length) occurred seemed to move slowly but consistently from east to west against the backdrop of the fixed stars. He gave a value for the annual precession of around 46

The first person to use the concepts of longitude and latitude in his geographical positionings

seconds of the arc, which is exceptionally close to the modern figure of 50.26 seconds, given the tools and data then available to him.

▸ THE DISTANCE OF THE MOON

From these observations, Hipparchus was able to make much more accurate calculations on the length of the year, producing a figure that was accurate to within six and a half minutes. He was also able to correctly determine the lengths of the seasons and offer more exact predictions of when eclipses would take place. He made observations of the sun's supposed orbit and attempted to do likewise with the more irregular orbit of the moon. Although partially successful, he could not make entirely accurate calculations. Using measurements and timings related to the earth's shadow during eclipses, other attempts were made to determine the size of the sun and moon and their distances from the earth. Again, while not entirely accurate, Hipparchus proposed that the distance of the moon from the earth was 240,000 miles. This is remarkably close to the modern figure.

▸ A CATALOGUE OF STARS

Perhaps Hipparchus' most important astronomical achievement was his plotting of the first known catalogue of the stars, despite warnings from some of his contemporaries that he was thus guilty of impiety. He was inspired to begin this work in 134 BC after allegedly seeing a 'new star' which prompted his speculation that the stars were not fixed as had previously been thought. He went on to record the position of 850 stars in the remaining years of his life, a significant achievement given the resources available to him. Moreover, he devised a scale for recording a star's magnitude or brightness: from the most visible – the first magnitude – to the faintest – the sixth. Though amended considerably, it is a scale still used today.

▸ DEVELOPING TRIGONOMETRY

Because of the accelerated developments Hipparchus was making in astronomy, he was required to break new ground in other disciplines, particularly mathematics, to facilitate his celestial observations and calculations. Most notably of all, he developed an early version of trigonometry. With no notion of sine available to him, he was required to construct a table of chords which calculated the relationship between the length of a line joining two points on a circle and the corresponding angle at the centre.

FURTHER INFLUENCE OF HIPPARCHUS

Although Hipparchus is considered to be one of the most influential astronomers of the ancient world, it is arguable that his most impacting achievements lay in the areas of mathematics and geography. The geographer and astronomer Ptolemy *cited Hipparchus as his most important predecessor and he is most often revered for his astronomical measurements and cataloguing. Yet as the attributed inventor of trigonometry, as well as being the first person to plot places on the earth's surface using longitude and latitude, his influence was long lasting and widespread.*

He was able to apply his work on the trigonometry of spheres to the planet from which he made his observations. Significantly, he was the first person to use longitude and latitude in his mathematical calculations to position where places were on the earth's surface. Like so many of Hipparchus's achievements, it is his further pioneering work that still resonates today.

ZHANG HENG

78–139AD

CHRONOLOGY • **78 AD** Zhang Heng born in Nan-yang, China • **123 AD** Corrects the calendar, bringing it into line with the seasons • **132 AD** Invents the first seismograph for measuring earthquakes • **138 AD** Zhang Heng's machine detects the location of an earthquake 500 kilometers away

Western science is often credited with discoveries and inventions which have been observed in other cultures centuries before. This can be due to a lack of reliable records, difficulty in discerning fact from legend, problems in pinning down a finding to an individual or group, or frequently, simple ignorance. No such excuses exist for the work of Zhang Heng, whose life and achievements are well recorded, and whose major invention was created some 1,700 years before European scientists 'invented' the same thing.

STUDYING THE EARTH

Zhang, a Chinese scholar in the East Han Dynasty, was a man of many disciplines, including astronomy, mathematics and literature. Yet his greatest achievement was in geography, inspired by one of the duties assigned to him in the course of his work as Imperial Historian! China regularly suffered from earthquakes and as part of his job Zhang was required to record when and where they occurred. Rather than accept the common superstition that the quakes were punishment from angry gods, Zhang believed that if he took a scientific approach to

He devised the world's first seismograph, which he named Di Dong Yi, Earth Motion Instrument

noting data about tremors, the Dynasty would be better equipped to predict, prepare for and deal with them. To this end he devised the world's first seismograph, an invention he named Di Dong Yi, or 'Earth Motion Instrument'.

▸ THE EARTH MOTION INSTRUMENT

The seismograph was large, at almost two metres in diameter, and made out of bronze. Eight thin copper rods were attached to a central shaft at one end and to a corresponding number of dragons' heads at the other. These heads pointed in the eight major directions of a compass (north, north-east, east, south-east and so on), and each contained a copper ball in its mouth. Underneath each dragon was an open-mouthed copper frog. When a tremor occurred, the copper ball fell out of the mouth of the dragon nearest to the direction from which the earthquake came and into the frog's mouth, which in turn rang a bell alerting the royal household. A story is recorded that in 138 AD a copper ball had dropped to the west. Zhang recounted his finding to the emperor, but for two days nothing unusual happened and there were no reports of activity elsewhere. Sceptics were left to question the validity of Zhang's machine. Finally, though, messengers arrived on horseback reporting a severe earthquake 500 kilometres to the west. Zhang was vindicated.

▸ OBSERVING THE STARS

Fabled to be a man with intense powers of concentration, Zhang was also able to employ his abilities to excellent effect in astronomy. Through his observations he correctly deduced that the sun caused the illumination of the moon, and that lunar eclipses were caused by the earth's shadow passing over its surface. He mapped the night sky in fine detail, recording 2500 'brightly shining' stars in 124 constellations, 320 of which were named. He estimated that in total, including the 'very small,' there were 11520 stars. In addition, Zhang wrote a number of books on astronomy, the most famous being *Lin Xian*. In another, *Hun-i-chu*, he outlined his perception of the universe and the earth's position within it. 'The sky is like a hen's egg,' he wrote, 'and is as round as a crossbow pellet; the earth is like the yolk of the egg, lying alone at the centre. The sky is large and the earth is small.'

Zhang Heng then, in common with his Greek predecessors, believed that the earth was spherical and at the centre of the universe. This drove him to create possibly the first three-dimensional model of the cosmos: a bronze celestial orb which turned by water-power. Each year, making a single complete rotation, it showed how the stars' positions changed.

FURTHER ACHIEVEMENTS

Zhang undertook other work which had a lasting impact. He improved the previous figure of π from 3, the traditional figure in use by the Chinese, to ÷10 or 3.162, closer to the number of 3.142 used today. Zhang also performed calculations involving time, notably correcting the Chinese calendar in 123 AD to harmonise it with the seasons.

Zhang's seismograph is recognized by the world as an instrument that was well ahead of its time. To this day no one has been able to reproduce it. He constructed the first accurate odometer, or 'mileage cart'.

He is considered one of the four great painters of his era. Zhang also produced over twenty famous literary works.

PTOLEMY

90–168AD

A NOTE ON DATES For all that is known of Ptolemy's work, very little is known about his life. Of Greek descent, he was born and lived in Alexandria, Egypt. It is thought that he rarely, if ever, left Alexandria, ironic for a man who mapped the world. Instead, he obtained his geographical knowledge from the accounts of sailors and Roman visitors.

Claudius Ptolemy's work in astronomy and geography had a profound impact on man's perception of the world and universe from the second century AD until the Renaissance. His genius lay in his ability to distil and summarise the important findings of his predecessors, then add to or provide 'scientific' proof of their theories from an all-encompassing viewpoint. Ptolemy's texts were written with such authority that later generations struggled for more than a thousand years to convincingly challenge his theses.

PTOLEMY'S ALMAGEST

The work for which Ptolemy was most revered is his thirteen-volume *Mathematical Collection*, later more commonly referred to as the *Almagest*. It provided for the first time a definitive compilation of everything that was known and accepted in the field of astronomy up to that point. In particular, the work of **Hipparchus** was used as the starting point of many of Ptolemy's developments and it is largely through the records of the latter that Hipparchus's theories have been passed down to us today. In addition, as the starting point for his

The 'Ptolemaic' astronomical system would not be rivalled until Copernicus, 1,400 years later

arguments, Ptolemy used the **Aristotelian** notion that the earth was at the centre of the universe, with the stars and planets rotating in perfect circles around it. He then set about attempting to justify this interpretation through astronomical observation and mathematical speculation. The result was the 'Ptolemaic System', a mathematically 'proven' interpretation of the universe that would not be rivalled until 1543 by **Copernicus.**

In order to explain his observations in the context of a geocentric model of the universe – which would ultimately be proved wrong – Ptolemy had to introduce some complex explanations and calculations to arrive at a convincing result. Most notably, in explaining planetary and star motion he argued for a system of 'deferents', or large circles, rotating around the earth, and eighty 'epicycles', or small circles, which circulated within the deferents. He also examined theories of 'movable eccentrics'. These proposed just one circle of rotation, with its centre offset slightly from the earth, as well as 'equants' – imaginary points in space which also helped define the focal point of the rotation of the celestial bodies. Ptolemy needed to employ these complex theories because he did not know that the planets actually moved in elliptical orbits, not the perfect circles he supposed. As a result, his predictions for some of their movements continued to be inaccurate but they were the best explanations available at the time, and for many centuries afterwards.

▸ GEOGRAPHY

Almost as significant in its impact on the world was Ptolemy's *Geography*. For the first time, a detailed mathematical explanation for calculating lines of longitude and latitude was offered. and again, this built on the work Hipparchus had begun. It allowed Ptolemy to undertake groundbreaking research into the projection of the earth's sphere onto a plane, leading to the drawing of a scaled map of the known world which would resonate for as long as *Almagest*. Although there were many errors on this map, such as the equator being too far north and Asia stretching too far to the east, its importance to later generations cannot be underestimated. Most notably, it has been argued that because Asia appeared much closer to Europe on Ptolemy's map than it actually was (assuming the earth was a sphere and one 'looked' westwards), it was this that encouraged Christopher Columbus to sail west in the hope of finding a shorter route to Asia, and to accidentally discover America.

▸ ASTROLOGY

Still frequently read and referred to today, Ptolemy's other major text is his *Tetrabiblos*, possibly the founding work on the then 'science' of astrology. Although this more properly belongs in the category of 'pseudo-science', Ptolemy does at least suppose that the influence of the stars on humans may be due to some sort of radiation.

FURTHER INFLUENCE OF PTOLEMY

Ptolemy wrote on a number of other subjects, and two works in particular were of some importance. His final text is the Optics, *regarded by many as his most successful. In this work, Ptolemy gives a statement of various elementary principles of optics, which he then sets out to demonstrate. After setting out the principles of reflection, Ptolemy then proceeds to examine the refraction of rays of light passing through water, providing tables for various angles of incidence, tables which are clearly based upon empirical observation.*

GALEN OF PERGAMUM

130–201AD

CHRONOLOGY

• **129 AD** Galen born in Pergamum (now Bergama in Turkey)
• **148–157AD** Travels and studies in Corinth and Alexandria
• **157AD** Takes the post of surgeon to the Pergamum gladiators • **c. 161 AD** Becomes physician to the emperors Marcus Aurelius and Commodus • **1628** William Harvey's system of blood circulation becomes the first viable alternative to Galen's

Unlike many of the other entrants in this book, Galen is not famous for any single achievement, but more for the sheer volume of medical thought which he presented. That accomplishment in itself may not necessitate an inclusion, but the fact that his works on medical science became accepted as the only authority on the subject for the following 1400 years does.

UNDISPUTED FOR A MILLENNIA

The question, therefore, is why? Some commentaries suggest the answer is simply because Galen's studies were so all-encompassing that there was very little left for those following him to dispute. Another is the readiness with which the Arab, Christian and Jewish authorities accepted his work, lending it a weight which might have made it difficult for others to challenge. A third explanation could be that Galen not only incorporated the results of his own findings in his texts, but also compiled the best of all other medical knowledge that had gone before him into a single collection, such as that of

Galen's studies were so comprehensive, those following him had very little to dispute

Hippocrates, for example. In particular, Galen readily adopted Hippocrates' 'four humours' approach to the body, and this was one of the main reasons it endured for so long.

▸ A METICULOUS INVESTIGATOR

That is not to say Galen was at all lacking in original material and thinking. He was meticulous and methodical in his approach to his own medical investigations, above all in anatomy. Many important dignitaries came to the shrine of Asklepios, the god of healing in Galen's home town, to seek cures for ailments. Thus Galen was able to observe first hand the symptoms and treatments of diseases. After spells in Smyrna (now Izmir), Corinth, and Alexandria studying both philosophy and medicine, which he considered inextricably linked, and including work on the dissection of animals, he returned to Pergamum in 157. There he took up a four-year appointment as a physician to gladiators, giving him further first hand experience in practical anatomical medicine.

▸ PHYSICIAN TO EMPERORS

All of this was excellent preparation for his transfer to Rome. Here he spent most of the rest of his career and became the esteemed physician to emperors Marcus Aurelius, Lucius Verus, Commodus and Septimius Severus. This position not only brought him prestige, but it allowed him the freedom to undertake detailed research and dissection in the quest for the improved knowledge it provided. Galen was not permitted to scrutinise human cadavers, so he dissected animals, predominantly Barbary apes because of the characteristics they shared with man. His most influential conclusions concerned the central operation of the human body. Sadly they were only influential in that they limited the search for accurate information for the next millennia and a half.

Galen believed that blood was formulated in the liver, the source of natural spirit. In turn, this organ was nourished by the contents of the stomach which was transported to it. Veins from the liver carried blood to the extremes of the body where it was turned into flesh and 'used up', thus requiring more food on a daily basis to be converted into blood. Some of this blood passed through the heart's right ventricle, then seeped through to the left ventricle and mixed with air from the lungs, providing vital spirit which regulated the body's heat and blood flow. Using the arteries, a portion of this blood was then transported to the brain where it blended with animal spirit. This created movement and the senses. The combination of these three spirits managed the body and contributed to the make up of the soul. It was for this reason that Galen missed the idea of a single, integrated system of the circulation of the blood, a result which was not conclusively proved until 1628 by William Harvey.

Galen's Legacy

Although some of Galen's deductions were wrong, his surviving 129 volumes are a phenomenal contribution to his subject and offered a platform from which Renaissance physicians could begin their critical progress. It was Galen who first introduced the notion of experimentation to medicine.

Many of the anatomical errors made by Galen were due due to the fact that he could only operate on animals – human dissections were out of favour at the time. Galen became a doctor supposedly because his father had a dream in which Asklepios, the god of healing, appeared to him.

AL-KHWARIZMI

800–850

CHRONOLOGY • **c. 786** Al-Khwarizmi born in Khwarizm, now Khiva, in Uzbekistan • **813** Caliph al-Ma'mun, the patron of Al-Khwarizmi, begins his reign in Baghdad • **c. 820** 'House of Wisdom' founded in Baghdad by al-Ma'mun • **833** Death of al-Ma'mun

One of the greatest scientific developments of all time was the introduction of 'Arabic' numerals into mathematics. The man often credited with their invention was Al-Khwarizmi, an Arabian mathematician, geographer and astronomer. Yet strictly speaking, the concept was neither invented by Al-Khwarizmi, nor was it Middle Eastern in origin.

▸ NUMERICAL NOTATION

What Al-Khwarizmi did do for Arabic numerals, though, was introduce them to Europe, which is why many Western textbooks subsequently acknowledged the development as his. The notation actually finds its roots in India around 500 AD and the naming of the numerical scheme employing the figures, now called the 'Hindu-Arabic' system, acknowledges this. The method of using only the digits 0–9, with the value assigned to them determined by their position (e.g. the '1' in '100' has a different value than the '1' in '10' because of its location in relation to other digits), as well as introducing a symbol for zero for the first time, completely revolutionised mathematics. Without it many of the develop-

The name 'Al-Khwarizmi' became 'algorithm', meaning 'rule of calculation', in the West

ments of later times, and what have become norms in the modern world, would have been impossible. Al-Khwarizmi observed this system, then clearly explained how it worked in his text *Calculation with the Hindu Numerals.* When translated later into Latin, it was widely adopted by the West and ultimately the entire world. Even today the numerical system is perhaps the only truly global 'language'.

▸ THE HOUSE OF WISDOM

Al-Khwarizmi does have a much more original claim, however, to writing the first book on algebra, and indeed, introducing the word into our language. He was afforded the opportunity to develop such texts as a patron to Caliph Al-Ma'mun in Baghdad, who ruled the huge Muslim empire extending from the Indian subcontinent to the Mediterranean. Al-Ma'mun's father, Caliph Harun al-Rashid, had been keen to facilitate the development of academic disciplines in his kingdom, and Al-Ma'mun had continued to back his father's goals, founding his 'House of Wisdom' to this end. This academy housed a library, including translations of important Greek texts, and also established astronomical observatories. Al-Khwarizmi repaid the investment with his work *Calculating by Completion and Balancing.*

▸ A PRACTICAL GUIDE TO ARITHMETIC

In this treatise, Al-Khwarizmi set out to provide a practical guide to arithmetic using, where applicable, calculations later described as algebraic. In so doing he introduced quadratic equations, although he described them in words and did not use the symbolic algebra (e.g. $x^2+3x=10$) we more commonly understand today. The two key concepts he outlined were the ideas of 'completion' and 'balancing' of equations. Completion is the method of expelling negatives from an equation by moving them to the opposite side (e.g. $4x^2=54x-2x^2$ becomes $6x^2=54x$). Balancing, meanwhile, is the reduction of common positive terms on both sides of the equation to their simplest forms (e.g. $x^2+3x+22=7x+12$ becomes $x^2+10=4x$).

▸ THE 'FATHER OF ALGEBRA'

It is not clear whether Al-Khwarizmi was familiar with the works of Euclid, despite the fact that one of his colleagues at the House of Wisdom had translated *The Elements* into Arabic.

Although Al-Khwarizmi was clearly building on the work of others before him, such as Diophantus and Brahmagupta, his was a much closer expression of modern elementary algebra, and this is the reason he is sometimes referred to as 'the father of algebra.'

Indeed, the Arabic title of *Calculating by Completion and Balancing* is *Hisab al-jabr w'al-muqabala* and it is from the 'al-jabr' in this heading that the word 'algebra' descends.

FURTHER ACHIEVEMENTS

Through carelessness of pronunciation the name 'Al-Khwarizmi' became referred to in the West as 'algorismi', then 'algorism', and ultimately 'algorithm', which is where we get the word meaning 'a rule of calculation' today.

Al-Khwarizmi undertook other work in mathematics, such as writing tables of sines and tangents. He also performed many astronomical observations and was a keen investigator of geography. In particular, he expanded on Ptolemy's use of longitude and latitude in plotting the positions of places around the world, developing a series of maps more accurate than those of his predecessor.

JOHANNES GUTENBERG

1400–1468

CHRONOLOGY • **1420** Gutenberg moves from Mainz, Germany, to Strasbourg, France • **1450** Returns to Mainz, and sets up his printing press using moveable type • **1450–56** Prints a number of books, a calendar, and a letter of Papal indulgence • **1456** prints his famous 42-line Bible • **1465** Made courtier to the Archbishop of Mainz

Johannes Gutenberg was born and spent much of his life in the German town of Mainz. His family background was in minting and metal-working, an ideal foundation for his training as an engraver and goldsmith. These skills enabled him to craft the first individual metal letter moulds, matrices, which were the core of his achievements in printing. Hand-held block printing, however – a laborious process of carving whole pages of 'fixed' text out of wooden slabs, reproducing copies using dyes – had been used for many decades before the German inventor appeared.

MOVEABLE TYPE

What Gutenberg mastered was the idea of placing individual metal letters into temporary mounts, which could then be dismantled or 'moved' once a page of text had been completed and reused to produce other pages. In comparison to the slow engraving and single use of wooden blocks, the theoretically infinite number of different sides which could be made out of a set of metal characters, together with the speed at which a template could be created, revolutionised printing and the spread of the printed word.

The development closest to the impact of the numerical system in the history of science

It is thought that Gutenberg began experimenting with the creation of metal letter casts towards the end of the 1430s during a period living in Strasbourg. It was probably not until sometime between 1444 and 1448, however, that he finally perfected the development of the moveable type printing press. Certainly, it is known he borrowed money from a relative in 1448 on his return to Mainz most probably to fund his printing business. The invention itself consisted of an adapted wine press with the plate of metal characters at the bottom upon which a piece of paper was laid and the top of the press lowered from above to force the imprint. The matter of developing a suitable dye for this machine was no easy task, either, but it is thought Gutenberg finally found the answer in an amalgamation of linseed oil and soot.

▸ THE FORTY-TWO LINE BIBLE

No works survive bearing Gutenberg's name, but the earliest printed piece attributed to him is a Calendar for 1448. Much more famous, with around forty-eight of the original two hundred copies still in existence, is the first Bible printed using moveable type, known as the forty-two line Bible' because of the number of lines to a page. It is believed Gutenberg and his assistants made the copies between 1450 and 1456.

In the later years of his life, Gutenberg lived off the patronage of the Archbishop of Mainz, an offer thought to be an acknowledgement of his ground-breaking achievement. Others, however, have been less willing to recognise Gutenberg as the inventor of moveable type and instead claim the inventor of printing to be a certain Laurens Janszoon Coster (c. 1370–1440).

Very little is known about this Dutchman and, like Gutenberg, no printed fragments bearing his name survive, but a legend persists that he carved individual letters into wood for the entertainment of his grandchildren. To amuse them further he used dye to print words and sentences onto paper whereupon he realised the possibilities for these moveable pieces. A block printer by trade, it is thought that Coster began using the wooden characters to speed up his printing processes probably by bringing together a combination of block and moveable type in the production of texts. The evidence to support these claims is limited at best. Even if true, what is notable is the superior quality of Gutenberg's metal casts and press – they are almost as important as the idea of moveable type itself.

Some sources credit the Chinese with inventing moveable type printing, using characters made of wood, in the early fourteenth century. It is almost certain, however, that Gutenberg developed his ideas independently and was unaware of any similar developments which may or may not have taken place on the other side of the world more than a century before him.

The Legacy of Gutenberg

The one development in the history of science which probably comes closest to matching the impact of the Hindu-Indian numerical system is the invention of the printing press using moveable type. Although not specifically a scientific achievement, its emergence provided one of the key tools in helping to begin the revolutionary progress of the subject in Europe, giving academics the opportunity to share scientific knowledge widely and cheaply.

By the end of the fifteenth century tens of thousands of books and pamphlets were already in existence and the stage was set for the imminent explosion of scientific ideas.

LEONARDO DA VINCI

1452–1519

CHRONOLOGY • **1469** Leonardo apprenticed to the studio of Verrocchio in Florence • **1482** Works for the Duke of Milan • **1502** Returns to Florence to work for Cesare Borgia as his military engineer and architect • **1516** Journeys to France on invitation of Francis I • **1519** Dies in Clos-Luce, near Amboise, France

It is perhaps something of an indulgence to include Leonardo Da Vinci in a book of scientists who changed the world, not least because most of his work remained unpublished and largely forgotten centuries after his death. His, however, was undoutedly one of the most brilliant scientific minds of all time; arguably the biggest handicap preventing him from profoundly changing the world was the era in which he lived.

The genius of Leonardo's designs for his inventions so far outstripped both his contemporaries' intellectual grasp and contemporary technology that they were rendered literally inconceivable to anyone but him. If Leonardo could have teleported to **Edison's** time, with his access to nineteenth century technology, one can only speculate how much more he may or may not have achieved than even Edison himself. But even in his own time, Leonardo's achievements were notable.

▸ RENAISSANCE MAN

Leonardo is celebrated as the Renaissance artist who created such masterpieces as the Last Supper (1495–97) and the Mona Lisa (1503–06), yet

The term 'Renaissance Man' could have been coined specifically for Leonardo

much of his time was spent in scientific enquiry, often to the detriment of his art. The range of areas Leonardo examined was breathtaking. It included astronomy, geography, palaeontology, geology, botany, zoology, hydrodynamics, optics, aerodynamics and anatomy. In the latter field in particular, he undertook a number of human dissections, largely on stolen corpses, to make detailed sketches of the body. Irrespective of the breadth of his studies, however, perhaps the most important contribution Leonardo made to science was the method of his enquiry, introducing a rational, systematic approach to the study of nature after a thousand years of superstition. He would begin by setting himself straightforward scientific queries such as 'How does a bird fly?' Next he would observe his subject in its natural environment, make notes on its behaviour, then repeat the observation over and over to ensure accuracy, before making sketches and ultimately drawing conclusions.

▸ AERODYNAMICS

Moreover, in many instances he could then directly apply the results of his enquiries into nature to designs for inventions for human use. For example, his work in aerodynamics led him to make sketches for a number of flying machines, – which, potentially, could have flown – including a primitive helicopter, some five hundred years before the invention became reality! He even envisaged the need for his flying machines to have a retractable landing gear to improve their aerodynamics once airborne. In 1485 he designed a parachute, three hundred years before it became an actuality, and included calculations for the necessary size of material to safely bring to ground an object with the same weight as a human. He also had an excellent understanding of the workings of levers and gears, enabling him to design bicycles and cranes.

▸ HYDRODYNAMICS

Leonardo's studies in hydrodynamics led to numerous sketches on designs for waterwheels and water-powered machines centuries before the industrial revolution. In addition, he sketched humidity-measuring equipment as well as a number of primitive diving suits, mostly with long snorkel devices to provide a supply of air.

▸ MILITARY INVENTIONS

During his work for the Duke of Milan between 1482 and 1499, Leonardo prepared an array of designs for weaponry such as catapults and missiles. Even in this arena, however, he could not help but create sketches of weapons that lay way ahead of their time such as hand-grenades, mortars, machine-type guns, a primitive tank and, most audaciously, a submarine!

Leonardo's Influence

If this were a book of scientists who 'could have' changed the world, Leonardo Da Vinci would be at the top of the list. But regardless of the fact that many of the designs for his potentially world-changing creations were never published, his methodical approach to science marks a significant and symbolic stepping-stone from the Dark Ages into the modern era.

Hoping to secure employment with the Duke of Milan, he wrote to him that his areas of expertise included: the construction of bridges and irrigation canals, the designing of military weapons and architecture, as well as painting and sculpture. To add to the list, Leonardo is also credited with being the first ever person to conceive of a bicycle!

NICOLAS COPERNICUS

1473–1543

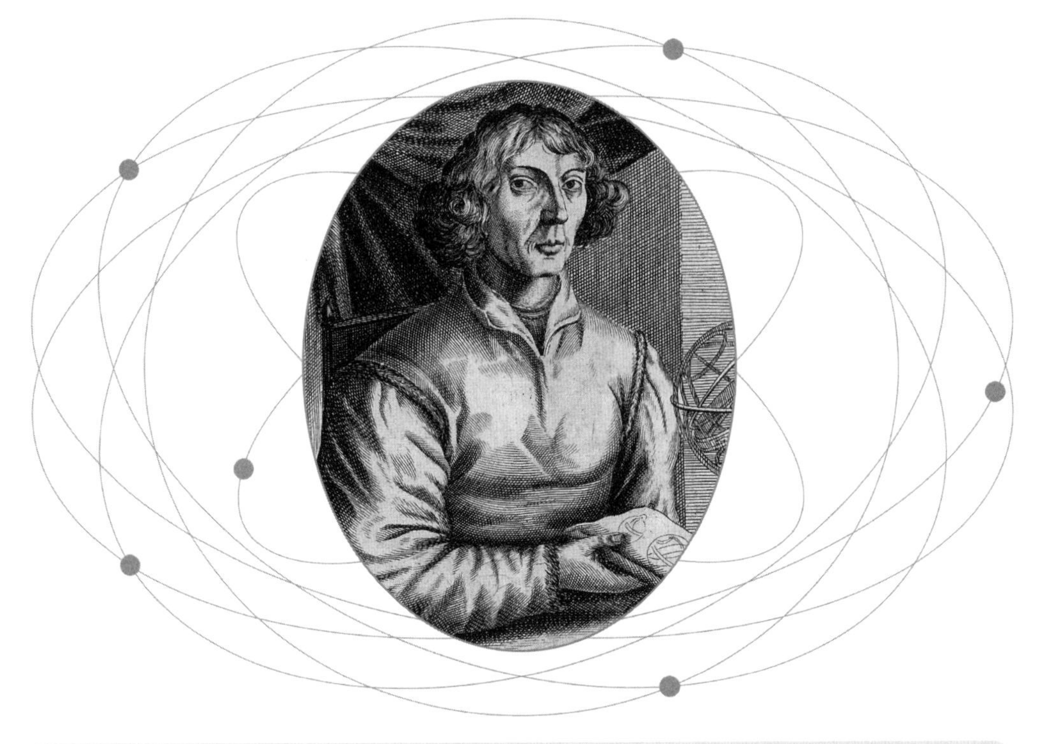

CHRONOLOGY • **1491** Copernicus enters the University of Krakow • **1510–14** The revolutionary *Commentariolus* is circulated • **1543** *De revolutionibus orbium coelestium* (*On the Revolution of the Celestial Spheres*), his definitive work, is published while he is on his deathbed, but is banned by the Catholic Church. The ban is not lifted until 1835.

For all the impact the idea the planets might revolve around the sun, not the earth, would have on astronomy and science, arguably its biggest challenge would be to religion. The explanation of an earth inhabited by human beings, made in God's image as the most superior of all creatures, at the centre of a cosmos around which everything else revolved, suited the Christian Church's interpretation of the universe and mankind's position within it. It was a concept which dated back to **Aristotle**, was given observational legitimacy by **Ptolemy** and authority by Christendom. The Catholic religion still opposed the heliocentric model of planetary motion nearly three centuries after it was first published. And yet ironically its author, Nicolaus Copernicus, was himself a man of the Church.

A MAN OF FAITH

Indeed, it was Copernicus's faith which had led him to question Ptolemy's accepted geocentric model of the universe in the first place. Why would God create a hugely complicated system of equants, epicycles and eccentrics, as Ptolemy had proposed, to explain the planets' motion around

Copernicus literally used the Church to advance his studies, observing the stars from a belltower

the earth when it would be far more simple, logical and graceful to have them all revolving around the sun? It was a theory Copernicus spent many years contemplating while studying in Krakow and then Italy, and continued to develop as he returned to Poland to take up a post as canon in Frauenberg Cathedral. He even used his position within the Church to quite literally advance his studies, using a cathedral tower to quietly and solitarily observe the stars.

▸ THE EARTH CIRCLES THE SUN

Gradually Copernicus became more convinced of his proposition that a fixed sun was at the centre of planetary motion, with the earth rotating around it once a year. Between 1510 and 1514 he drafted *Commentariolus*, his initial exposition of the theory. In order to have any credence, the idea also required that the earth itself was not fixed in position as had previously been thought, but revolved on its axis once every twenty-four hours. This would also explain the apparent movement of the stars and sun across the sky. Perhaps because of his position within the Church, fearing a backlash, or perhaps because he was a perfectionist and recognised that his ideas were not fully developed, Copernicus resisted publishing *Commentariolus*, circulating it instead only among friends.

▸ CHURCH OPPOSITION

Copernicus continued to work on his ideas for the next twenty years and though his final work was largely completed by 1530 he continued to resist pleas by his friends to publish. Word of Copernicus' theories was already spreading across Europe and it is thought that even the Pope himself knew of them but offered no initial resistance to the idea of a heliocentric model. Indeed, it was not until 1616 that the Church banned the text Copernicus eventually published for its 'blasphemous' content, although that sanction subsequently remained in place until 1835, long after the 'Copernican system' had been widely accepted by most others.

▸ A CRITICAL RECEPTION

On The Revolutions of Celestial Spheres was finally published in 1543. But as powerful and revolutionary as Copernicus's theories were, the text was rejected by many academics. This was partially because the author had undermined the simplicity of his initial ideas by clinging onto the Aristotelian belief that planetary motion took place in perfect circles. As we now know this not to be true, it meant Copernicus had been forced to introduce his own system of epicycles and other complex motions to fit in with observational evidence, thereby producing as equally complicated an explanation as the geocentric one he had initially rejected for its lack of simplicity. It was not until Johannes **Kepler** offered the solution that the planets moved in an elliptical, not circular, motion in 1609 that the simplicity Copernicus had been seeking was offered and the rest of his model could be vindicated.

A Man of Contradictions

Copernicus was brought up by his maternal uncle Lucas, the Bishop of Ermeland, and took a doctorate in canon law at the University of Ferrara in 1503. By this time he had become a canon of Frauenburg. Throughout his life Copernicus struggled to come to terms with the conflict between his mathematics and his religious faith. Indeed, one of the main reasons he did not publish his works was through fear of contradicting the Bible.

ANDREAS VESALIUS

1514–1564

CHRONOLOGY • **1514** Born in Brussels, Belgium • **1537** Appointed Professor of Anatomy and Surgery, Padua University • **1543** Publishes first anatomically accurate medical textbook, *De humani corporis fabrica* (*On the Structure of the Human Body*) • **1543** Joins Hapsburg court where he serves as physician to the Emperor Charles V and King Philip II of Spain • **1564** Dies on a pilgrimage to the Holy Land

It takes a brave person to challenge the accepted authority on any subject, especially one which has endured without dispute for some 1400 years, and more especially when the person raising the objection is only twenty-eight years old and has only relatively recently graduated. That is just the task Andreas Vesalius took upon himself, however. For many of his contemporaries, there was nothing about this confrontation to consider as 'brave': instead, they described him as anything from a liar to a madman.

CHALLENGING GALEN

The authority Vesalius dared to challenge was that of **Galen**, the celebrated Roman physician who wrote what had been considered the definitive work on human anatomy. Such was his clout that when dissections of humans began to be permitted in Europe from the fourteenth century for research and tuition purposes, lecturers would simply read directly from Galen as the cadaver was cut by a butcher or assistant. Yet what was somehow lost sight of in all of this reverence was the fact that Galen himself had never actually

Vesalius encouraged the hands-on approach, revolutionising the teaching of anatomy

dissected a human body, forbidden as this had been by Roman religious laws. Academics before Vesalius, however, still considered Galen as the authority on the subject, with any advance on his texts regarded as impossible.

▸ A NEW APPROACH

Vesalius's approach was completely different. Born and raised in Belgium, to a family with a distinguished background as doctors to royalty, Vesalius was a keen dissector of animals from a young age. He went on to study medicine at institutions around Europe, notably the universities of Louvain, Paris and then Padua, where he was appointed Professor of Anatomy and Surgery at the age of 24. He insisted on performing the dissection of human bodies himself during lectures to students, rejecting the traditional clean-handed, textbook method.

▸ HUMAN ANATOMY

Although schooled in the Galenic tradition like all other medical students, Vesalius began questioning its teachings towards the end of the 1530s. From 1540 onwards, having been granted an ample number of human corpses to dissect, mostly from the local executioners, Vesalius was convinced. Galen's findings, he argued, did not reflect the human anatomy, but that of apes. This had led to numerous errors based on assumptions Galen had made on similarities between the two.

▸ THE DEFINITIVE TEXT

In 1543, Vesalius published his masterwork *De Humani Corporis Fabrica Libri Septem* or *The Seven Books on the Structure of the Human Body*. It was the first definitive work on human anatomy actually based on the results of methodical dissections of humans and, as such, was the most accurate work on the subject ever written. Furthermore, it was beautifully and clearly illustrated with woodcut drawings, probably prepared at the artist Titian's studios, and was excellently structured and organised. Its publication outdated all that had gone before and the text became the guide upon which future teachers would base their lectures. It was some time before its wisdom was widely accepted, however, due to the hostility which Vesalius often endured for challenging Galen. For example, Vesalius stated he could find no evidence for the 'pores' in the heart which allowed blood to seep from the right to the left ventricle, a key foundation of the Galenic tradition and one which was resolutely defended by many of his contemporaries.

Vesalius spent much of the rest of his life after the *Fabrica* in the service of kings, firstly as the physician to Charles V, the Holy Roman Emperor, then to Phillip II of Spain. He left Spain in 1564 on a pilgrimage to Jerusalem, but died on the return journey.

Further Achievements

In spite of his premature death, Vesalius left behind a revolutionary legacy to anatomy students. It was only after his publications that both anatomy and medicine in general were first treated as sciences in their own right. By his reasoned critical approach to Galen, he had broken the reverence ascribed to the former 'master' and created a model for independent, rational investigation for his successors in the development of medical science.

Vesalius also changed the organisation of the medical school classroom, and actively encouraged the participation of medical students in dissection lectures.

WILLIAM GILBERT

1540–1603

CHRONOLOGY • **1569** Receives degree at Cambridge university • **1600** Publishes *De magnete, magnetisque corporibus, et de magno magnete tellure* (*On the Magnet, Magnetic Bodies, and the Great Magnet Earth*), the first great English scientific work. • **1600–03** Serves as physician to Queen Elizabeth

William Gilbert has often been considered one of the first great English scientists and arguably the first great physicist of the modern era. His principle area of study related to magnetism in which he made groundbreaking revelations. For all the fame the subject of his observations brought him, however, his method of enquiry is equally, if not more, significant.

▸ DANGEROUS TIMES

Living in the time of Shakespeare and Elizabeth I, for whom he acted as physician from 1600–03, England was still largely a place of superstition and religious fervour. Rational scientific enquiry was rare with some of the few earlier European attempts at it, such as the observations of Leonardo **Da Vinci**, unknown to Gilbert. He was, however, familiar with the work of **Copernicus** with whom he passionately concurred, a potentially dangerous sentiment in an era when elsewhere in Europe others such as Giordano Bruno and later **Galileo** were being persecuted (and in the case of Bruno, executed) for sharing the same opinion.

Gilbert's first work, De Magnete, *is considered to be one of the first scientific texts*

▸ NEW METHODS

Given this background, then, Gilbert's approach to his work is all the more remarkable. In an unheard of manner, he cast aside all prior speculation on his subject, including that of the 'authorities' from antiquity, and resolved to only make deductions based on proof. Although that approach seems perfectly natural to the modern reader, it was a rational mode of enquiry which religion and superstition had hitherto largely made impossible. Gilbert's work was instrumental as a model for the scientific revolution.
By the same token, his principle work, *De magnete, magnetisque corporibus, et de magno magnete tellure* or *On the Magnetic, Magnetic Bodies, and the Great Magnet Earth* (1600), is considered to be one of the first truly scientific texts. It was the result of years of painstaking observations and experiments which Gilbert had undertaken to learn more about magnetism and electricity, a term he popularised, and to systematically dispel common myths. For example, it had been believed garlic could destroy the accuracy of a compass needle, one of many folk tales Gilbert sought to redress.

▸ FROM EXPERIMENT TO CONCLUSION

What he did prove through his repeated experiments was that a spherical magnet would force a small compass needle to point north or south pole-ward according to where it was positioned near the sphere, and also 'dip' downward towards its surface. This mimicked the behaviour of a normal compass needle when used under ordinary circumstances in the wider world. From these results, he deduced that the earth itself was effectively a large magnet, with a magnetic 'bar' running through the centre of it (causing the compass to 'dip'), which contained north and south poles at its extremities. Although, these revolutionary findings were not confirmed beyond doubt for several hundred years, they were a vital discovery in beginning to truly comprehend the physics of the earth and even the wider universe beyond.

▸ INVISIBLE FORCES

Indeed, Gilbert went on to reason that magnetism played a part in holding the planets in their orbits. This established the concept of invisible forces and explained much of the behaviour of the universe, which Galileo and **Newton** would go on to exploit. He also correctly surmised that the earth's atmosphere was not very deep at all and the vast majority of space between planets was a vacuum. Through further observations involving experiments with amber, which was known to cause static electricity, he suggested that there might be some kind of link between electricity and magnetism, a theory which equally was not conclusively proven for several centuries.

FURTHER ACHIEVEMENTS

As well as his insistence upon modern methodology in scientific practice, Gilbert introduced a number of terms into the English language including: magnetic pole, electric force and electric attraction. A term of magnetomotive force, the gilbert, is named after him, and it was he who first popularised the term electricity.

Gilbert also disproved many commonly held beliefs about magnetism, including the belief that a diamond can magnetise iron.

As a further contribution to the study of magnets and magnetism, he proved that the earth acts as a bar magnet with magnetic poles.

FRANCIS BACON

1561–1626

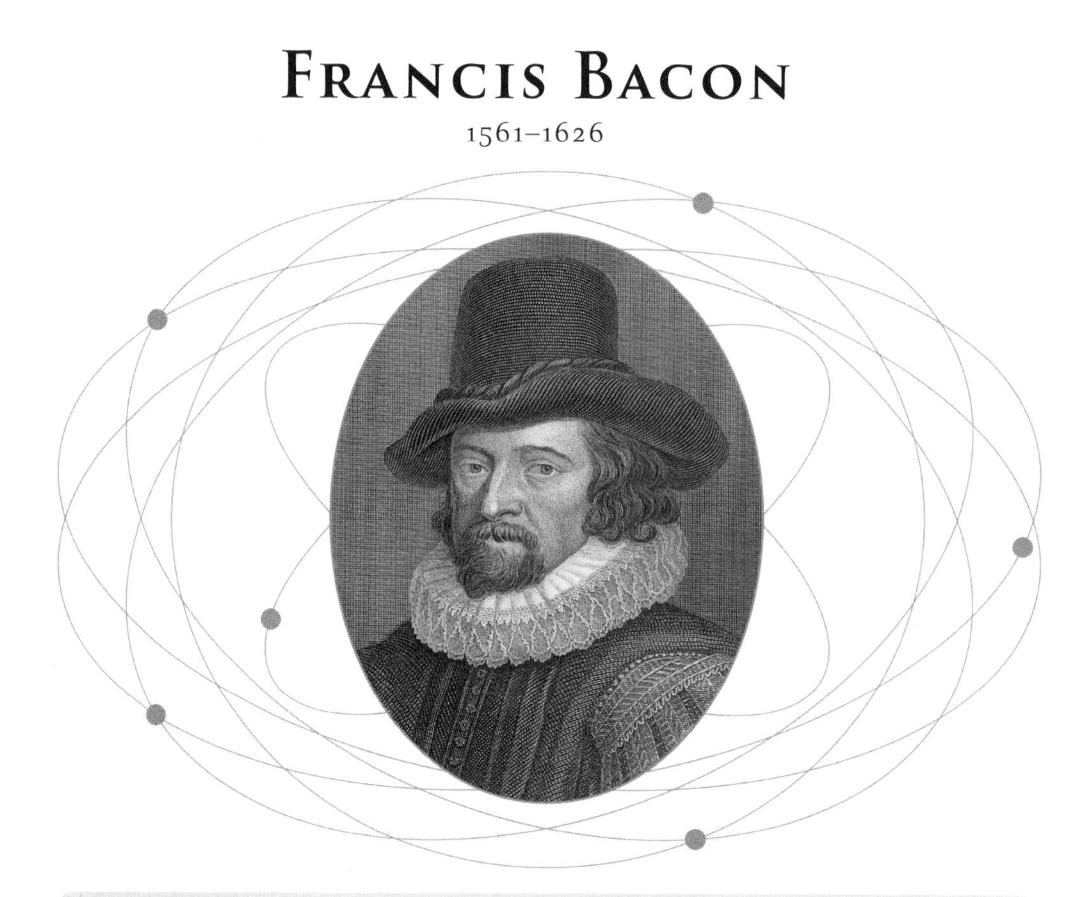

CHRONOLOGY • **1561** Bacon born in London • **1594** Bacon receives MA from Cambridge • **1605** Accession of James I of Scotland to the throne of England • **1607** James I appoints Bacon King's Counsel • **1613** Appointed Attorney General • **1620** His *Novum Organum* insists that the correct method for science is experimentation • **1621** Bacon's legal career ends in scandal and ignominy

If William **Gilbert** gently suggested a new rational approach to experimental science in his famous book *De magnete*, then Francis Bacon stood on the rooftops and bellowed out its arrival to the world. Although not strictly a scientist himself, Bacon was responsible for crystallising the methodology behind the scientific revolution which would go on to change the world so drastically. Ironically, he knew little about Gilbert's book, but nonetheless implored academics to introduce a systematic approach to their studies, following Gilbert.

AN EARLY START

Bacon proved his academic capability from a very young age, entering Trinity College, Cambridge, at only twelve years old. At the age of twenty-three he became Member of Parliament for Dorset by which stage he had also qualified as a barrister. He would go on to have a prestigious career in the Royal Court of James I, rising to Lord Chancellor of England. Having risen to such heights, however, his fall from grace in 1621 was all the more great when he was convicted of taking a bribe while a judge,

Bacon stood on the rooftops and bellowed out the arrival of the rational method

and was stripped of his office and power. Both during and after his legal career, Bacon undertook academic studies in philosophy and science. In the history of science, he is notable for two texts in particular. *The Advancement of Learning* in 1605 signalled his dissatisfaction with the limits of and approaches to knowledge to date, and foresaw a future where the work of the ancient masters would be far surpassed. The *Novum Organum*, of 1620, advanced this sentiment, boldly challenging the **Aristotelian** view and approach to the world. Aristotle himself had written a text called *Organum* or *Logical Works*, and Bacon's 'new' approach to the work of his predecessor suggested in the title alone an alternative direction to scientific study.

▸ A CRITICISM OF METHOD

In the text itself, Bacon strongly criticised Aristotle's 'deductive' method of science, involving formulating abstract ideas and 'logically' building upon them step-by-step to find 'truths', without thorough consideration of whether the theoretical foundation in itself was ever valid. Alternatively then, Bacon argued for 'inductive' reason where the only 'certain' statements that should ever be made were based on repeated observation and proof collected from the natural world. Rather than rely on superstition or accept unquestioningly the flawed solutions of the ancient academics as had largely been the case for two thousand years, Bacon implored scientists to only draw conclusions from exactly what was 'known'. Gathering the data from which to induce these certainties involved a rational, systematic, scientific approach using Bacon's 'Tables of Comparative Instances', which basically provided a methodology for eliminating irrelevancies when examining any given question, and pinpointing the proven facts.

▸ AN EARLY FINISH

Novum Organum was only one part of what Bacon had envisaged as a six-part work outlining his new approach, to be called *Instauratio*. This was never finished, owing to the premature death of the scholar from bronchitis, but his plan indicated that as well as outlining the new method of scientific enquiry, he had hoped to reclassify the sciences into new divisions, assemble a collection of scientific 'facts', provide proven examples using his new method and prepare and espouse a new philosophy based on the practical success of his approach.

Further Achievements

Even without the planned additional material that Bacon never got around to writing, his work went on to profoundly influence the science of the future. In particular, the science of the seventeenth century, as the scientific revolution found a framework within which to operate. In many instances it continues to do so today.

Bacon cautioned those trying to practise his new method, urging them to repudiate four kinds of intellectual idol:

- *Perceptual illusions –' idols of the tribe'*
- *Personal biases – 'idols of the cave'*
- *Linguistic confusions – 'idols of the market-place'*
- *Dogmatic philosophical systems – 'idols of the theatre'*

Only once we have abandoned metaphysical baggage, said Bacon, can we approach the scientific method in the correct manner.

GALILEO GALILEI

1564–1642

CHRONOLOGY • **1564** Galileo Born in Pisa, Italy • **1581** Studies medicine at Pisa, but fails to complete the course • **1583** Observes swinging lamps in Pisa Cathedral and notes that the time for the swing is the same no matter what the amplitude • **1586** Invents a hydrostatic balance for the determination of relative densities • **1610** Designs and constructs a refracting telescope. Publishes observations in *Sidereus nuncius* (*Starry Messenger*) • **1632** *Dialogue Concerning the Two Chief World Systems* published. This leads to Galileo being forced by the Church to recant his Copernican views. He is put under house arrest

In both his life and through the imprisonment which he was forced to endure in the years leading up to his death, Galileo more than any other figure personified the optimism and struggle of the scientific revolution. He was responsible for a series of discoveries which would change our understanding of the world, while struggling against a society dominated by religious dogma, bent on suppressing his radical ideas.

▸ A MATHEMATICIAN

Although he was initially encouraged to study medicine, Galileo's passion was mathematics, and it was his belief in this subject which underpinned all of his work. One of his most significant contributions was not least his application of mathematics to the science of mechanics, forging the modern approach to experimental and mathematical physics. He would take a problem, break it down into a series

'Nevertheless, it turns!'

GALILEO, AFTER BEING FORCED TO RENOUNCE HIS HELIOCENTRIC VIEW OF THE EARTH

of simple parts, experiment on those parts, and then analyse the results until he could describe them in a series of mathematical expressions.

One of the areas in which Galileo had most success with this method was in explaining the rules of motion. In particular, the Italian rejected many of the **Aristotelian** explanations of physics which had largely endured to his day. One example was Aristotle's view that heavy objects fall towards earth faster than light ones. Through repeated experiments rolling different weighted balls down a slope (and, legend has it, dropping them from the top of the leaning tower of Pisa!), he found that they actually fell at the same rate. This led to his uniform theory of acceleration for falling bodies, which contended that in a vacuum all objects would accelerate at exactly the same rate towards earth, later proved to be true. Galileo also contradicted Aristotle in another area of motion by contending that a thrown stone had two forces acting upon it at the same time; one which we now know as 'momentum' pushing it horizontally, and another pushing downwards upon it, which we now know as 'gravity'. Galileo's work in these areas would prove vital to Isaac **Newton's** later discoveries.

▸ THE PENDULUM

Galileo's earliest work involved the study of the pendulum, inspired by observing a lamp swinging in Pisa cathedral. Following further experiments, he concluded that a pendulum would take the same time to swing back and forth regardless of the amplitude of the swing. This would prove vital in the development of the pendulum clock, which Galileo designed and was constructed after his death by his son.

▸ THROUGH THE TELESCOPE

One of the inventions Galileo is often mistakenly credited with today is the invention of the telescope. This is not true; there had been a numerous early prototypes mostly developed in Holland before him, and a Dutch optician called Hans Lippershey applied for a patent on his version in 1608. Galileo did, however, develop his own far superior astronomical telescope from just a description of Lippershey's invention, and quickly employed it to make numerous discoveries. A strong supporter of the **Copernican** view of planetary motion, Galileo's initial findings published in the *Sidereal Messenger* (1610) provided the first real physical evidence to back up this interpretation. As well as discovering craters and mountains in the moon, sunspots and the moonlike phases of Venus for the first time, he also noted faint, distant stars which supported the Copernican view of a much larger universe than **Ptolemy** had ever considered. More importantly, he discovered Jupiter had four moons which rotated around it, directly contradicting the still commonly held view, including that of the Church, that all celestial bodies orbited earth, 'the centre of the universe.'

GALILEO AND COPERNICUS

Galileo's Dialogue Concerning the Two Chief World Systems – Ptolemaic and Copernican *in which the Ptolemaic view was ridiculed, attracted the attention of the Catholic Inquisition when it was published in 1632. Threatened with torture, Galileo renounced the Copernican System. His work was placed on the banned 'Index' by the Church where it remained until 1835, and he was subject to house arrest for life. But the tide of scientific revolution Galileo had helped instigate proved too powerful to hold back.*

JOHANNES KEPLER

1571–1630

CHRONOLOGY • **1600** Kepler works in Prague with Tycho Brahe, the imperial mathematician • **1601** On Brahe's death, Kepler inherits his position • **1609** Publishes *Astronomia Nova*, containing two of his laws of planetary motion • **1611** Publishes *Dioptrics*, which has been called the first work of geometrical optics • **1619** Publishes *Harmonices mundi* (*Harmonics of the World*). It contains his third law of planetary motion

The German mathematician, Johannes Kepler, while probably not as well remembered as **Copernicus**, was one of the key reasons why the Polish astronomer's theories finally became widely accepted. What Copernicus had started in suggesting a heliocentric model of the solar system, i.e. that the planets actually rotated around the sun, Kepler finished in providing the arithmetical and observational proof to support such a thesis.

TYCHO BRAHE

Kepler himself ultimately owed much of his success to the most famous astronomer of the second half of the sixteenth century, Tycho Brahe, a Dane. Brahe had become aware of Kepler's potential after reading a paper he had written while at university in Tübingen. After Kepler had been forced to leave his post as a mathematics lecturer at Graz in Austria, Brahe invited him to become his assistant in Prague under the patronage of Rudolph II, ruler of the Holy Roman Empire. Kepler took up the post in

Kepler one day 'saw a new light break' as he realised the orbits of the planets were ellipses

1600 and formulated a productive, if somewhat stormy, relationship with Brahe. One of the reasons the two argued was because Brahe rejected outright the Copernican view of the universe, which Kepler held in such high regard. The Dane had formulated his own alternative and rather obscure view on the rotation of the planets, which never caught on. Although history would subsequently prove Brahe to be wrong, his importance to Kepler, and astronomy in general, was that he was a brilliant observer of the skies and kept excellent records. When Brahe died in 1601, Kepler not only inherited his position of imperial mathematician in Rudolph's court but, crucially, his astronomical notes.

▸ THE ORBIT OF MARS

Using Brahe's records from the previous twenty years, Kepler set about trying to calculate and explain the orbit of Mars. Unfortunately, because he shared Copernicus's view that the planets orbited in perfect circles, the German struggled for the next eight years to produce a satisfactory conclusion. One day, he 'awoke from sleep and saw a new light break' as suddenly he realised that the planets did not rotate in perfect circles at all. They orbited around an ellipse, that is, a (flattened) circle with two 'centres' very close together. At a stroke, this would provide the simple mathematical explanation which had eluded Copernicus and **Ptolemy** when trying to predict the movement of the planets.

▸ KEPLER'S LAWS

In 1609, Kepler published his findings in *Astronomia Nova* or *New Astronomy*, which crystallised two 'laws' that would have a vital influence on our understanding of the universe. In a later book of 1619, *Harmonices Mundi* or *Harmonies of the World*, he added another important rule. These three together made up 'Kepler's Laws of Planetary Motion'. The first formalises his earlier discovery that the planets rotate in elliptical orbits with the sun at one of the centres, or focus points. The second states that all planets 'sweep' or cover equal areas in equal amounts of time regardless of which location of their orbit they are in. This is important because, as the sun is only one of two centres in a planet's orbit, a planet is nearer to the sun at some times than at others, yet it still 'sweeps out' the same area. What this means is that a planet must speed up when it is nearer the sun and slow down when it is further away. Kepler's third law finds that the 'period' (the time it takes to complete one full rotation – a year for the earth for instance) of a planet squared is the same as the distance from the planet to the sun cubed (in astronomical units). This allows distances of planets to be worked out from observing their cycles alone.

As well as providing the credible solution to predicting planetary motion that had previously proved so difficult, Kepler's findings would later act as the stimulus for questions that would lead to Isaac **Newton's** theory of gravity.

Further Achievements

Kepler's last major work was Tabulae Rudolphinae, Ruldolphine Tables *(1627) which were a painstakingly developed series of tables widely used in the next century to help calculate the positions of planets. He also made important discoveries in the field of optics, proposing the ray theory of light.*

He published a work of science fiction, A Dream, or Astronomy of the Moon *in 1634*

At the time only two new stars visible to the naked eye had been discovered since antiquity. The second was observed by Kepler in 1604.

WILLIAM HARVEY

1578–1657

CHRONOLOGY • **1609** Harvey appointed as physician at St Bartholomew's Hospital, London • **1618** Physician to James I • **1628** Publishes *On the Motion of the Heart and Blood in Animals* • **1651** Publishes *Essays on the Generation of Animals* • **1661** Marcello Malpighi uses his microscope to prove Harvey's assumptions regarding anastomoses

If Johannes **Kepler** thrust astronomy into the modern world by 'completing' the work of Nicolaus **Copernicus** – who himself had confronted that of **Ptolemy** – then William Harvey was surely his anatomical equivalent. What **Galen** had begun and **Vesalius** had challenged, Harvey credibly launched into the modern arena with perhaps the most significant theory in his field of biology, before or since. What he postulated and convincingly proved was that blood circulated in the body via the heart – itself little more than a biological pump.

▸ A NEW THEORY

Galen had concluded that blood was made in the liver from food which acted as a kind of fuel which the body used up, thereby requiring more food to keep a constant supply. Vesalius, for all his corrections of Galen's work, added little to this theory. So it was left to the Englishman William Harvey, physician to King's James I and later Charles I, to prove his theory of circulation through rigorous and repeated experimentation on the 'royal' stock of animals over two decades. In the first instance, he had believed the heart could simply not produce the quantities of blood

To Harvey's mind, the blood was not used up, as Galen believed, but recycled around the body

required to support Galen's 'refuelling' theory. To Harvey's mind then, the only sound alternative was that blood was not used up but was recycled around the body. His dissections led him to correctly conclude that the arteries took blood from the heart to the extremities of the body, able to do so because of the heart's pump-like action. The veins, with their series of one-way valves, brought the blood back to the heart again. This rejected Galen's accepted explanation of how the body functioned.

Harvey published his findings in the 720-page *Exercitatio Anatomica de Motu Cordis et Sanguinis in Animalibus* or *Anatomical Exercise on the Motion of the Heart and Blood in Animals* at the Frankfurt Book Fair in 1628. He had, however, been lecturing on his theories of circulation since as early as 1616 but had taken a long time to publish his work. Rather like Copernicus, he was something of a perfectionist, partially explaining why he delayed for so long, but equally he feared a backlash against his theories for challenging Galen head on.

▸ DIVIDED OPINION

And rightly so. Although he initially received support from some academics, an equal number reacted with outrage and ridiculed his ideas. One of the areas where Harvey's work was weakest, which the author himself acknowledged but had been unable to solve, was that he struggled to offer a proven explanation for how the blood moved from the arteries to the veins. He speculated that the exchange took place through vessels too small for the human eye to see, which was confirmed shortly after his death with the discovery of capillaries by Marcello Malpighi with the recently-invented microscope. Harvey, though, had had no such luxury and even lost patients at his London practice as a result of the criticism directed towards him. By the time of his death, however, he had answered most of his detractors' objections and his conclusions became increasingly accepted, even before Malpighi's final proof.

▸ REPRODUCTION

In 1651, Harvey published another notable work, this time in the area of reproduction. *Exercitationes de Generatione Animalium* or *Essays on the Generation of Animals* included conjecture which rejected the 'spontaneous generation' theory of reproduction in mammals which had hitherto persisted. Instead, he suggested the only plausible explanation was that female mammals carried eggs which were somehow spurred into reproduction through interaction with the male's semen. While he did not foresee the egg itself being fertilised in the sense we now understand reproduction, his belief that the egg was at the root of all life was convincing, and gained acceptance long before the observational proof some two centuries later.

A Modern Methodology

Harvey's significance comes not only from his discoveries, but also his methodology. As William **Gilbert** *had begun in physics, and Francis* **Bacon** *had subsequently implored in all aspects of life, Harvey was the first to take a rational, modern, scientific approach to his observations in biology, sewing the seeds for a methodolgy that we can accept today. He cast aside the prejudices of his predecessors and only 'induced' conclusions based on the results of experiments which he could repeat identically again and again. It was a model which gained popularity following Harvey's success, and continues to be employed.*

JOHANN VAN HELMONT

1579–1644

CHRONOLOGY • **1579** Van Helmont born to a wealthy, noble Brussels family • **c. 1621** Put under house arrest by the Church • **1648** *Ortus Medicinae* (*Origin of Medicine*), his collected papers, published by his son

While the sciences of physics, astronomy and anatomical biology were taking increasingly large strides towards the modern era, chemistry, as we know the subject today, was still lagging somewhat behind. Whether this was down to chance or because the other sciences had as their subjects more obviously observable phenomena, readily accessible for scrutiny, such as stars, motion or the human body, areas of chemistry were not immediately subject to the rational approach typical of the rest of the scientific revolution. Instead they had to wait for the posthumous publications of a slightly eccentric Flemish chemist, Jan Baptista van Helmont, for the transformation to begin.

▸ A KEEN OBSERVER

Born of a wealthy family, van Helmont had the luxury of being able to turn down paid employment and instead keep a rather solitary working life locked up in his own private laboratory conducting experiments. Van Helmont retained a belief in some areas of superstition such as the healing of wounds by treating the weapon that made them, although he insisted, contrary to the teaching of the Church, that this was an entirely

Van Helmont introduced the word 'gas', from a Flemish pronunciation of the Greek 'chaos'

natural phenomenon, and in no way miraculous. This attitude, predictably, soon led him into conflict with the Church, with the result that he spent much of his life under house arrest.

▸ THE PHILOSOPHER'S STONE

Another belief from the world of alchemy that Van Helmont retained was the existence of the Philosopher's Stone. Indeed, faith in this famous jewel's reality was one of the driving forces behind the 'science' of alchemy. To the alchemists the Stone was the elixir of life in solid, material form, and was capable, so the story went, of transmuting base metals into gold. Although imaginary, the pursuit of this fantastic gem has resulted in many important chemical discoveries.

Van Helmont still achieved enough to be considered by some to be the 'father' of modern biochemistry. In particular, he was the first to employ a calculated approach to his subject, most notably through the application of scientific measurement to the results of his experiments. He made use of a chemical balance and meticulously monitored his observations.

▸ WATER, WATER EVERYWHERE

For example, in one famous experiment, van Helmont planted and measured the growth of a willow tree over five years, during which time it gained 164 pounds. The scientist did this to 'prove' his belief that almost all matter was chiefly made up of water. While he rejected the **Aristotelian** belief in the 'four elements' (plus aether), van Helmont followed another early Greek scientist, Thales, in the fact that he was convinced of the dominance of water, even more so after his experiment. For five years the tree had only been fed on rainwater and the limited supply of soil in which it had been planted. Van Helmont found that the decrease in the soil's mass was only a few ounces, leading him to conclude that the tree was almost entirely composed of the water it had consumed. What he had failed to realise, ironically, given that he was the discoverer of the gas, was that about fifty per cent of the increased weight came from carbon dioxide in the air.

▸ SPLITTING AIR

Not that van Helmont failed to notice the existence of the gas in his other experiments, however. Indeed, the chemist's most important discovery was that gases other than 'air' existed at all. He realised that different elements gave off different gases when heated and was able to identify four distinct ones. He named them gas carbonum, gas sylvester of two variants, and gas pingue. These are the gases we now know as carbon dioxide, carbon monoxide, nitrous oxide and methane. In fact, van Helmont introduced to the world the term 'gas' itself, from the Flemish pronunciation of the Greek word 'chaos'.

Further Achievements

As well as chemistry, Van Helmont undertook studies in nutrition, digestion and physiology, applying the same scientific methodology to much of his work. He was, however, unable to divorce his fascination with the mystical from the majority of his studies, thereby discrediting much of their value when viewed from a modern perspective. More importantly, he recognised the law of the 'indestructibility of matter', realising, for instance, that metals dissolved in acid were only 'concealed' and could be regained in equal amounts. Van Helmont's son finally published the scientist's collected works in 1648 under the title Ortus Medicinae *or* Origin of Medicine.

RENÉ DESCARTES

1596–1650

CHRONOLOGY • **1596** Descartes born in La Haye, France • **1616** Graduated in law from the University of Poitiers • **10 November 1619** Descartes first begins to ponder the principles that would form his later work • **1637** *Discours de la Méthode* (*Discourse on Method*) published • **1637** *La Geométrie* (*Geometry*), published as an appendix to *Discours de la Méthode* • **1641** *Meditations on First Philosophy* published

René Descartes has been described as the first truly 'modern' mathematician and philosopher. Certainly his systematic, logical approach to knowledge was revolutionary, dominating philosophy for the next three centuries. Even more importantly, from the perspective of this book at least, it led to a new breakthrough which would greatly impact the future of mathematics and science.

Descartes initially gained a degree in law and spent a number of years in the military before eventually settling in Holland in 1628 where he composed all of his great works. In 1649 he accepted a post as personal tutor to Queen Christina of Sweden. A lifelong late riser and lover of a warm bed – where Descartes claimed to have undertaken his most profound thinking – he succumbed to the harsh Swedish weather. Within months he had contracted pneumonia and died.

▸ A REVELATION OF PHILOSOPHY

Three decades earlier, on the night of 10 November 1619, while campaigning with the

'Give me matter and motion and I will construct the universe'

army on the Danube, Descartes' life had changed forever when his influential journey began. He later claimed to have had a number of dreams on that date which formulated the principles behind his later work. In particular, it left him certain that he should pursue the theory that all knowledge could be gathered in a single, complete science and set about putting in place a system of thought by which this could be achieved. In turn, this left him to speculate on the source and truth of all existing knowledge. He began rejecting much of what was commonly accepted and vowed only to recognize facts which could be intuitively taken to be true beyond any doubt.

The full articulation of these processes came in Descartes' 1641 work *Meditations on First Philosophy*. The book is centred around his famous maxim 'Cogito, ergo sum' or 'I think, therefore I am,' from which he pursued all 'certainties' via a method of systematic, detailed mental analysis. This ultimately led him to a very detached, mechanistic interpretation of the natural world, reinforced in his 1644 metaphysical text the *Principia Philosophiae* or *Principles of Philosophy*, in which he attempted to explain the universe according to the single system of logical, mechanical laws he had earlier envisaged and, although largely inaccurate, would have an important influence even after **Newton's** more convincing explanations later in the century. He also regarded the human body as subject to the same mechanical laws as all matter, distinguished only by the mind which operated as a distinct, separate entity.

MATHEMATICAL CERTAINTIES

Descartes passionately believed in the logical certainty of mathematics and felt the subject could be applied to give a superior interpretation of the universe. It is through this reasoning that his greatest legacy to mathematics and science came. In his 1637 appendix to the *Discourse*, entitled *La Geométrie*, Descartes sought to describe the application of mathematics to the plotting of a single point in space. This led him to the invention of what are now known as Cartesian Coordinates, the ability to plot a position according to x and y (that is, perpendicular) axes (and in a 3D environment by adding in a third 'depth' axis). Moreover, this method allowed geometric expressions such as curves to be written for the first time as algebraic equations (using the x, y and other elements from the graph).

The bringing together of geometry and algebra was a significant breakthrough and could, in theory at least, predict the future course of any object in space, given enough initial knowledge of its physical properties and movement. It is from his mathematical interpretation of the cosmos that Descartes would later claim, 'Give me matter and motion and I will construct the universe.'

The 'Cogito'

Perhaps the most famous of philosophical maxims, 'Cogito, ergo sum', is best translated as 'I am thinking, therefore I am'. It was the result of a form of a thought experiment by Descartes, in which he resolved to cast doubt on any and all of his beliefs, in order to discover which he was logically justified in holding. He argued that although all his experience could be the product of deception by an evil demon (a more modern version has a brain in a vat, fed information by an evil scientist, an idea used in the film The Matrix*), the demon could not deceive him if he did not exist. That he can doubt his existence proves that he in fact exists.*

BLAISE PASCAL

1623–1662

CHRONOLOGY • **1642–44** Pascal invents and produces a calculating device to aid his father; in effect this is the first digital calculator • **1647** on the summit of the Puy de Dome he proves that air pressure decreases at higher altitudes • **1655** abandons his studies and enters the Jansenist retreat at Port-Royal • **1670** Pascal's Wager

Perhaps one of the lesser noted benefits of being a child prodigy is that if you die at an early age, you have still had sufficient time to fulfil your potential! Blaise Pascal, a Frenchman who passed away at just thirty-nine, was one such example. Although his time on earth was unfortunately cut short by poor health, and his contributions to mathematics and science severely limited by his abandonment of his studies in favour of religious devotion in 1655, he still had a significant influence within both fields of endeavour.

MEASURING PRESSURE

During his twenties Pascal spent a large amount of time undertaking experiments in the field of physics. The most important of these involved air pressure. An Italian scientist, Evangelista Torricelli (1608–47), had argued that air pressure would decrease at higher altitudes. Pascal set out to prove this by using a mercury barometer. He took initial measurements in Paris and then, at the 1200m-high Puy de Dome in 1646, accompanied by his brother-in-law, he confirmed in no uncertain terms that Torricelli's speculation was true.

By the age of twenty-one, Pascal had finished what was effectively the first pocket calculator

▸ PASCAL'S LAW

More significantly, though, his studies in this area led him to develop Pascal's Principle or Law, which states that pressure applied to liquid in an enclosed space distributes equally in all directions. This became the basic principle from which all hydraulic systems derived, such as those involved in the manufacture of car brakes, as well as explaining how small devices such as the car jack are able to raise a vehicle. This is because the small force created by moving the jacking handle in a sizeable sweep equates to a large amount of pressure sufficient to move the jack head a few centimetres. Applying the lessons of his studies in a practical way, Pascal went on to invent the syringe and, in 1650, the hydraulic press.

▸ CHILD PRODIGY

In spite of these developments, however, Pascal is probably better remembered for his work in the area of mathematics. It was here that he showed his genius from an early age. For example, having independently discovered a number of Euclid's theorems for himself by the age of eleven, he went on to master *The Elements,* the great mathematician's definitive text, by twelve. When he was sixteen he published mathematical papers which his older contemporary **Descartes** at first refused to believe could have been written by one so young. In 1642, still only nineteen, Pascal began work on inventing a mechanical calculating machine which could add and subtract. He had finished what was effectively the first digital calculator by 1644 and presented it to his father to help him in his business affairs.

▸ THEORY OF PROBABILITY

It was not until later in his short life, around 1654, that Pascal jointly made the mathematical discovery which would have the most impact on future generations. It had begun with a request by an obsessive gambler, the Chevalier de Méré, for assistance in calculating the chance of success in the games he played. Together with Pierre de Fermat, another French mathematician, Pascal developed the theory of probabilities, using his now famous Pascal's Triangle, in the process. As well as its obvious impact upon all parts of the gambling industry, the importance of understanding probability has had subsequent application in areas stretching from statistics to theoretical physics.

The SI unit of pressure – the pascal – and the computer language, Pascal (named in honour of his contribution to computing through his invention of the early calculator), are named after him in recognition of two of his main areas of scientific success.

Seven of the calculating devices that he produced in 1649 survive to this day.

PASCAL'S WAGER

Like many of his contemporaries, Pascal did not separate his science from philosophy, and in his book Pensees, he applies his mathematical probability theory to the perennial philosophical problem of the existence of God. In the absence of evidence for or against God's existence, says Pascal, the wise man will choose to believe, since if he is correct he will gain his reward, and if he is incorrect he stands to lose nothing, an interesting, if somewhat cynical argument.

ROBERT BOYLE

1627–1691

CHRONOLOGY • **1638** Boyle takes up studies in Geneva, after four years at Eton College • **1644** Retires to his estate in Dorset, to pursue his own studies • **1654** moves to Oxford, where he meets and befriends Robert Hooke • **1659** Boyle and Hooke carry out experiments with vaccuum • **1662** Formulates his famous chemical dictum, Boyle's Law

Born in the modern day Republic of Ireland as the fourteenth child of the richest man in what was then part of Great Britain, Boyle enjoyed all the privileges of an aristocratic education. Schooled at Eton and then privately, he continued his studies as he undertook a long European tour from 1639 to 1644. Eventually he returned to an inherited estate in Dorset, England, largely avoiding the worst excesses of the Civil War. Here he began his scientific studies. In 1656 he moved to Oxford where, with the philosopher John Locke, and the architect Christopher Wren, he formed the Experimental Philosophy Club. He also met Robert **Hooke**, who became his assistant and the two formed a productive partnership. It was together with Hooke that Boyle began making the discoveries for which he became famous.

BOYLE'S LAW

Chief amongst these was the expression of what is now known as Boyle's Law (also independently discovered by the French scientist Edme Mariotte) which established a direct relationship between air pressure and volumes of gas. By using

Boyle's Law: pressure is inversely proportional to volume at constant temperature

mercury to trap some air in the short end of a 'J' shaped test tube, Boyle was able to observe the effect on its volume by adding more mercury. What he found was this: if he doubled the mass of mercury (in effect, doubling the pressure), the volume of the air in the end halved; if he tripled it, the volume of air reduced to a third, and so on. As long as the mass and temperature of the gas were constant, his law concluded that the pressure and volume were inversely proportional.

▸ THE VACCUUM PUMP

This experiment was the culmination of a series of other tests involving air and its effects. They had begun shortly after Boyle had moved to Oxford and progressed rapidly when Robert **Hooke** had constructed an air pump upon his request. The pump was able to create the best man-made vacuum to date and through experiments involving bells, animals and candles, Boyle was able to draw a number of important conclusions. He found that sound could not travel through a vacuum and required air in order to do so. Air was required for respiration and combustion, and not all of the air was used up during breathing and burning processes. In addition, he proved **Galileo's** proposal that all matter fell at equal speed in a vacuum.

▸ THE SCEPTICAL CHEMIST

In 1661, Boyle published *The Sceptical Chemist* which criticised the **Aristotelian** view of a universe composed of only four elements (earth, water, air and fire), plus aether in wider space. The text helped pave the way to our current view of the elements. Although he did not describe elements exactly as we understand them today, he believed that matter consisted at root of 'primitive and simple, or perfectly unmingled bodies' which could combine with other elements to form an infinite number of compounds. This was an extension of his support for early atomic theory, believing in what he described as tiny 'corpuscles'. In spite of an interpretation which does not entirely correspond with the modern view, his importance was in promoting an area of thought which would influence the later breakthroughs of Antoine **Lavoisier** (1743–93) and Joseph **Priestley** (1733–1804) in the development of theories related to chemical elements.

The Influence of Robert Boyle

Robert Boyle made a number of contributions to the history of science, but perhaps most significant is his claim to being the man responsible for the establishment of chemistry as a distinct scientific subject in its own right. Like his idol Francis **Bacon**, *he experimented relentlessly, accepting nothing to be true unless he had firm empirical grounds from which to draw his conclusions.*

Boyle's other important legacies were the creation of flame tests in the detection of metals, as well as tests for identifying acidity and alkalinity.

Boyle was also a founder member of the Royal Society, the longest running scientific society in the world. But it was his insistence on publishing chemical theories supported by accurate experimental evidence – including, for the first time, details of apparatus and methods used as well as failed experiments – which would have the most impact upon modern chemistry.

CHRISTIAAN HUYGENS

1629–1695

CHRONOLOGY • **1655** Huygens discovers Titan, Saturn's largest moon • **1657** Clock constructed according to his groundbreaking design • **1658** *Horologium* (*The Clock*) published • **1673** *Horologium Oscillatorium* (*The Clock Pendulum*) published • **1690** *Traité de la Lumière* (*Treatise on Light*) published

The Dutchman Christiaan Huygens is widely considered to be the second most important physical scientist of the seventeenth century. Son of the distinguished diplomat and scholar Constantjin Huygens, Christiaan was acquainted from an early age with notables such as René **Descartes**, a friend of the family.

Unfortunately for him, one of his key propositions on the behaviour of light contrasted directly with the first and most important, originally proposed by Isaac **Newton**. As a result Huygens' theory was largely ignored for over a century. His other achievements in time measurement did impact immediately, however, helping his science to progress in a way it otherwise would have struggled to do.

▸ AGAINST NEWTON

Isaac Newton articulated a particle theory of light, believing it to be made up of 'corpuscles'. This was a view he summarised in his 1704 text *Opticks* but had held for the preceding decades. He vigorously challenged anyone who tried to contradict this opinion, as both Leibniz and Robert **Hooke** (1635–1703) – who shared similar

His key idea on the behaviour of light went against Newton: it was ignored for a century

views to the Dutchman – were to find out. Huygens believed light actually behaved in a wave-like fashion, in a method which became known as the 'Huygens Construction,' which he outlined in his 1690 work *Treatise on Light* (although he had first expressed it in 1678). This opinion much more satisfactorily explained the way light reflected and refracted, and correctly anticipated that in a denser medium light would travel more slowly. Although the modern interpretation is that light can behave in both a particle and wave-like fashion depending on the situation, Huygens's view, when rediscovered and championed by Englishman Thomas Young (1773–1829), in the early nineteenth century, would eventually become the more commonly accepted version. Such was the dominance of Isaac Newton, however, that Huygens's theories were totally ignored for the whole of the eighteenth century, and they still faced fierce resistance in Young's time.

▸ THE PENDULUM CLOCK

Of much more immediate impact, though, were Huygens's breakthroughs in clock-making. Ever since the time of **Galileo** (1564–1642), scientists had been aware that a swinging pendulum could keep a regular beat and they had hoped to use this knowledge to create an accurate time-measuring device. They had been unsuccessful. Huygens realised that this was partly because a pendulum mimicking a circle's curve did not maintain a perfectly equal swing and in order to do this it actually needed to follow a 'cycloidal' arc. This discovery set him on the path to designing the first successful pendulum clock. He had it constructed in 1657 and announced his creation to the world in his 1658 book *Horologium* or *The Clock*. The invention was of monumental importance to the progress of physics, for without an accurate method of measuring time the progress of the subject over the following centuries would have been severely hampered.

Huygens backed up his practical findings with mathematical explanations describing a pendulum's swing in his 1673 work *Horologium Oscillatorium* or *The Clock Pendulum*. The text included a number of other dynamic explanations and anticipated the first of Newton's motion laws, the 'law of inertia', which states: an object moving in a straight line will continue to move in the same way indefinitely until it meets another force.

Huygens was an associate of Leibniz, whom he supported during his controversial bout with Isaac Newton over the law of gravity. Despite this, and despite Huygens' opinion that Newton's theory of gravity was incomplete without a mechanical explanation, as expressed in the *Principia*, Newton was a staunch admirer of the Dutchman.

ASTRONOMY AND LIGHT

As well as being an accomplished physicist, Christiian Huygens was also a keen astronomer and made some important contributions in this area. He devised a much-improved telescope and used it to make a number of findings, including the discovery of Saturn's biggest moon, Titan, in 1655. In addition, he observed and accurately explained Saturn's ring system.

Huygens' hypothesis that light is a wave was largely ignored at the time as it conflicted with Newton's theory which proposed that light had a particle structure. Both were in fact correct.

Huygens was one of the founding fathers of the French Academy of Sciences in 1666, and was granted a larger pension from that body than anyone else.

ANTON VAN LEEUWENHOEK

1632–1723

CHRONOLOGY • **1673** Van Leeuwenhoek begins correspondence with the Royal Society • **1674** First to observe protozoa • **1677** First to observe human spermatozoa • **1683** First to observe bacteria • **1684** First to observe red blood cells

Who said you had to be a full time scientist with money and an aristocratic background to make world-changing discoveries? Probably the vast majority of people who lived in the seventeenth century, when most scientists came either from the nobility – possessing the independent wealth to undertake research with no need for a job – or were funded by them through patronage. Not quite so for Anton van Leeuwenhoek, a humble Dutch draper, who, despite little formal education, went on to entertain kings and queens with his remarkable revelations.

HIS HOBBY

Born and remaining all his life in Delft in the Netherlands, van Leeuwenhoek became an apprentice linen-draper at the age of 16 and went on to open his own business in the town around 1654. In 1660 he took on a better-paid position in the town's law courts. This gave him greater means and more spare time to pursue the subject he would bring to impact on history: van Leeuwenhoek had developed a passion for microscopy, and by 1660 was devoting all the spare time he could get to producing lenses with a greater magnification than had ever been made before.

He discovered that when his faeces were 'a bit looser than usual', protozoa were observed

▸ THROUGH THE LOOKING GLASS

Van Leeuwenhoek kept secret his methods for producing the lenses during his entire ninety years. Even though his finest single, short focal length lenses could enlarge a specimen by up to three hundred times, it is believed he employed an additional technique, perhaps some form of illumination, to view the miniscule 'animalcules' he observed. Another major discovery was protozoa, effectively tiny one-celled plants, which he came across in water specimens in 1674. In more recent scientific studies protozoa would be linked to a number of tropical diseases including, most significantly, malaria and amoebic dysentery. In 1683, and perhaps even more importantly, van Leeuwenhoek observed bacteria for the first time. They were smaller than protozoa and were later linked to diseases such as cholera and tetanus, as well as their treatment.

▸ ANIMAL REPRODUCTION

In between these findings, van Leeuwenhoek discovered spermatozoa. The story goes that in 1677 his contemporary Stephen Hamm brought the microscopist a sample of human semen. Upon examination he discovered the short-lived sperm, reinforcing his opinion of their importance in reproduction by finding similar creatures in the semen of frogs, insects and other animals. He made detailed and exact observations of both fleas and ants, proving that the former were generated from eggs like any other insects, rather than arising spontaneously. He also showed that the eggs and pupae of ants were phenomena occurring at two entirely different stages. From this he correctly claimed their existence as evidence of his belief that the commonly held view of 'spontaneous generation' of insects and other small organisms was wrong.

Other important discoveries included the observation of red blood cells in 1684, providing further support for Marcello Malpighi's 1660 work on blood capillaries, which in itself had been so important in reinforcing William Harvey's speculation on the transfer of blood from the arteries to the veins. From ants to shellfish, van Leeuwenhoek also undertook a range of further studies, including observations of animal life .

▸ LANGUAGE BARRIER

In keeping with his lack of academic training, van Leeuwenhoek wrote up his findings in Dutch rather than the scholarly Latin, and published little directly. Instead he was introduced to the Royal Society of England via correspondence in 1673. For the rest of his life, van Leeuwenhoek would subsequently write regularly to the Royal Society in Dutch, outlining his latest discoveries, for he knew no English. In all, the society translated and printed some 375 entries in their publication *Philosophical Transactions* before van Leeuwenhoek's death.

Further Achievements

Van Leeuwenhoek's letters, and his subsequently assembled collected works, made the part-time scientist world famous and brought many noble visitors to Delft. Amongst those who came to see the animalcules first hand were James II of England and Peter the Great of Russia.

When van Leeuwenhoek died he left behind 247 complete microscopes, nine of which survive to this day. One of his microscopes had a resolution of 2 micrometers.

Examining his own faeces, he observed that 'when of ordinary thickness' there were no protozoa observed, but when 'a bit looser than ordinary' protozoa were observed.

ROBERT HOOKE

1635–1703

CHRONOLOGY • **1656** Hooke meets Robert Boyle at Oxford • **1659** Devises the pump, the most efficient creator of a vacuum at the time • **1662** Becomes the first Curator of Experiments to the Royal Society • **1665** *Micrographia* (*Small Drawings*) published • **1670** Discovers the law of elasticity

Perhaps one of the most 'underrated' scientists of the seventeenth century, Robert Hooke, an Englishman, experimented and made advances in a wide range of scientific areas. Yet because of this breadth of coverage, he seldom developed any of his concepts to their fullest extent. This explains why he rarely gained credit for them. Indeed, it is arguable that his role as a provider of ideas to others is his most important legacy.

▸ BOYLE'S ASSISTANT

The most obvious example of his contributions to others was the work he undertook with Robert Boyle at Oxford, where they met in 1656. Boyle, as the aristocrat, was clearly the dominant partner in the relationship, in social terms at least. Hooke, as his assistant, acted on Boyle's instructions, yet many of his creations were worthy inventions in their own right. The most obvious example is the air pump that he devised in 1659, the most efficient vacuum creator of its time. It enabled Boyle to go on to make many of his discoveries.

▸ PROVIDER OF IDEAS

Moreover, Boyle was responsible, albeit indirectly,

Hooke accused Newton of plagiarism, sparking a bitter lifelong feud between the two

for keeping Hooke in his position as jack of all sciences, master of none. The aristocrat had been influential in having Hooke elevated to the position of Curator of Experiments for the Royal Society in 1662. While the prestige of the role pleased Hooke, the job requirement of showing 'three or four considerable experiments' to the Society at each of its weekly meetings was almost certainly the factor that ensured Hooke would never have the time to develop any of his findings fully.

▸ A SOURCE OF IDEAS

Another scientist to whom Hooke felt he had provided source material was the Dutch physicist Christian **Huygens**. Huygens is credited with creating the influential wave theory of light, which he published in 1690. Yet as early as 1672, Hooke had explained his discovery of diffraction (the bending of light rays) by suggesting that light might behave in a wave-like fashion.

Isaac **Newton** vehemently argued against Hooke's theory of light, beginning a bitter feud which would continue for the rest of Hooke's life. Hooke also claimed to have discovered one of the most important theories credited to Newton, arguing that the latter had plagiarised his ideas from correspondence between the two during 1680. Certainly, Hooke's letters suggested some notion of universal gravitation and hinted at an understanding of what later became Newton's law of gravity. In spite of this, though, it is unquestionable that Newton's mathematical calculations and endeavours in proving the law give him a much stronger claim.

Hooke's countless experiments did, however, result in some other discoveries solely credited to him. He was, for example, the first to describe the universal law that all matter will expand upon heating. He is credited with the law of elasticity, discovered in 1670. Also known as Hooke's Law, it states that the strain, or change in size, placed upon a solid – when stretched – is directly proportional to the stress, or force, applied to it. Hooke was also the first person to use the word 'cell' in the scientific sense understood by us today, after observing the properties of cork under one of the powerful microscopes that he had developed. This word was used in his 1665 work *Micrographia* or *Small Drawings,* which also included many other advances such as Hooke's theory of combustion, as well as other discoveries of the microscope. These included crystalline structure of snow, and studies of fossils which led to the proposition that they were the remains of once living creatures. He suggested that whole species had lived and died out long before man, centuries before Charles Darwin came to the same conclusion.

Hooke also made discoveries in astronomy, locating Jupiter's Great Red Spot, and proposed that the huge planet rotated on its axis.

Further Achievements

Hooke's inventions were greatly influential. He either invented or significantly improved the reflecting telescope, compound microscope, dial barometer, anemometer, hygrometer, balance spring (for use in watches), quadrant, universal joint and iris diaphragm (later used in cameras). He also showed impressive vision, foreseeing the development of the steam engine and the telegraph system.

The inventions of the compound microscope and a balance spring for use in watches are also credited to him. Beyond this he was an accomplished architect who designed parts of London following the great fire of 1666.

SIR ISAAC NEWTON

1642–1727

CHRONOLOGY • **1670–71** Newton composes *Methodis Fluxionum* (*Method of Fluxions*), his main work on calculus which is not published until 1736 • **1672** New *Theory about Light and Colours* published. It is his first published work • **1687** *Philosophiae Naturalis Principia Mathematica* (*Mathematical Principles of Natural Philosophy*), known as the *Principia*, published • **1704** *Opticks* published

So many extensive books and articles have been written on the life and impact of Sir Isaac Newton over the last three centuries it is impossible to do his achievements justice in a short entry like this. He is quite simply one of the greatest scientists of all time.

A SLOW BEGINNING

His early years did not necessarily suggest, however, he would end up as such. Born and brought up in the quiet village of Woolsthorpe in Lincolnshire, England, and schooled in the nearby town of Grantham, he was not particularly noted for academic achievements as a child. Even on entry to Trinity College, Cambridge, he did not stand out until, ironically, the University was forced to close during 1665 and 1666 due to the high risk of plague. Newton returned to Woolsthorpe and began two years of remarkable contemplation on the laws of nature and mathematics which would transform the history of human knowledge. Although he published nothing during this period, he formulated and tested

'If I saw further than others, it is because I was standing on the shoulders of giants'

many of the scientific principles which would become the basis for his future achievements.

However, it would often be decades before he returned to his earlier discoveries. For example, his ideas on universal gravitation did not re-emerge until he began a controversial correspondence on the subject with Robert **Hooke** in around 1680. Furthermore, it was not until Edmond **Halley** challenged Newton in 1684 to find out how planets could have the elliptical orbits described by Johannes **Kepler**, and Newton replied he already knew, that he fully articulated his law of gravitation. Yet he had begun work on the subject back in the 1660s in Woolsthorpe after famously seeing an apple fall from a tree and wondering if the force which propelled it towards the earth could be applied elsewhere in the universe. After his declaration to Halley, Newton was forced to recalculate his proof, having lost his original jottings, and the result was published in Newton's most famous work *Philosophiae Naturalis Principia Mathematica* (1687). This law of gravitation proposed that all matter attracts other matter with a force related to the combination of their masses, but this attraction is weakened with distance, indeed, in inverse proportion to the square of their distances apart. This universal principle applied just as equally to the relationship between two small particles on earth as it did between the sun and the planets, and Newton was able to use it to explain Kepler's elliptical orbits.

▸ NEWTON'S LAWS OF MOTION

In the same work, Newton built on earlier observations made by **Galileo** and expressed three laws of motion which have been at the heart of modern physics ever since. The 'law of inertia', states that an object at rest or in motion in a straight line at a constant speed will carry on in the same state until it meets another force. The second stated that a force could change the motion of an object according to the product of its mass and it acceleration, vital in understanding dynamics. The third declares that the force or action with which an object meets another object is met by an equal force or reaction.

Aside from the wide ranging uses for the laws Newton outlined in the *Principia*, the important point is that all historical speculation of different mechanical principles for the earth from the rest of the cosmos were cast aside in favour of a single, universal system. It was clear that simple mathematical laws could explain a huge range of seemingly disconnected physical facts, providing science with the straightforward explanations it had been seeking since the time of the ancients. Newton's insistence on the use of mathematical expression of physical occurrences also underlined the standard for modern physics to follow.

FURTHER ACHIEVEMENTS

Newton achieved major breakthroughs in other areas too. His proof that white light was made up of all the colours of the spectrum was outlined in his 1672 work New Theory about Light and Colours. *In* Opticks *(1704), he also articulated his influential (if partially innacurate) particle or corpuscle theory of light. Another achievement significant to mathematics was his invention of the 'binomial theorem'.*

Newton had a practical side too, inventing the reflecting telescope in the 1660s. This new instrument bypassed the focusing problems caused by chromatic aberration in the refracting telescope of the type Galileo had created.

During his time as master of the Mint twenty-seven counterfeiters were executed.

EDMUND HALLEY

1656–1742

CHRONOLOGY • **1679** *Catalogue of the Southern Stars* published • **1682** Halley observes the comet that now bears his name • **1687** Encourages Newton to publish the *Principia* and finances the publication with his own personal funds • **1705** *Astronomiae Cometicae synopsis* (*A Synopsis of the Astronomy of Comets*) published • **1758** The comet returns as he predicted. From then on it is known as Halley's Comet.

It is frequently the case with the famous dead that they are remembered for a single discovery, action, theory or invention. Edmund Halley, the English astronomer and mathematician, is perhaps the greatest example of this phenomenon, renowned today for the discovery of the comet which bears his name. Yet more than almost any other scientist featuring in this book, his academic interests were broad and wide, with an impact far greater than the observation of a single cosmological boomerang.

HALLEY'S COMET

Not that his most celebrated achievement should be underestimated. Halley's Comet, as it became known for the first time when it reappeared sixteen years after the astronomer's death in 1758, exactly when he said it would, was the first comet whose return had been predicted. Halley had come to this conclusion after observing the body for himself in 1682. After further research he deduced that other comets, visible in 1531 and 1607, were so similar in characteristic to the one that he had seen that they were in fact probably a

Halley should be remembered for more than the observation of a single cosmic boomerang

single visitant simply returning at an interval of seventy-six years. The findings, which also calculated the orbits of twenty-three other comets, and published in 1705 as *A Synopsis of the Astronomy of Comets*, were seminal in the subsequent approach to the study of the subject.

▸ SOUTHERN SKIES

Halley's astronomical interests were not restricted to comets, however, and he contributed many other important studies. In 1718, he demonstrated that stars must have a 'proper' motion of their own by making comparisons between **Ptolemy's** catalogue and the position of the stars in his own time. He also observed the moon's full nineteen-year cycle and after doing so confirmed the theory of secular acceleration that he had originally predicted in 1695. In 1716, he proposed a way of calculating the Earth's distance from the Sun from transits of the planet Venus across the Sun's disc. One of his greatest celestial achievements was also one of his first. At the age of 20 he travelled on an East India Company ship to St Helena in order to map the stars in the southern hemisphere. He left Oxford University without completing his degree in order to do it. After two years of study on the remote island his publication of *The Catalogue of Southern Stars* in 1679 was not only the first accurate mapping of the southern skies, but also the first telescopically determined survey of the stars that he observed.

▸ NOT JUST ASTRONOMY

Closer to home, Halley was to gain credit in many other arenas. He is considered by some to be the founder of geophysics, beginning with his publication of a map in 1686 of the Earth's prevailing winds, and went on to prepare detailed maps of the tides and magnetic variation. He undertook work on the salinity and evaporation of lakes between 1687 and 1694, using his results to offer theories on the Earth's age. Halley developed a mathematical law which demonstrated the relationship of height to air pressure, allowing him to go on to make improvements in the design of the barometer. The population of the city of Breslau was the subject of the mortality tables he published in 1693, pioneering work in social statistics which later influenced the life insurance industry. The size of the atom, the optics of the rainbow and even the design of the diving bell did not escape the scrutiny of the man. Halley was definitely not just an astronomer.

As well as commanding the *Paramour*, a Royal Navy man of war, from 1698 to 1700, he was also a prolific mapmaker, showing prevailing winds, tides and magnetic variations in his cartography.

Halley's Comet will return to the skies in 2062.

HALLEY AND NEWTON

Despite his many achievements, it is arguable that perhaps the most important way Halley can be seen to have changed the world is in his friendship with Newton. He met him for the first time at Cambridge in 1684 and from then on would have an important role in the development and presentation of the theory of gravitation. He encouraged Newton to undertake his greatest work, the Principia, *in the first place. He went on to edit and proof-read the text, write the preface and perhaps most importantly of all, to finance its publication himself in 1687, when the Royal Society failed to do so.*

Had Edmund Halley not been born, his comet would still exist, albeit under a different name. Newton's Principia, *at least in the form the world knows it today, would almost certainly not.*

THOMAS NEWCOMEN

1663–1729

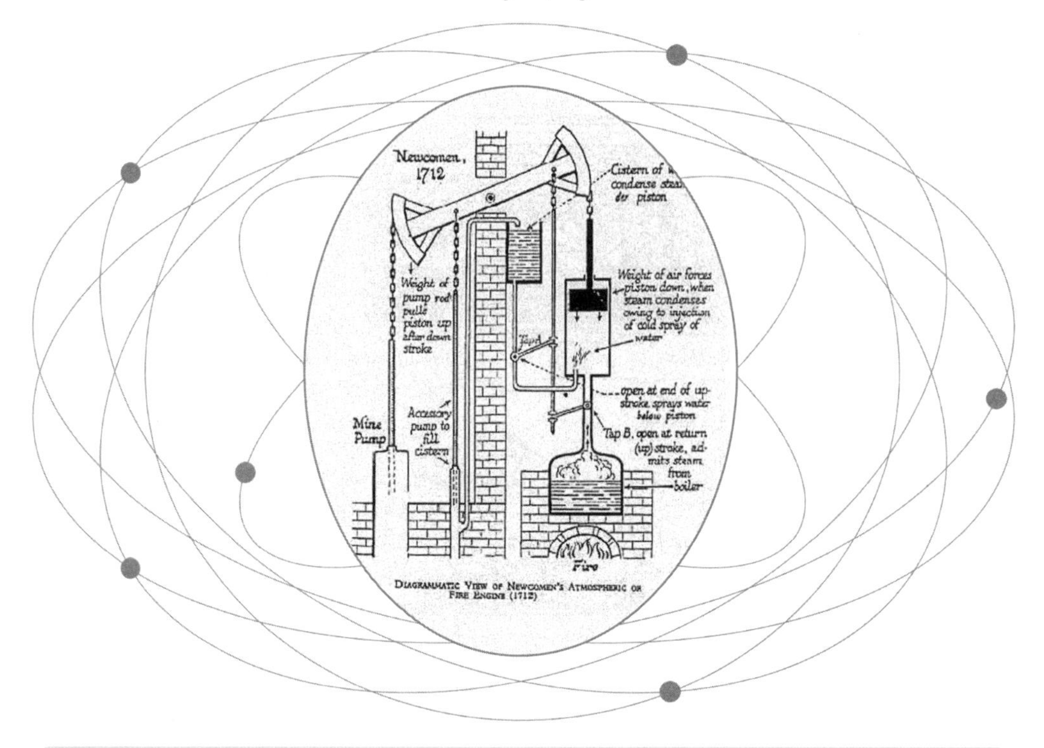

DIAGRAMMATIC VIEW OF NEWCOMEN'S ATMOSPHERIC OR FIRE ENGINE (1712)

CHRONOLOGY • 1663 Newcomen born in Dartmouth, Devon, England • **1705** Begins working to build a steam engine • **1712** Successfully builds and puts into use his improved steam engine • **1729** Dies in London

If the Industrial Revolution changed the world, then the man who made available the power source which facilitated the transformation must be heralded. This man was Thomas Newcomen, the English inventor of the world's first commercially successful low-pressure steam engine.

Of course, Newcomen had not set out to alter the development of society quite so dramatically. He had begun life as a simple ironmonger and smith in his hometown of Dartmouth in Devon, establishing a living in his chosen trade long before he began work on his ground-breaking invention. It was, however, his occupation which brought to his attention the problems which inspired his invention.

▸ THE PROBLEM

Many of Newcomen's most important customers owned mines. They described to him the problem they encountered when forced to dig deeper mines to meet the growing demand for natural resources such as coal, tin and iron ore. Owing to the depth, they were increasingly hindered by flooding. The solution was to pump this water out of the mines but with only horse or manpower available to do the job, it was an expensive and slow task.

Thomas Newcomen provided the power sources for the Industrial Revolution

▸ FAILED ATTEMPTS

The idea of using atmospheric pressure as a new kind of power source to be employed in carrying out repetitive, mechanical work like pumping had been known to engineers before Newcomen. It had been proved that when a vacuum was created, air, when given the opportunity, would rush into it with considerable force. But nobody had ever harnessed this discovery successfully into a practical power supply. In 1698, an English engineer called Thomas Savery (1650–1715) had made an attempt at it through his design and patent of the 'Miner's Friend', a high-pressure steam pump engine. Due to technological and practical limitations, however, it was never successfully employed for pumping.

▸ THE SOLUTION

It was against this context that Newcomen decided to begin work in 1705 on building a steam engine to take advantage of atmospheric pressure. By 1712, he had solved the problem and his engine was successfully constructed and used for pumping in South Staffordshire Colliery. The design involved heating water underneath a large piston which was encased in a cylinder. Steam that was released as a result of the heating forced the piston upwards. A jet of water was then released from a tank above the piston. The sudden cooling of the steam made it condense, creating a partial vacuum which atmospheric pressure then pushed down on, forcing the piston downwards again. The piston was attached to a two-headed lever, the other side of which was attached to a pump in the mineshaft. As it moved up and down, the lever moved likewise and a pumping motion was created in the shaft which could be used to eject flood water. The first engine could remove about 120 gallons per minute, completing about twelve strokes in that time, and had the equivalent of about 5.5 horsepower.

Even thought the engine was still not particularly powerful, was hugely inefficient to run, and burnt large amounts of coal, it would work reliably twenty-four hours a day and was far better than the previous alternatives. Consequently, even though each one cost an expensive £1000 to build, they were highly successful commercially and as a result more than a hundred were installed, chiefly in Britain's mines and factories, before Newcomen's death in 1829. Sales continued to increase across Britain and Europe for the next one hundred years. Even though more efficient engines were to follow, such as that invented by James Watt (1736–1819), its relative simplicity, reliability – and lower price tag than the competition – ensured that working engines continued to be used well into the twentieth century. By then, the Industrial Revolution had changed the world, and Newcomen's harnessing of steam and atmospheric pressure had been at its helm.

A FORGOTTEN GENIUS

The steam engine, although originally developed by Newcomen for use in mines, went on to become one of the cornerstones of the Industrial Revolution. It was quickly developed by engineers like James Watt and Richard Trevithick into the steam locomotive, eventually going on to power the ocean-going ironclads which cut sailing times to and from Britain so drastically. Today, the credit for the steam engine is usually given to James Watt while the name Thomas Newcomen remains shrouded in obscurity. And although he undoubtedly changed the world, there is no single known portrait of Newcomen in existence.

DANIEL FAHRENHEIT

1686–1736

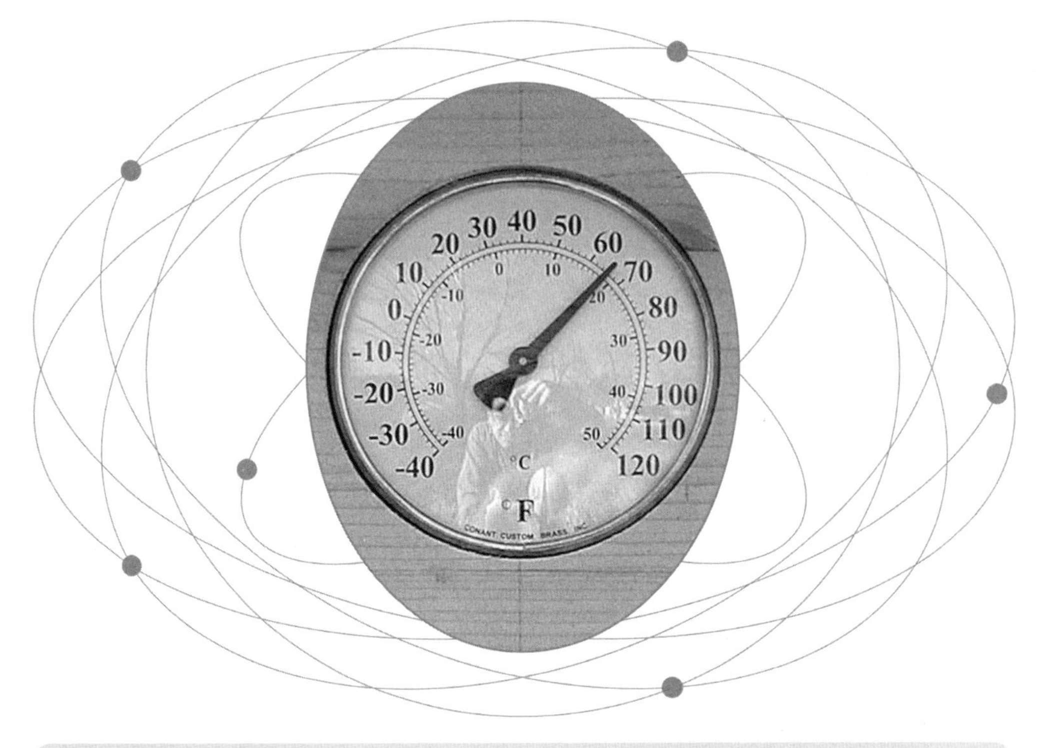

CHRONOLOGY • c. 1701 Fahrenheit arrives in Amsterdam, Holland, apprenticed to merchants • **1709** Develops a superior alcohol thermometer • **1714** Invents the first mercury thermometer • **1715** Develops the Fahrenheit temperature scale

Daniel Fahrenheit spent most of his working life in the Netherlands. He was born in the Polish city of Danzig, now Gdansk, the oldest of five children. At the age of fifteen his parents died after eating poisonous toadstools and while his four siblings were sent to foster homes he became apprenticed to merchants who sent him to Amsterdam. Here he became interested in thermometry, particularly the primitive thermometers invented in Florence around 1640. Fahrenheit borrowed money against his inheritance to develop the idea and left his apprenticeship – this forced him to go on the run.

The measurement and description of temperature is so commonplace today that it is virtually impossible to imagine a world without it. Yet as late as the start of the eighteenth century, scientists were still struggling to find a reliable device which would accurately measure temperature, and a uniform scale by which to describe the limited measurements they could make.

▸ PRIMITIVE THERMOMETERS

Galileo (1564–1642) had, in fact, been the first to create a primitive kind of thermometer. He had used his knowledge that air expanded when heated and contracted when cooled to build his

The Fahrenheit scale, perhaps unsurprisingly, is named for its discoverer, Daniel Fahrenheit

instrument. By placing a cylinder tube in water, he noticed that when the air was hotter it pushed the level of the water in the device downwards, just as it rose when the air cooled. He soon realised the reading was unreliable, however, because the volume, and therefore the behaviour, of the air also fluctuated according to atmospheric pressure. Gradually, scientists began using other, more stable substances to improve the accuracy of the reading, with alcohol being introduced as a possible substitute later in the century.

▸ THE MERCURY THERMOMETER

Fahrenheit eventually made the thermometer reliable and accurate enough for the purposes scientists required. A producer of meteorological instruments, he first achieved progress in 1709 with the development of an alcohol thermometer far superior to any that had previously been created. However, it was in building upon Guillaume Amontons' (1663–1705) work on the properties of mercury that Fahrenheit truly took the measurement of temperature into another domain. He invented the first successful mercury thermometer in 1714, particularly useful in its application across a wide range of temperatures.

▸ THE FAHRENHEIT SCALE

In 1715 he complemented his breakthroughs in instrument-making with the development of the now famous Fahrenheit temperature scale. Taking zero degrees Fahrenheit (0°F) to be the lowest temperature he could produce (from a blend of ice and salt), he used the freezing point of water and the temperature of the human body as his other key markers in its formulation. In his initial calculations this placed water's freezing point at 30°F and the body's at 90°F. Later revisions changed this to the now well-known 32°F for water and 96°F for humans. The boiling point of water worked out to be 212°F, meaning that there were a hundred and eighty incremental steps between freezing and boiling. The scale became widely used, particularly in English speaking countries where it largely remained dominant until the 1970s. The fahrenheit scale is still in common use throughout the USA.

▸ FAHRENHEIT'S SUCCESSORS

Across much of the scientific community Fahrenheit's range has been superseded by the Celsius scale. This metric scale, with water freezing at 0°C and boiling at 100°C, was begun by the Swede Anders Celsius (1701–44), and revised by his countryman Carolus Linnaeus (1707–78) into the scale we know today (Celsius originally had the scale the other way round, with boiling point at 0°C and freezing at 100°C!). The Celsius scale is also known as the Centigrade scale, from the Latin meaning 'one hundred steps'. A conversion can be made from degrees Fahrenheit to degrees Celsius by subtracting thirty-two and multiplying this figure by five, then dividing by nine.

THE INFLUENCE OF FAHRENHEIT

The Kelvin scale is more suitable for scientific purposes and the Celsius scale is neater, based as it is on decimals. The advantage of using the Fahrenheit scale is that it is designed with everyday use in mind, rarely needing, for example, negative degrees. Fahrenheit, using his glass-blowing skills to create his thermometer, found that the boiling points of different liquids varied according to fluctuations in atmospheric pressure; the lower the pressure, the lower the boiling point of water, a useful fact for anyone wishing to make tea at high altitude!

BENJAMIN FRANKLIN

1706–1790

CHRONOLOGY • **1716** Due to financial difficulties, Franklin leaves school at the age of ten • **1751** *Experiments and Observations on Electricity, made at Philadelphia in America* published • **1752** Famously flies a kite in a thunderstorm • **1776** One of five men who drafted the Declaration of Independence.

Benjamin Franklin had a rare genius. Unlike most entrants in this book, whose outstanding talents are generally restricted to the scientific, the American Franklin was brilliant in a wide range of arenas. In a five-year period between 1747 and 1752, he contributed more to science than most scientists would achieve in a lifetime of dedicated study. Yet, during other periods of his life, he operated in, and conquered, completely different fields. He was a master printer and publisher, a successful journalist and satirist, an inventor, a world famous ambassador and, probably most notably of all, a politician at a vital time in American history. Indeed, Franklin was one of the five signatories of the Declaration of Independence from Great Britain in 1776 and was a key participant in the later drafting of the American Constitution.

▸ STUDYING ELECTRICITY

Yet Franklin merits an entry in this book for his achievements in physics alone – he was a pioneer in understanding the properties and potential benefits of electricity. Although the phenomenon of electricity had been noted since the time of the

Franklin's legacy, in addition to his many inventions, was essentially one of learning

ancients, very little was known about it from a scientific perspective, and many considered the extent of its usefulness to be limited to 'magic' tricks. At around the age of forty, however, Franklin became fascinated by electricity and began to experiment with it, quickly realising it was a subject worthy of scientific study and research in its own right. So, he sold his printing interests and dedicated himself for the next five years to understanding it.

▸ FLYING A KITE

Although Franklin wrongly believed electricity was a single 'fluid' (this was in itself an advance on earlier theories which posited the idea of two different fluids), he perceived this fluid to somehow consist of moving particles, now understood to be electrons. More importantly, he undertook important studies involving electrical charge and introduced the terms 'positive' and 'negative' in explaining the way substances could be attracted to or repelled by each other according to the nature of their charge. He also believed these charges ultimately cancelled each other out so that if something lost electrical charge, another substance would instantly gain the amount being cast away. His work on electricity climaxed in his now famous kite experiment of 1752. Believing lightning to be a form of electricity, and in order to prove it, Franklin launched a kite into a thunderstorm on a long piece of conducting string. Tying the end of the string to a capacitor, he was vindicated when lightning did indeed charge it, proving the existence of its electrical properties. From these results, and realising the potential of a device that could deflect the harmful effects of lightning strikes away from buildings and property, he developed the lightning conductor.

Franklin had also published his text *Experiments and Observations on Electricity, made at Philadelphia in America* in 1751, which went on to inspire future scientists in the study and development of the uses of electricity.

▸ A PROLIFIC INVENTOR

From 1753 the time Franklin dedicated to science reduced dramatically due to his taking up a new post as deputy postmaster general and, later, political and ambassadorial roles. He did, however, leave a legacy of other inventions from the wide range of experiments conducted throughout his life, including: an iron furnace 'Franklin' stove (still in use today), bifocal spectacles, the street lamp, the rocking chair, the harmonica, an odometer and watertight bulkheads for ships. Franklin also came up with the idea of Daylight Saving Time and was the first to charter the Gulf Stream from observations made by sailors. A man of many talents, Benjamin Franklin was a successful inventor, politician, printer, oceanographer, ambassador, journalist, satirist and, of course, scientist.

The Legacy of Benjamin Franklin

Franklin's legacy, in addition to the many inventions such as lightning conductors, bifocal lenses and street lamps, was one of learning. He established one of the first public libraries, as well as one of the first universities: Pennsylvania, in America. On a broader societal level, he established the modern postal system, set up police and fire fighting departments and established the Democratic Party.

He certainly lived up to his own quotation, 'If you would not be forgotten as soon as you are dead and rotten, either write things worth reading, or do things worth the writing.'

JOSEPH BLACK

1728–1799

CHRONOLOGY • **1746–50** Black studies chemistry at Edinburgh University • **1754** Presents thesis on the cycle of reactions in chemistry • **1756–66** Professor of Medicine and lecturer in chemistry at Glasgow University • **1757** Discovers the concept of latent heat • **1766** Appointed Professor of Chemistry at Edinburgh, a post he holds until his death in 1799

Joseph Black was born in Bordeaux, France, the son of a wine merchant, and educated at Belfast and later Glasgow universities. At Glasgow he came under the tutelage of the renowned William Cullen, although their relationship soon became that of professor and assistant, rather than that of teacher and student. Cullen was an innovator in the fields of chemistry and the classification of diseases, identifying four major divisions, although he was better known for his unorthodox teaching methods and inspirational lecturing style. His lectures certainly seem to have inspired the young Black, who was to put chemistry onto a sound, scientific footing, although he never published his works during his lifetime.

▸ THE REDISCOVERY OF CARBON DIOXIDE

Although Jan Baptista **van Helmont** had identified the existence of gases distinct from air more than a century before Joseph Black became prominent, little work had been undertaken to build on these observations in the intervening

Latent heat: the ability of matter to absorb heat while remaining at the same temperature

hundred years. It is for this reason the Scottish scientist is often credited as the discoverer of carbon dioxide, which he called 'fixed gas', even though van Helmont had clearly been aware of its existence. What is true, however, is that Black was the first to fully understand and quantify the properties of carbon dioxide, and in so doing laid one of the key foundations of modern chemistry.

▸ THE IMPORTANCE OF METHOD

Black's insistence on the importance of quantitative experiments was another notable step towards the setting of the standard for the new era of chemistry. He employed such methods in producing the results of his most significant text, *Experiments upon Magnesia Alba, Quicklime, and some other Alcaline Substances* (1756). He outlined the cycle of chemical changes in what has become one of the defining experiments in the teaching of the science. Black observed how limestone (Calcium Carbonate) produced quicklime (Calcium Oxide) and fixed air (Carbon Dioxide) when heated. He then mixed the quicklime with water to produce slaked lime (Calcium Hydroxide), and by combining the output of that with fixed air he was able to make limestone again (plus water). Anticipating the discoveries of Antoine **Lavoisier** a century later, Black concluded from his experiments that carbon dioxide was a distinct gas from 'normal' (atmospheric) air, as well as one of its constituents in small quantities. He also demonstrated that removing carbon dioxide from the limestone made the latter more alkaline, with the effects reversed on the addition of carbon dioxide again, observing therefore the gas's acidic properties. Black was also able to prove carbon dioxide was made by respiration, the burning of charcoal and through fermentation, but that the gas would not allow a candle to burn in it nor sustain animal life.

▸ THE PHYSICS OF HEAT

Black later turned his attention to physics and here too he made some elementary discoveries now at the core of the subject. Through meticulous experimentation and measurement of results, the scientist discovered the concept called 'latent heat', or the ability of matter to absorb heat without necessarily changing in temperature. The best example of this principle is that of the transformation of ice into water at 0°C. This requires heat to form water, although the liquid formed is still at the same temperature. The same principle is true in the process of transforming water to steam and, indeed, all solids to liquids and liquids to gases. Through this work, Black made the important distinction between heat and temperature. As well as its more general application since, the experiments in latent heat became important almost immediately, as one of Black's friends was James **Watt**, who benefited from these discoveries during his development of the condensing steam engine.

FURTHER ACHIEVEMENTS

Black also articulated many other findings involving heat. In particular, he formulated the theory of specific heat, that is, the theory which states that different quantities of heat are required to bring equal weights of different materials to the same temperature. Out of this work came the development of calorimetry, an accurate way of measuring heat for the first time and still used in an amended form today, as well as a device to be employed to this end, the calorimeter.

HENRY CAVENDISH

1731–1810

CHRONOLOGY

• **1731** Cavendish is born in Nice, France, to an aristocratic family • **1753** Leaves Cambridge University without taking a degree • **1798** Publishes his estimate of the density of the earth, an estimate almost precisely what it is now believed to be • **1871** The endowment of the famous Cavendish Laboratory was made to Cambridge University, by Cavendish's legatees.

If ever a person were to fit the stereotypical image of a wacky, eccentric scientist, Henry Cavendish would be that man. Born of the English aristocracy and inheritor of a huge sum of money mid-way through his life, Cavendish used his wealth to indulge his unusual behaviour. He built private staircases and entrances to his homes in London so he would not have to interact with his servants, and only communicated with them through written notes. He never spoke to women, doing all he could to avoid having to look at them, and only usually appeared in public for the purposes of attending scientific meetings. His love of solitude did, however, offer him plenty of time to work on the experiments which would advance science, in spite of his equally eccentric approach to the publication of his work.

▸ PROMPTED BY CURIOSITY

Cavendish's main motivation was not scientific acclaim, but curiosity, and it is because of this that he failed to put many of his discoveries into print. He conducted meticulous experiments in

Some of Cavendish's discoveries are considered to be half a century ahead of their time

both physics and chemistry, but it is largely for his work in chemistry that he is best remembered, since he did publish a number of papers in this field.

Of the most famous were his 1766 *Three Papers Containing Experiments on Factitious Airs (gases made from reactions between liquids and solids)*. In these he demonstrated how hydrogen (inflammable air) and carbon dioxide (fixed air) were gases distinct from 'atmospheric air'. Although Joseph **Black** was making similar discoveries with fixed air, Cavendish is credited with being a pioneer in distinguishing and understanding inflammable air. He managed to develop reliable techniques for weighing gases and, in further experiments undertaken around 1781, he discovered that inflammable air, mixed with what we now know as oxygen (from atmospheric air) in quantities of two to one respectively, formed water. In other words, water was not a distinct element but a compound made from two parts hydrogen to one part oxygen (now famously expressed in chemistry as H_2O). Due to his typical tardiness in publication (he did not declare his findings until 1784), his claim to this discovery became confused with similar observations subsequently made by Antoine **Lavoisier** (1743–94) and James **Watt** (1736–1819). The important point is that water was proved not to be a distinct element – a view held since the time of **Aristotle**. In the same paper, Cavendish also explained his discovery that air (whose composition remained constant from wherever it was sampled in the atmosphere) was composed of approximately one part oxygen to four parts nitrogen. In these experiments – performed to decompose air by 'exploding' it with electrical sparks – he also found that there was always a residue of about one per cent of the original mass which could not be broken down further. This 'inert' gas would not be studied again for a century, when it was named argon. In the same series of experiments, Cavendish also discovered nitric acid, by dissolving nitrogen oxide in water.

▸ AHEAD OF HIS TIME

Potentially, Cavendish could have been remembered as a great physicist as well, since some of his experiments and discoveries were considered to be more than half a century ahead of their time. Almost all of his work in this arena remained unpublished until the late nineteenth century however, when his notes were found. The scientist James Clerk **Maxwell** (1831–79) dedicated himself to publishing Cavendish's work, a task he completed in 1879. But by then Cavendish's potential breakthroughs, significant at the time, had been surpassed by history. In particular, Cavendish had undertaken significant work with electricity, anticipating laws later named after their 'discoverers' Charles **Coulomb** (1736–1806) and Georg Ohm (1789–1854), as well as some of Michael **Faraday**'s (1791–1867) later conclusions. In the absence of any other appropriate device and in keeping with his eccentric tendencies, he even resorted to measuring electrical current by grabbing electrodes and estimating the degree of pain it caused him!

THE DENSITY OF THE EARTH

One physical experiment for which Cavendish was acclaimed in his time (and which is now named after him) was working out the density of the earth. The experiments involved a torsion balance and the application of Newton's theories of gravity. In 1798 he concluded that the earth's density was 5.5 times that of water, a figure almost identical to modern estimates.

JOSEPH PRIESTLEY

1733–1804

CHRONOLOGY • **1766** Priestley meets Benjamin Franklin, who awakes his scientific curiosity • **1767** *The History and Present State of Electricity* published • 1771 Discovers that a plant will replenish enough air in a jar to sustain a burning candle • 1774 Discovers oxygen independently of Karl Scheele (who announces his discovery in 1777)

Joseph Priestley was not first and foremost a scientist, yet he became one of the most important British experimental chemists of the eighteenth century. Trained as a Unitarian minister and with an active interest in politics, philosophy, history and languages, it was not until he met Benjamin **Franklin** in 1766 that his scientific fascination was awakened. The consequent journey this would take him on made Priestley famous, even though he remained an amateur in science for the rest of his life and dedicated most of his time to teaching and preaching.

▸ BEGINNINGS IN ELECTRICITY

Physics, not chemistry, was the subject of Priestley's first scientific endeavour, however. Encouraged by Franklin, and given the use of his books, Priestley wrote *The History and Present State of Electricity* (1767) which, as well as acting as a summary of everything known about electricity to that point, included some of Priestly's own contributions such as the discovery that graphite conducts electricity.

▸ THE ALLURE OF CHEMISTRY

Physics did not sustain his long-term scientific

Priestley's scientific work virtually ceased after his enforced emigration to the United States

interest, however, and Priestley became more intrigued with chemical experiments. On taking up a new ministerial post in Leeds in 1767, he gained access to a virtually unlimited supply of 'fixed air' (carbon dioxide) with which to begin his work. This he obtained from the gas released through fermentation at his local brewery. Among other things, this led to Priestley's creation of soda water (carbon dioxide in water). Little did he know the consequences this would have for the future development of the soft drinks industry! The episode stimulated the Englishman's interest in working with gases, leaving him determined to add to the mere three that were then known: 'fixed air' (carbon dioxide), 'inflammable' air (hydrogen) and atmospheric air. By improving the design of a piece of apparatus called the 'pneumatic trough', filling it with mercury and then heating solids floating in it, he was able to isolate and capture gases above the mercury. Soon Priestley had discovered four new gases that we now know as nitrous oxide (laughing gas), nitrogen dioxide, nitric oxide and hydrogen chloride.

▸ DISCOVERING GASES

Arguably Priestley's most successful period, however, came while under the patronage of Lord Shelburne at his estate in Calne, Wiltshire, from 1773–80. Ostensibly employed as a librarian and teacher to Shelburne's children, Priestley was largely given the freedom to pursue his scientific studies as he wished. In due course he discovered nitrogen, carbon monoxide, sulphur dioxide, ammonia and silicon tetrafluoride. But it was his discovery of the gas now known as oxygen for which Priestley is most famous.

▸ THE DISCOVERY OF OXYGEN

Priestley stumbled across oxygen in 1774 while heating mercury oxide and seeing that it greatly enhanced the burning of a candle's flame, prolonged the life of mice and was 'five or six times as good as common air.' (Karl **Scheele** had also discovered oxygen independently in 1772 but did not publish his results until 1777). Priestley's findings were in addition to his 1771 discovery that a plant will replenish the 'air' in a jar sufficient to burn a candle again after the candle has burnt itself out. As well as his later observation of the significance of sunlight in the growth of plants, this was an important foundation for other scientists in subsequent research into photosynthesis. Yet despite this, Priestley did not realise the true impact of his discovery and it was left to Antoine **Lavoisier** (1743–1794), whom he told of his findings in 1775, to establish the central place oxygen has in the fields of chemistry and biology. Instead, Priestley named his gas 'dephlogisticated' air, in keeping with the accepted theory that all flammable substances contained the elusive substance 'phlogiston' which was central to the combustion process and released during it.

Afterword

Priestley also undertook other experimental work involving the density, diffusion and heat conductivity of gases, as well as the impact of electrical discharges upon them.

His scientific work virtually ceased, however, following his emigration to Pennsylvania, USA in 1794. This was a move Priestley felt forced to make after his Birmingham laboratory was subjected to mob violence as a result of his vocal political support for the French Revolution, which he saw as an antidote to the corruption of an un-Godlike society.

JAMES WATT

1736–1819

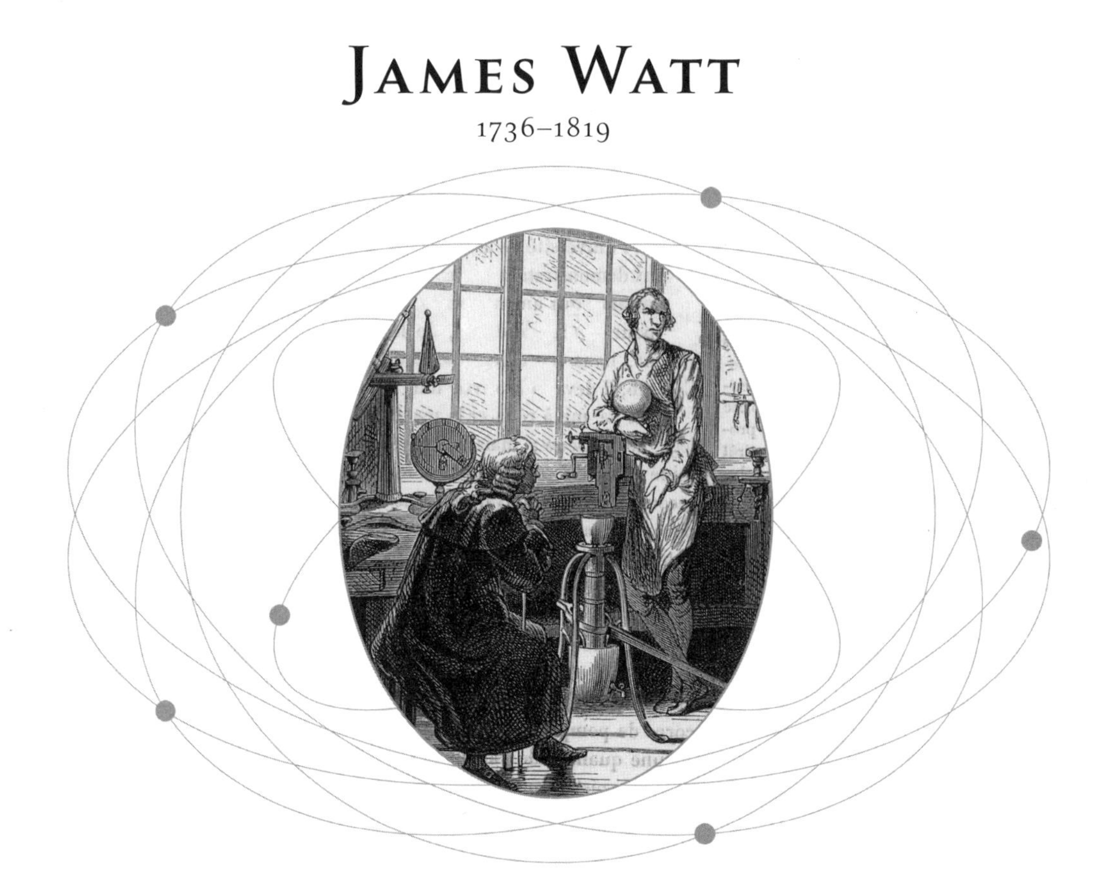

CHRONOLOGY • **1764** Watt finds a model of Newcomen's steam engine to be inefficient • **1765** Has the idea that kick-starts the Industrial Revolution • **1768** He produces the first prototype of his new engine •**1788** Invents the 'centrifugal governor', a mechanism that automates speed control • **1790** Perfects the 'Watt Engine'

James Watt is often mistakenly perceived by many people to have been the inventor of the first steam engine. In reality Thomas **Newcomen** had achieved this nearly a quarter of a century before Watt was even born. Watt's engines, however, had the wider impact. Newcomen's machines had been restricted to the world of mining, Watt's were used across all industries. If Newcomen is remembered as the inventor of a power source which changed the world, it is Watt who made its potential available, and provided the catalyst for the Industrial Revolution in the process.

▸ A HAPPY ACCIDENT

As with all the best tales of discovery and invention, the occurrence which began the chain of events leading to Watt's engine was nothing more than a happy accident. In 1764, Watt was asked to repair a scale model of Newcomen's engine which had been used by the University of Glasgow for teaching purposes. The close examination of the model Watt undertook in the process of fixing it made him realise it was hugely inefficient. The biggest weakness Watt identified was in the heating and cooling of the engine's cylinder during every stroke. This

Watt's development of the rotary engine brought mechanisation to industry

wasted unnecessary amounts of fuel, as well as time, in bringing the cylinder back up to steam producing temperature which limited the frequency of strokes.

Consequently, Watt began pondering on improvements to the design of Newcomen's engine. It is said that the Scotsman hit upon his solution in 1765 while wandering through Glasgow Green, and today a memorial stone marks this spot as the place where the idea was born which truly sparked the Industrial Revolution. He had realised the key to improved efficiency lay in condensing the steam in a separate container, thereby allowing the cylinder and the piston to remain always hot.

▸ WATT'S PARTNERS

By 1768, Watt had constructed a fully functioning prototype of his new engine, at which point he entered into a business partnership with John Roebuck to finance and sell the production of the machine. Shortly afterwards, the partnership took out a patent for the engine under the title 'A New Invented Method of Lessening the Consumption of Steam and Fuel in Fire Engines,' and began selling it to colliery owners. Unfortunately, in 1772, Roebuck went bankrupt, although this later gave Watt the opportunity to enter into a more fruitful partnership with businessman Matthew Boulton in 1775.

'Boulton & Watt' immediately applied to the British Parliament for a new patent allowing the company to be the sole makers and sellers of Watt's engines in the country for the next twenty-five years. The success of the application gave the business a virtual monopoly in steam engine production, guaranteeing its financial success and the individual wealth of Watt himself by the time he retired in 1800.

The patent did not stop the inventor from trying to make continuous improvements to his engine, however, and it was not until 1790 that he had finally perfected the 'Watt Engine'. In the intervening years, he made the breakthrough of modifying the steam engine to work in a rotary-motion. The up-down action of his and Newcomen's engines before then had been fine for pumping water out of mines, but of little use elsewhere. With a circular, rotary-motion however, other industries could make use of steam power for driving machines. For example, in the cotton industry, Richard Arkwright was the first to realise the engine could be used to spin cotton, and later in weaving. Flour and paper mills were other early adopters and in 1788 steam power was used to paddle marine transportation for the first time. In that same year, Watt developed the 'centrifugal governor' to regulate the speed of the engine to keep it constant, itself an important foundation in the science of automation.

WATT'S INFLUENCE

Watt is also credited with a number of other inventions including the rev. counter and early letter copying press. More significantly, he was the first to coin the term 'horsepower' which he used when comparing how many horses it would require to provide the same pull as one of his machines. In 1882 the British Association also named the 'watt' unit of power in his honour, further cementing the inventor's fame.

Watt's steam engine was the driving force behind the Industrial Revolution and his development of the rotary engine in 1781 brought mechanisation to several industries such as weaving, spinning and transportation.

CHARLES DE COULOMB

1736–1806

CHRONOLOGY • **1777** De Coulomb publishes a paper outlining the principles behind the construction of an extremely sensitive torsion balance • **1781** Elected to the French Académie des Sciences • **1785** Publishes the principle that becomes known as Coulomb's Law • **1802** Appointed as an inspector of public instruction

Charles Augustin de Coulomb came from a family eminent in the law, in the Languedoc region of France. After being brought up in Angoulême, the capital of Angoumois in southwestern France, Coulomb's family moved to Paris., where he entered the Collège Mazarin. Here he studied language, literature, and philosophy, and he received the best available teaching in mathematics, astronomy, chemistry and botany, before going on to study engineering . The study of electricity was gaining important new ground throughout the eighteenth century, but scientists were still only just beginning to understand how it behaved and could be manipulated, and more importantly, how it could be used. Newton's law of gravitation had been a remarkable breakthrough in comprehending the way in which the universe worked, so imagine the impact of the discovery that an identical principle could be applied to electrical forces. Coulomb is credited with this finding.

▸ A FINE BALANCE

Before Coulomb could prove such a phenomenon, however, he would be forced to invent an

Coulomb believed electricity and magnetism to be two distinctly separated 'fluids'

extremely sensitive torsion balance for taking measurements in his electrical experiments. He outlined the principles behind this instrument in a paper of 1777, and went on to construct a balance which could sense force down to as little as 1/100000th of a gram. The appliance itself consisted of a large, glass cylinder with degrees marked around its circular edge, and a wax covered straw inside it, suspended parallel to the ground on a piece of silk thread. An electrically charged ball could be fixed to one end of the straw, balanced by a counter-weight at the other end. By holding a second charged ball at various distances from the one in the cylinder, Coulomb could measure the impact of the electrical force by the degree in which the sphere suspended on the straw rotated.

▸ COULOMB'S LAW

Henry **Cavendish** (1731–1810) had also hinted at the law, but had never published. In its simplest form, Coulomb's Law states that the force between two electrically charged bodies is linked to the square of the distance between them in an inversely proportional relationship. So, for example, by tripling the distance between the charges, the force would decline by nine times. Equally, the force is directly proportional to the product of the charges. In other words, Newton's law of gravitation was mirrored in electricity. Coulomb's law was published in 1785 in one of a series of seven papers the Frenchmen wrote between that year and 1791 outlining the results of his observations. In these ongoing experiments Coulomb also found a similar principle linking the relationship of magnetic forces, leading to speculation by others that perhaps gravity, magnetism and electricity had some kind of interconnected relationship. Coulomb himself, however, had no time for such conjecture, believing electricity, and magnetism in particular, to be two separate 'fluids'. It was left to Hans Christian Oersted (1777–1851), André-Marie **Ampère** (1775–1836) and most notably Michael **Faraday** (1791–1867) to enunciate the phenomenon of electromagnetism.

▸ A MILITARY ENGINEER

While Coulomb is mostly remembered for his electrical studies, he also made discoveries in other areas. The Frenchman spent much of his life as an engineer in the army, working on various French West Indian possessions, as well as a significant amount of this time designing and overseeing the building of fortresses. Hardly surprising, then, that much of his early scientific speculation was linked to engineering theory, such as the concept of a 'thrust line', still used today in construction. The SI unit of electric charge, one unit of which is shifted when a current of one ampere flows for one second, is named after him: the coulomb.

Friction Burns

Coulomb is credited by many commentators with the invention of the science of friction. During his work as a military engineer the issue of friction frequently arose. It was this that inspired him to devote several years of study to the subject. The end result was an articulation of Coulomb's Law of Friction, which outlined a proportional relationship between friction and pressure, and it was this work for which he was elected to the mechanics section of the Académie des Sciences in 1781.

JOSEPH MONTGOLFIER

1740–1810

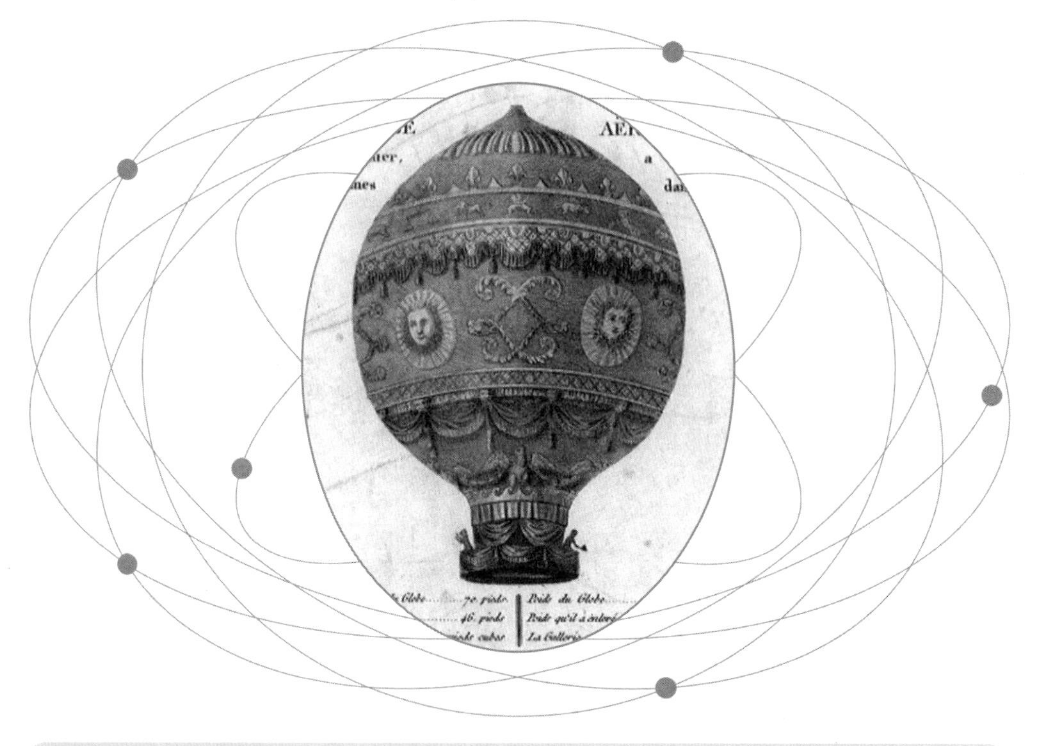

CHRONOLOGY • **1782** The Montgolfiers begin their quest to see if balloons containing heated air can be used to lift humans • **4 June 1783** First public demonstration of their hot air balloon at Annonay • **19 September 1783** Their balloon carries a sheep, duck and rooster to King Louis XVI • **21 November 1783** The first manned flight in history takes place.

The Montgolfier brothers, Joseph and Etienne, were part of a family of sixteen siblings who grew up near Lyons where their father owned a paper-making factory. They had noticed that when paper was burnt on an open fire the hot air would often force the burnt pieces upwards. Around 1782 the brothers began to examine whether this fact could somehow be used in the quest to get humans airborne. They began experimenting, not only with hot air, but also with hydrogen gas (which their countryman Jacques-Alexandre-César Charles later made use of in producing the first hydrogen balloon) and steam, to see which would raise small paper models the most efficiently. While they struggled to make progress with hydrogen and steam, towards the end of 1782 they succeeded in making paper envelopes filled with hot air rise to the ceiling of their residence.

FIRST FLIGHT

From this initial success, the brothers were spurred on to devise a large-scale balloon. By 4

From burning paper, the brothers were spurred on to devise a large scale balloon

June 1783 they were ready. They had produced a cloth sphere with a paper lining some twelve yards across. Inflating the balloon using heat from a fire burning wool and hay, they released it in front of spectators at Annonay, near Lyons. The flight was a success: their invention rose some 2000 yards into the sky in an ascent lasting ten minutes.

▸ A ROYAL AUDIENCE

Word of the brothers' achievement reached Paris where King Louis XVI requested that a display of the balloon should take place at Versailles. The Montgolfiers agreed and upped the ante by not only building a newer, larger balloon, but this time sending it airborne with a sheep, duck and cock as cargo. As well as providing an impressive demonstration, this would also provide proof that living creatures could survive the ascents without adverse effects. The launch took place on 19 September 1783 and although the flight was still short, the balloon travelled an impressive two miles at a height of over 1500 feet. More importantly, the animals touched down unscathed.

The foundations had thus been laid for the achievement of a dream which had burned within mankind for a long time: flight. The volunteers who would make the journey into the unknown were friends of the Montgolfiers, Jean-François Pilâtre de Rozier and François Laurent, the Marquis d'Arlandes. Joseph and Étienne went away again and constructed an even larger balloon capable of carrying human weight. The new balloon was built with a furnace to enable the voyagers to maintain their altitude.

▸ MANNED FLIGHT

The day history was made was 21 November 1783. The human guinea pigs were launched into a twenty-five minute flight across Paris. Even though a large crowd had assembled to watch the launch, the Marquis D'Arlandes later commented, 'I was surprised at the silence and the absence of movement which our departure caused among the spectators, and believed them to be astonished and perhaps awed at the strange spectacle.' Throughout the duration of the flight, the pair rose only a few hundred feet, borne on a hay-fuelled fire, but still gained enough height to avoid the rooftops which sat precariously below. Although the journey was short, they covered several miles and eventually descended into a field on the outskirts of the city, unhurt and triumphant. Man's attainment of the sky saw the beginning of a new era.

Using animals as test pilots was an idea put into practice again when Sputnik II was launched into space by the Russians on 3 November 1957. On board was a dog named Laika, the first canine to enter the cosmos. Unlike the animals that the Montgolfiers used, however, Laika was not destined to survive her journey.

THE MONTGOLFIER LEGACY

For as long as man has observed the birds, he has always dreamt of flying.

From the early Greek fantasies of Icarus's wings, to Leonardo's designs for helicopters and other flying machines, it had been something long talked of, but rarely thought possible. The tale of Icarus was held up as a stern warning to those who sought to better nature. Then came the Frenchman Joseph-Michel Montgolfier and his brother Jacques-Étienne (1745–99). They observed one simple natural phenomenon and sought to take advantage of it. The result was the realisation of the 'unachievable'.

KARL WILHELM SCHEELE

1742–1786

CHRONOLOGY • **1772** Scheele discovers oxygen, two years before Joseph Priestley, but does not publish his findings until 1777 • **1774** Discovers chlorine • **1775** Elected to the Stockholm Royal Academy of Sciences • **1777** *Chemical Observations and Experiments in Air and Fire* published

One of the few challengers to Joseph **Priestley**'s greatness in eighteenth century experimental chemistry had more than just a love of scientific testing in common with him. Like Priestley, the Swede Karl Wilhelm Scheele was only an amateur scientist, had little interest in theoretical science and shared in the claim to the discovery of oxygen. In fact Scheele's claim to be the first is more convincing than Priestley's, given that he made the discovery almost two years before his British contemporary!

▸ DESPITE THE OBSTACLES

Scheele's achievements are all the more remarkable because, unlike Priestley, he did not benefit from a good education, beginning his first full-time apprenticeship at fourteen. Moreover, he was of poor means and throughout his life was forced to conduct most of his experiments restricted by limited apparatus and a lack of laboratory space. His day job was initially as an apothecary and he practised in Gothenburg, Malmo, Stockholm and Uppsala in the years up to 1775. He then moved to Köping to run a pharmacy where he remained for the rest of his life.

Scheele was a man who, quite literally, would die for his science

In the same year as his final move, he was also elected to the Stockholm Royal Academy of Sciences. This recognition brought with it some prestige and consequent well paid job offers overseas, but Scheele preferred to remain in Köping where his limited working hours left plenty of free time to experiment at his leisure without obligation.

▸ OXYGEN BEFORE PRIESTLEY

In 1772 Scheele's scientific efforts resulted in his most significant achievement, the discovery of oxygen, two years before Priestley. He made his 'fire air' from a variety of sources. These including the heating of compounds such as mercury oxide, nitric acid and potassium nitrate. Furthermore, he encompassed his discovery into a more general theory of the atmosphere and combustion, and concluded that the former was composed of only two gases. The first, 'vitiated air,' or nitrogen, suppressed combustion; the other, 'fire air', facilitated it. In common with Priestley, however, he did not appreciate the full significance of his finding, and interpreted it only within the boundaries of the 'phlogiston' theory of his day. But, unlike Priestley, he did not get the credit for it (and even now is often forgotten), because he did not publish his findings until 1777 in his only text *Chemical Observations and Experiments in Air and Fire*. By that point, Priestley was already well known as the discoverer of 'dephlogisticated' air.

▸ A GREEN GAS

The discovery of chlorine was another instance of Scheele not appreciating the significance of what he had achieved. He isolated the green gas in 1774, but it was not until work by others in the early years of the nineteenth century that it was found to be an element in its own right.

▸ A CATALOGUE OF DISCOVERIES

Scheele was obsessed with the discovery of new chemical elements and compounds. From 1770 onwards he identified a remarkable number of substances. These included manganese dioxide, silicon tetrafluoride, barium oxide, copper arsenite, glycerol and hydrogen fluoride, sulphide and cyanide. He also recognised an array of new acids which included citric, arsenic, hydrocyanic lactic, tartaric, prussic and tungstic varieties.

▸ A MARTYR TO HIS CAUSE

Scheele's breakthroughs were not without sacrifice. It is highly likely that the toxicity of many of the chemicals he was working with and his processes for identifying them (including smell and taste tests!) contributed to his relatively early death at the age of forty-three. Scheele was quite literally a man who would die for his science.

An Ardent Scientist

Scheele was possibly the most prolific discoverer of new substances that the world has ever seen. This feat is even more remarkable when you consider that he accomplished his achievements in spite of very poor schooling. Taking into account the breadth of his scientific output, his lack of space and adequate laboratory equipment can only reinforce our view of what must have been an iron determination to produce results.

The Swede is also credited with demonstrating the effect that light has on silver salts, a phenomenon that would later became the basis for photography.

ANTOINE LAVOISIER

1743–1794

CHRONOLOGY • **1784** Lavoisier meets the English chemist Joseph Priestley in Paris • **1788** names oxygen • **1789** *Traité élémentaire de chimie* (*Elementary Treatise on Chemistry*) published • **1794** Lavoisier guillotined in Paris

Despite possible claims to the title by other scientists, the Frenchman Antoine Lavoisier is regarded by most as the true founder of modern chemistry. Although he often undertook similar work to that of Henry **Cavendish** (1731–1810), Joseph **Priestley** (1733–1804) and Karl **Scheele** (1742–86), it was the interpretation of his findings which distinguished Lavoisier. His conclusions led to the restructuring of chemistry into a format which laid the foundations for the modern era, arguably achieving an impact comparable to that of **Newton** (1642–1727) in physics. For this it would be reasonable to assume the scientist could have expected accolades and awards from his countrymen. Instead, they chopped off his head.

▸ THE CONSERVATION OF MATTER

Lavoisier's early studies involved experiments concerning the weight loss or gain in substances when heated. By burning matter such as lead and phosphorus in closed vessels, and accurately weighing them before and after heating, he was able to observe the containers did not gain or lose any mass at all during combustion. This ultimately led to his conclusion of the law of conservation of matter. Lavoisier suggested matter

France's most outstanding scientist of his day, Lavoisier met a tragic end on the guillotine

was simply rearranged on heating and nothing was actually added or destroyed overall, hence the equal weight of the vessel before and afterwards. This in itself called into question the 'phlogiston' theory, the commonly held belief all combustible material contained a mysterious element which was released (and lost) on heating.

▸ COMBUSTION THEORY

While the overall weight of the vessel remained the same during Lavoisier's experiments, he made the further interesting discovery that the solids being heated could in fact gain mass. He observed such a reaction, for example, in 1772 when burning phosphorus and sulphur. The only logical conclusion, therefore, was the weight gain of the solid had been caused by some kind of combination with the air trapped in the container. This idea was given further impetus when Lavoisier met Joseph Priestley in Paris in 1774, and the latter explained his discovery of 'dephlogisticated' air. While Priestley failed to realise the impact of this new gas, maintaining a belief in the phlogiston theory, Lavoisier repeated the Englishman's experiments to see if this was the source of the weight gain in some solids during heating. By 1778, he had definitively concluded that not only was Priestley's dephlogisticated air the gas from the atmosphere which was combining with the matter but, moreover, it was actually essential for combustion to take place at all. He renamed it oxygen ('acid producer' in Greek) from the mistaken belief the element was also evident in the make up of all acids. He also noted the existence of the other main component of air, the inert gas nitrogen which he named 'azote'.

▸ MODERN CHEMISTRY

The Frenchmen summarised his new order for chemistry in his 1789 book *Traité élémentaire de chimie* or *Elementary Treatise on Chemistry*, sounding the death bell for the phlogiston theory and beginning modern chemistry. By this point, Lavoisier had also concluded oxygen was equally vital in the process of respiration, performing a similar kind of role in the body as it did during carbon combustion, and was at the basis of all animal life. In addition, the text included a list of known elements to date, identification work he had begun with a number of other French chemists in the mid-1880s. This founded in turn the naming process of chemical compounds which remains to this day. Lavoisier had already outlined one such combination in 1783, proving water was a combination of hydrogen and oxygen, becoming involved in confusion with James Watt (1736–1819) and Cavendish over who had made the discovery first.

Regarded today as the father of modern chemistry, Lavoisier's name is still used in the title of the modern chemical naming system.

AN UNHAPPY END

Despite the importance of Lavoisier's scientific achievements, they were not enough to save his life in the aftermath of the 1789 French Revolution. Lavoisier was a prominent figure in French public life and, most significantly, ran a tax-collecting firm. Those running such companies were considered to be enemies of the revolution. A prominent member of the revolutionary leadership, Marat, who had earlier attempted to forge a career in science and had had his work criticised by Lavoisier, used this as a pretext to try him. It ultimately led to the guillotine and a tragic end to the life of France's most outstanding scientist.

COUNT ALESSANDRO VOLTA

1745–1827

CHRONOLOGY • **1775** Volta invents the electrophorus to produce and store static electricity • **1778** Discovers methane gas • **1780** Volta's friend Luigi Galvani discovers that dead frogs' legs twitch when touched by two different metals • **1800** Creates the Voltaic pile, the first battery, which revolutionises the study of electricity

Whilst the study of electricity had begun to make progress through the work of Benjamin **Franklin** (1706–90) and others, there was still no reliable way of storing and producing a regular electric current. This had hampered ongoing experimentation in the subject, limiting the scope and usefulness of investigations. One scientist fascinated by electricity and determined to overcome this hurdle was Alessandro Volta.

The Italian aristocrat was born in Como, Lombardy, into a family where most of the male line had entered into the priesthood. Science was clearly Volta's calling, however, and in 1774 he became first teacher and shortly afterwards professor of physics at the Royal School in his hometown. Within a year, he had developed his first major breakthrough in the field of electricity with his invention of the 'electrophorus', used in the production and storage of static electricity.

▸ DANCING FROGS

This device brought Volta recognition within his field and in 1779 he was offered the chair of physics at Pavia University, a post he accepted

To gauge the strength of current from his batteries, Volta used his tongue

and would go on to hold for the next quarter of a century Here he continued his electrical investigations, becoming particularly absorbed in the work of Luigi Galvani (1737–98) during the 1780s. Volta's countryman had made a strange discovery during dissection work. He had found that simply by touching a dead frog's legs with two different metal implements, the muscles in the frog's legs would twitch. Through various other experiments, Galvani wrongly concluded it was the animal tissue which was somehow storing the electricity, releasing the substance when touched by the metals.

▸ THE VOLTAIC PILE

Volta, however, was not convinced the animal muscle was the important factor. He set about recreating Galvani's experiments and concluded, controversially at the time, the different metals were the important factor in the production of the current. Indeed, Volta and Galvani had been friends before the former began criticising the deductions his peer had made concerning the importance of the animal tissue. To make matters worse, it was Galvani himself who had sent Volta his papers on the subject for Volta's review and, he hoped, support. Instead, a bitter dispute broke out concerning whose analysis was correct. Although Galvani would not live long enough to see Volta's ultimate rebuttal of his work, the argument was already swinging in the latter's favour by the time of Galvani's death, and he ended his days a disillusioned man.

▸ DRY AND WET BATTERIES

To back up his theory, Volta had begun putting together different combinations of metals to see if they produced any current, even going as far as to use his tongue as an indicator of current strength from the shock they produced! It was, in fact, an important test because he deduced the saliva from his tongue was a factor in aiding the flow of the electric current. Consequently, Volta set about producing a 'wet' battery of fluid and metals. His decisive solution came in 1800, with the 'Voltaic pile', a stack of alternating silver and zinc disks interspersed with brine-soaked cardboard layers. By attaching a copper wire to the ends of this device and closing the circuit, Volta found it produced a regular, flowing electric current. He had created the first battery.

▸ IMPRESSING NAPOLEON

The invention radically improved the study of electricity, facilitating further breakthroughs in the subject by other scientists such as William Nicholson and Humphrey Davy (1778–1829), who made discoveries using electrolysis, and later aided the work of Michael Faraday (1791–1867). Napoleon, who at that time controlled the territory in which Volta lived, invited the scientist to demonstrate his invention in Paris in 1801. He was so impressed that he made Volta a count, and later a senator, of Lombardy, and awarded him the Legion of Honour medal.

Further Achievements

The volt, the SI unit of electric potential, is named after the Italian. A volt is defined as the difference of potential between two points on a conductor carrying one ampere current when the power dissipated between the points is one watt.

Volta was also the first to isolate methane gas, an achievement made in 1778.

EDWARD JENNER

1749–1823

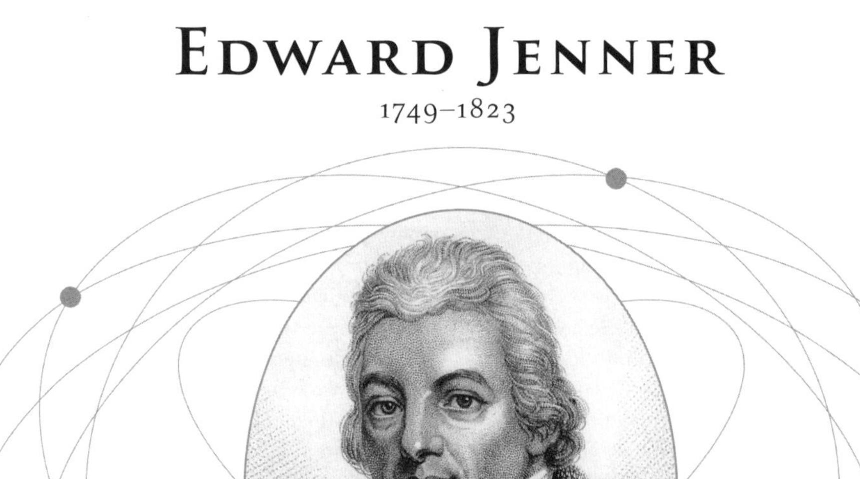

CHRONOLOGY • **14 May 1796** Jenner diagnoses a milkmaid with cowpox. He extracts matter from her pustules to infect an eight-year-old boy • **1 July 1796** Attempts to infect the boy with smallpox. He doesn't contract the disease and becomes the first person to be intentionally vaccinated against it • **1798** *An Inquiry into the Causes and Effects of the Variolae Vaccinae* published

The rapid development of science in the eighteenth century not only changed people's understanding of the world but often affected their ordinary lives. Few scientists would impact all levels of society in the manner of Edward Jenner, who developed the first ever vaccine.

Jenner was a skilled physician, having completed his training as a surgeon in London between 1770 and 1772 under the guidance of the noted surgeon, John Hunter, before returning to his home village of Berkeley in Gloucestershire, England to begin work as a medical practitioner. As well as becoming a successful doctor, Jenner was a keen observer of nature, in particular of bird migration habits and the behaviour of the cuckoo. He was also interested in medical experimentation, undertaking work on chemical treatments for certain diseases and investigating the causes of angina by human dissection.

▸ THE SMALLPOX PREDICAMENT

His big breakthrough came, however, with his experimental work on smallpox. The disease was

Smallpox was a ruthless, endemic disease, killing one in five of its victims

a ruthless, endemic killer in Jenner's time, ending the lives of one in five of those it infected. Survivors, meanwhile, were often left blind or badly disfigured. There was no known cure for the disease and attempts at prevention were limited to a crude form of inoculation known as variolation. This involved an otherwise healthy person deliberately infecting themselves with smallpox from someone who had a mild form of the disease by taking matter from the patient's sores and placing it into open wounds. As smallpox couldn't be caught for a second time, people hoped by this method to give themselves a mild version of the disease and become immune to it for life. Unfortunately this process was risky, with those inoculating themselves frequently contracting a fatal version of the disease.

▸ FIGHT DISEASE WITH DISEASE

In 1796 Jenner attempted a different approach to prevention. He had heard through anecdotal accounts and observed through local outbreaks that milkmaids who had earlier contracted cowpox from their cattle rarely, if ever, seemed to contract smallpox afterwards. Cowpox was a relatively mild disease caught from cow's udders, far less dangerous than smallpox. In May of 1796, one such milkmaid, Sarah Nelmes, came to Jenner's surgery with cowpox. He extracted some matter from his patient's pustules and persuaded a local farmer to allow him to infect his eight-year-old son, James Phipps, with cowpox in an experiment aimed at preventing him from ever contracting smallpox. The boy, as expected, contracted a mild form of cowpox and quickly recovered. Shortly afterwards Jenner attempted to infect Phipps with a lethal dose of smallpox. No disease was contracted. The doctor repeated the attempt on the boy a few months afterwards, and later on other subjects, but smallpox did not develop. The 'vaccination', as Jenner called it (from the Latin for cowpox, *vaccinae*), had been a success. Jenner did not understand the scientific reasons for this but the practical implications were clear.

▸ THE SUCCESS

Jenner communicated his results to the Royal Society in 1797 but they were declined for publication so in 1798 he printed a paper privately: *An Inquiry into the Causes and Effects of the Variolae Vaccinae.* Although the work initially provoked controversy and vaccination was not trusted by everyone its success soon became self evident. From the early 1900s it was acknowledged as the best method for preventing smallpox. The British government agreed, awarding Jenner substantial grants for his work. In 1840 the previous preventative for smallpox, variolation, was outlawed and compulsory vaccination of infants was introduced in 1853. The impact was dramatic, with deaths by smallpox reduced from 40 per 10,000 in 1800 to 1 in 10,000 by 1900. In 1980, with no reported cases at all, the disease was officially declared extinct.

FURTHER ACHIEVEMENTS

Jenner never sought to enrich himself on the back of his discovery and during the time he spent promoting its use he suffered financially as his private practice was neglected.

He adhered the advice given to him by his mentor John Hunter, 'why think? – why not try the experiment?'.

Jenner was the first to observe that the newly hatched cuckoo, not the adult, is responsible for removing the other eggs from the nest.

JOHN DALTON

1766–1844

CHRONOLOGY • **1793** *Meteorological Observations and Essays* published • **1801** Dalton states his Law of Partial Pressure • **1803** Outlines his atomic theory in a lecture • **1808** *A New System of Chemical Philosophy* published

For much of his life, the primary interest of the English Quaker, John Dalton, was the weather. Living in the notoriously wet county of Cumbria, he maintained a daily diary of meteorological occurrences from 1787 until his death, recording in total some 200,000 entries. Yet, it was his development of atomic theory for which he is most remembered.

▸ DIFFERENT ATOMS

It was around the turn of the nineteenth century that Dalton started to formulate his theory. He had been undertaking experiments with gases, in particular on how soluble they were in water. A teacher by trade, who only practised science in his spare time, he had expected different gases would dissolve in water in the same way, but this was not the case. In trying to explain why, he speculated that perhaps the gases were composed of distinctly different 'atoms', or indivisible particles, which each had different masses. Of course, the idea of an atomic explanation of matter was not new, going way back to **Democritus** of Abdera (c.460–370 BC) in ancient Greece, but now Dalton had the discoveries of recent science to reinforce his theory. On further examination of his thesis, he realised that not only would it

Dalton's atomic theory was to transform the basics of chemistry and physics

explain the different solubility of gases in water, but would also account for the 'conservation of mass' observed during chemical reactions as well as the combinations into which elements apparently entered when forming compounds (because the atoms were simply 'rearranging' themselves and not being created or destroyed).

▸ ATOMIC THEORY

Dalton publicly outlined his support for this atomic theory in a lecture in 1803, although its complete explanation had to wait until his book of 1808 entitled *A New System of Chemical Philosophy*. Here, he summarised his beliefs based on key principles such as: atoms of the same element are identical; distinct elements have distinct atoms; atoms are neither created nor destroyed; everything is made up of atoms; a chemical change is simply the reshuffling of atoms; and compounds are made up of atoms from the relevant elements. In the same book he published a table of known atoms and their weights, although some of these were slightly wrong due to the crudeness of Dalton's equipment, based on hydrogen having a mass of one. It was a basic framework for subsequent atomic tables, which are today based on carbon (having a mass of 12), rather than hydrogen. Dalton also wrongly assumed elements would combine in one-to-one ratios (for example, water being HO not H_2O) as a base principle, only converting into 'multiple proportions' (for example, from carbon monoxide, CO, to carbon dioxide, CO_2) under certain conditions. Although the debate over the validity of Dalton's thesis would continue for decades, the foundations for the study of modern atomic theory had been laid and with ongoing refinement were gradually accepted.

Prior to atomic theory, Dalton had also made a number of other important discoveries and observations in the course of his work. These included his 'law of partial pressures' of 1801, which stated that a blend of gases exerts pressure which is equivalent to the total of all the pressures each gas would wield if they were alone in the same volume as the entire mixture.

Dalton also explained that air was a blend of independent gases, not a compound. He was the first to publish the law later credited to and named after Jacques-Alexandre-César Charles (1746–1823). Although the Frenchman had been the first to articulate the law concerning the equal expansion of all gases when raised in equal increments of temperature, Dalton had discovered it independently and had been the first to print.

Dalton also discovered the 'dew point' and that the behaviour of water vapour is consistent with that of other gases, and hypothesised on the causes of the aurora borealis, the mysterious Northern Lights. His further meteorological observations included confirmation of the cause of rain being due to a fall in temperature not pressure.

FURTHER ACHIEVEMENTS

John Dalton began teaching at his local school at the age of 12. Two years later he and his elder brother purchased a school where they taught roughly 60 children.

His paper on colour blindness, which both he and his brother suffered from, and which was known as daltonism for a long while, was the first to be published on the condition. Dalton is also largely responsible for transforming meteorology from being an imprecise art based on folklore to a real science; how much more precise it is nowadays is perhaps debatable!

ANDRÉ-MARIE AMPÈRE

1775–1836

CHRONOLOGY • **1775** Ampère born in Lyons, France • **1799** Marries, and begins teaching mathematics in Lyons • **1802** Accepts a professorship at the École Centrale in Paris • **1808** Appointed inspector general of the newly formed university system by Napoleon • **1809** Appointed Professor of Analysis at the École Polytechnique in Paris • **1825** Announces his empirical law of forces (Ampère's Law)

The Dane, Hans Christian Oersted (1777–1851), was the first to show that an electric current could deflect a magnetic compass needle, thereby proving the long sought-after link between electricity and magnetism. The Englishman Michael **Faraday** (1791–1867) was the first to make real practical use of Oersted's discovery while it was left to the Frenchman André-Marie Ampère to explain the theory behind it. In so doing, he founded the science of electromagnetism, a branch of science which would have profound importance in helping to shape the modern world.

▸ MATHEMATICS

Ampère was first and foremost a brilliant mathematician and it was his skill in this subject which would facilitate his work in electromagnetism. He excelled in mathematics from an early age and he took up his first teaching post in the subject in 1799 in Lyons. In 1802, he became a Professor of Physics and Chemistry at Bourg-en-Brasse reverting back to mathematics as a

The founder of the science of electromagnetism, Ampère did not have the happiest of lives

professor at the Parisian École Polytechnique in 1809. Napoleon had also made him Inspector General of the university system in 1808. While his professional career advanced fairly effortlessly, Ampère did not have the same fortune in his private life. He had lost his father to the guillotine in 1793 in the aftermath of the French Revolution, and his beloved first wife died shortly after their son was born in 1803. He later remarried, but without happiness.

▸ ELECTRODYNAMICS

It was against this backdrop that Ampère later undertook his ground-breaking work in electromagnetism, after seeing a demonstration of Oersted's discovery in 1820. Ampère was by no means a prolific scientist, dipping in and out of the subject at will, but once his interest was stimulated in an area he could work extremely quickly. Within just seven days of learning about the link between electricity and magnetism, he had already conducted experiments and began making his own presentations on the phenomenon. Over the following months, he started to formulate mathematical explanations behind the relationship. It only later became referred to as electromagnetism; Ampère at this time had christened it electrodynamics.

▸ AMPERE'S LAW

In particular, Ampère became interested in the impact that one electric current could have upon another. He had noted two magnets could affect each other and wondered, given the similarities between electricity and magnetism, what effect two currents would have upon each other. Beginning with electricity run in two parallel wires, he observed that if the currents ran in the same direction, the wires were attracted to each other, and if they ran in the opposite direction, they were repelled. He then experimented with other shapes of wires to judge the impact, interpreting all the results mathematically in a bid to find an encompassing explanation for electromagnetism. This ultimately resulted in Ampère's law of 1827, another addition to the succession of 'inverse-square' laws begun with Newton's law of universal gravitation. This showed that the magnetic force between two electricity-carrying wires was related to the product of the currents as well as the inverse-square of their distance apart. This meant that if the distance between the wires were doubled, the magnetic force would reduce by a factor of four.

Further Achievements

The Solenoid

Ampère also introduced some important devices, inventing the solenoid, a cylindrical coil of wire which becomes an electromagnet when electricity is passed through it. The solenoid is most commonly used in devices such as bells and valves, where mechanical motion is required.

The Galvanometer

Ampère exploited Oersted's work, devising an early galvanometer which measured electric current flow via the degree of deflection upon its magnetic needle.

The Ampere

In addition, and perhaps most famously of all, the SI unit of electric current, the ampère, is named after him.

AMEDEO AVOGADRO

1776–1856

CHRONOLOGY • **1776** Avogadro born in Turin, northern Italy, of a long line of lawyers • **1800** Takes up the study of mathematics and physics • **1809** Becomes Professor of Physics at the Royal College at Vercelli • **1811** Proposes theory of volume of gas • **1834–50** Appointed Professor at Turin for a second time • **1856** Dies in Turin • **1860** Stanislao Cannizzaro 'rediscovers' Avogadro's theory and is a forceful advocate at a large chemical scientists' conference

Imagine working enthusiastically in scientific research for most of your life and your contribution to the advancement of the subject being a single achievement or theory. Indeed, many scientists invest years of effort to find themselves in this position and are rightly recognised for their one addition. Now imagine, however, your only theory of note being almost completely ignored in your lifetime and for half a century after you originally proposed it, even though it was the one idea scientists had needed for fifty-years to advance their field! Amedeo Avogadro, who found himself in exactly this position, would have had every right to die a frustrated man.

▸ COMBINATION OF ATOMS

The theory the Italian scientist formulated in 1811 involved a way of integrating the apparently irreconcilable hypotheses of Joseph Louis **Gay-Lussac** (1778–1850) and John **Dalton** (1766–1844). Indeed, the latter had actively

Avogadro's theory, despite its crucial nature, was completely ignored during his lifetime

sought to discredit Gay-Lussac's law of combining volumes. Gay-Lussac had observed that gases always combined in simple, consistent ratios of whole numbers such as 2:1 or 2:3 (and never in fractions), under the same temperature and pressure conditions. Dalton struggled to accept this because he believed, as a base case, that gases would seek to combine in a one atom to one atom ratio (hence believing the formula of water to be HO, not H_2O). Anything else would contradict Dalton's theory on the indivisibility of the atom, which he was not prepared to accept.The reason for the confusion was because at that time the idea of the molecule was not understood. Dalton believed that in nature all elementary gases consisted of individual atoms, which is true, for example, of the inert gases. This is not the case, however, for other gases which naturally exist, in their simplest form, in combinations of atoms called molecules. In the case of hydrogen and oxygen, for example, their molecules are made up of two atoms, described in chemical notation as H_2 and O_2 respectively. Avogadro realised a comprehension of molecules would explain Gay-Lussac's ratios while at the same time not contradicting Dalton's theories on the atom. For example, by this method Gay-Lussac's ratio for water could be explained by two molecules of hydrogen (making four 'atoms') combining with one molecule of oxygen (or two 'atoms') to result in two molecules of water ($2H_2O$). When Dalton had considered water previously, he could not understand how one 'atom' of oxygen could divide itself (thereby undermining his indivisibility of the atom theory) to form two particles of water. The answer, that oxygen existed in molecules of two and therefore the atom did not divide itself at all, was exactly what Avogadro proposed.

▸ AVOGADRO'S LAW

He built on this principle to famously suggest that at the same temperature and pressure, equal volumes of all gases have the same number of molecules. This became known as 'Avogadro's Law.' The principle in turn allowed a very simple calculation for the combining ratios of all gases, merely by measuring their percentages by volume in any compound (which in itself facilitated simple calculation of the relative atomic masses of the elements of which it was made up).

▸ REDISCOVERY OF AVOGADRO

Few scientists (André-Marie **Ampère** (1775–1836) was an exception) accepted Avogadro's speculation, partly due to lack of experimental evidence, until the Italian Stanislao Cannizzaro 'rediscovered' it and vehemently backed its suggestions at a large conference of chemical scientists in 1860. The law was consequently accepted by many present, immediately clearing up the confusion of the previous fifty years concerning atoms and molecules, and the calculation of the relative atomic and molecular masses of elements.

THE INFLUENCE OF AVAGADRO

As we have seen, Avogadro stands out in our collection of great scientists for having produced only a single theory, which was not recognised until after his death. He merits inclusion in this collection due to the contribution his theory made to molecular biology.

Avogadro was finally vindicated in 1860, four years after his death. In his honour, Avogadro's constant, which expresses the number of particles in a single 'mole' of any substance, currently given as 6.022 1367(36)$\times 10^{23}$, was named after him.

JOSEPH GAY-LUSSAC

1778–1850

CHRONOLOGY • **1816** Serves as joint editor of the *Annales de chimie et de physique* • **1818** Becomes a member of the government gunpowder commission • **1829** Appointed Director of the Paris Mint Assay Department • **1831** Elected to the Chamber of Deputies • **1848** Resigns from his various appointments in Paris and retires to the country

The Frenchman Joseph-Louis Gay-Lussac may not have died for his science but in 1809 he came close. Having prepared large quantities of sodium and potassium following the first successful isolation of the elements by the Englishman Humphry Davy, Gay-Lussac began using them in other chemical experiments, one of which went spectacularly wrong, blew up his laboratory and left him temporarily blinded. Such were the dangers facing the early chemical investigators. But if the risks were high, so were the rewards, as Gay-Lussac's enduring fame testifies.

▸ THE LAWS OF GAS

Although he went on to make many original contributions to chemistry, inheriting Antoine Lauren **Lavoisier's** (1743–94) mantle as the outstanding French scientist of his day, Gay-Lussac's first major contribution was not his own. Instead, in 1802, he brought to the world's attention a chemical law discovered by his countryman Jacques-Alexandre-César Charles (1746–1823) fifteen years earlier, but which his friend had chosen not to publish. Together with **Boyle's** law, the principle now known as Charles' law (although sometimes also named after Gay-

Gay-Lussac's experiments were as noted for their spectacular nature as for their results

Lussac because of his popularisation of it) completed the two 'gas laws'. It stated that a fixed amount of any gas expands equally at the same increments in temperature, as long as it is also at consistent pressure. Likewise, for a decline in temperature, all gases reduce in volume at a common rate, to the point at about -273°C, where they would theoretically converge to zero volume. It is for this reason that the Kelvin temperature scale later fixed its zero degree value at this point. While the law was not Gay-Lussac's own, his experimental proof was more accurate than Charles'. This helped it to gain acceptance when it was finally published (ironically at around the very time John **Dalton** made the same discovery).

One principle which is entirely attributable to Gay-Lussac is his 1808 articulation of the law of combining volumes. Having confirmed by experimental evidence in 1805 that water was made up of one part oxygen and two parts hydrogen (H_2O) and having further proceeded to break down numerous other compounds, Gay-Lussac noted that gases always combined with each other in simple, small numerical ratios (such as 2:1 or 2:3), and never in fractions. At the time the reason for this was not properly understood. Indeed, John Dalton (1766–1844) sought to discredit Gay-Lussac's conclusion because it appeared to conflict with his own theories on the indivisibility of the atom. In 1811 Amedeo **Avogadro** (1776–1856) would provide a framework for both men's theories to work in parallel through his distinction of atoms and molecules (although this too was largely ignored until Stanislao Cannizzaro (1826–1910) rediscovered and articulated it in 1860).

▸ A VOYAGE OF CHEMICAL DISCOVERY

Gay-Lussac would spend much of the remainder of his working life engaged in relentless chemical experimentation, either uncovering new compounds and elements, or greatly improving scientific understanding of the properties of other recently discovered substances. Much of this work was carried out in conjunction with his compatriot Louis Thenard. Together they discovered boron and undertook research into the 'new' element iodine, giving the chemical the name by which we now know it. In 1815 they were the first to create the compound cyanogen, discovering that this was the first in a series of related compounds called the cyanides. The duo conclusively proved Lavoisier's assumption to be wrong– that all acids had to contain oxygen.

Later work included detailed investigations into the properties and reactivity of nitrogen and sulphur, as well as research into the process of fermentation. In addition, Gay-Lussac also worked on modernising chemical experimentation techniques and is credited with creating a precise method of volumetric analysis.

Further Achievements

Gay-Lussac was as famous for the spectacular nature of his investigative work as for the results it produced.

As well as blowing up his laboratory, he had earlier gained fame for dangerous ascents in balloons, conducted in the name of scientific research. He first took to the skies in 1804 with Jean Baptiste Biot (1774–1862), and later on his own (attaining a then world-record height of 23,000 ft/7km) to investigate the air's composition and magnetic force at these altitudes. His results demonstrated that there was no change in either value from measurements taken on the ground.

CHARLES BABBAGE

1791–1871

CHRONOLOGY • **1815** Babbage helps to found the Analytical Society • **1823** Begins work on the machine later known as his Difference Engine No.1 • **1828–39** Lucasian Professor of Mathematics at Cambridge University, a post previously held by Isaac Newton and later Stephen Hawking • **1833** Work on the Difference Engine is abandoned as Babbage runs out of money • **1991** Doron Swade and his team at London's Science Museum complete the construction of a working No.2; it has functioned ever since

On his death in 1871, the *Times* mocked Charles Babbage, his work was virtually unknown to the wider public and those that had heard of it were generally unappreciative. The Reverend Richard Shipshanks, Babbage's harshest critic, wrote in 1854 with unconscious irony, that for all the public money invested in Babbage's work, 'We should at least have had a clever toy'. It had all begun so well for Babbage. Educated at Cambridge, he proved himself to be a brilliant mathematician, graduating in 1814 and receiving his MA three years later. In 1822, he began designing what was to become the world's first automatic calculator and after a meeting with the Chancellor of the Exchequer in June 1823 he obtained £1500 to fund the creation of his far-sighted vision. It would become known as the Difference Engine No.1 and would dominate the next ten years of Babbage's life.

The mathematician was driven to attempt the building of such a revolutionary concept and

Babbage's machine 'should be used to calculate the time at which it would be of use' Robert Peel

then persisted with it for so long because of his frustration with the alternative: books of mathematical tables written by teams of number-crunchers to help with complicated calculations. Due to human error they were inevitably prone to mistakes. Babbage was a champion of machines and the scientific approach, taking his enthusiasm to the point of eccentricity. He believed that if a mechanical solution to producing complex calculations could be devised, accuracy would always be assured.

▸ THE ANALYTICAL ENGINE

The government's patience finally ran out in 1834. Ironically this happened at the moment, and in part because of, Babbage's greatest conception: the first programmable computer. He called it his Analytical Engine. It was much more than a calculator, rather an all-purpose computing machine similar in concept to the modern computer. His design envisaged 'programs' written using loops of punched cards. It included a reader able to process the instructions they contained, a 'memory' which could store the results, 'sequential control' and other logical features which would become components of twentieth century computers. Babbage approached the government for more money to build his new machine, even though the Difference Engine was still unfinished. He argued that it would be cheaper and more beneficial to build the Analytical Engine than to make the necessary changes to finish the first machine. With the original still unfinished, the government was reluctant to fund another ambitious project – it had already swallowed up £17,000 of public money. Babbage persisted in making appeals to the government for extra capital. The scepticism reached the highest echelons of government. Robert Peel once commented wryly, when he was Prime Minister, that Babbage's machine should be employed to 'calculate the time at which it would be of use'.

▸ THE DIFFERENCE ENGINE NO.2

In 1842 the government confirmed it was definitely pulling the plug on Babbage's original project (even though no real work had been undertaken on it for nearly a decade) and there would be no money for his new project. Babbage persisted into the 1850s in trying to raise funds for his Analytical Engine but it never got beyond the design stage. By then, the mathematician had also designed the Difference Engine No.2, a much simpler model which used only a fraction of the 25,000 parts the first engine had employed. It was also slightly smaller than the first at six and a half feet tall rather than eight. Again, however, there was no funding from the government and it failed to progress beyond the design stage. To commemorate the bicentenary of his death in 1991, a team from the London Science Museum finally built a working replica of the No.2 based on Babbage's original plans.

FURTHER ACHIEVEMENTS

Babbage founded a number of societies, including the Royal Astronomical Society in 1820. He also made advances in the theory of algebra, and was responsible for or played a part in numerous other inventions, from the speedometer and the locomotive cowcatcher to the standard railway gauge and uniform postal rates. In addition, Babbage was the influence behind an Act of Parliament introduced to curtail the rights of street musicians; his Passages from the Life of a Philosopher *(1864) includes an interesting diversion on the social ill of street noise.*

MICHAEL FARADAY

1791–1867

CHRONOLOGY • **1821** Faraday creates the first electric motor • **1823** Accidentally liquefies chlorine • **1831** Discovers the principle that will lead to the creation of the electric generator, the transformer and the dynamo • **1833** States the basic laws of electrolysis • **1845** Discovers the Faraday effect

Michael Faraday is regarded as one of the greatest experimental scientists of all time. Even Albert **Einstein** (1879–1955) considered him to be one of the most important influences in the history of physical science. Yet the man whose discoveries and inventions, amongst them the electric motor, electric generator and the transformer, were to have such a profound impact on modern life, might not have entered the scientific arena at all but for certain fortuitous events in his youth. The first was his apprenticeship at a bookbinder's when he was thirteen. Here his interest in science and in particular electricity was stimulated upon reading pages from the books he was tasked to bind. The second fortunate incident was his appointment as assistant to the renowned chemist Sir Humphrey Davy (1778–1829), who had remembered the young Faraday attending his lectures. The temporary post soon turned permanent and shortly afterwards Davy took Faraday with him on a grand European tour which gave the young man the rare opportunity to meet and learn from many of the leading physicists and chemists of the day.

Faraday was considered by Einstein to be one of the most influential physical scientists of all

Much of Faraday's early work as a scientist in the 1820s was not in physics, the area which ultimately led to his breakthrough inventions, but in chemistry. In 1823, he became the first person to liquify chlorine, albeit accidentally, while he was conducting another experiment. He quickly deduced how the new form of chlorine had been obtained and applied the process, which made use of pressure and cooling, to other gases. By employing his talent as an outstanding analyst of his own chemical experiments, he also went on to discover benzene in 1825.

▸ THE ELECTRIC MOTOR

Yet it is physical science, in particular his work involving electricity, for which Faraday is best remembered today. As early as 1821, he was able to create the first electric motor after discovering electromagnetic rotation. He had developed Hans Christian Oersted's (1777–1851) 1820 discovery that electric current could deflect a magnetic compass needle. Faraday's experiment proved that a wire carrying an electric current would rotate around a fixed magnet and that conversely, the magnet would revolve around the wire if the experiment were reversed. From this work, Faraday became convinced that electricity could be produced by some kind of magnetic movement alone but it took ten further years before he successfully proved his hypothesis. In 1831, by rotating a copper disk between the poles of a magnet, Faraday was able to produce a steady electric current. This discovery allowed him to go on to produce electrical generators, the transformer (also invented independently at around the same time by an American, Joseph Henry) and even the dynamo: inventions which can truly be claimed to have changed the world!

▸ ELECTRICAL FIELDS

The reason Faraday was able to make such advances was because from early in his career he had rejected the concept of electricity as a 'fluid', an idea that had been accepted up until that time, and instead visualised its 'fields' with lines of force at their edges. He believed that magnetism was also induced by fields of force and that it could interrelate with electricity because the respective fields cut across each other. Proving this to be true by producing an electric current via magnetism, Faraday had discovered electromagnetic induction. He was encouraged by this and went on to explore the idea that all natural forces were somehow 'united'. He then focused on how light and gravity were related to electromagnetism. This led to the discovery of the 'Faraday effect' in 1845 which proved that polarised light could be affected by a magnet. James Clerk **Maxwell** proved that light was indeed a form of electromagnetic radiation, and eventually provided the mathematical expression for Faraday's law of induction.

THE LAWS OF ELECTROLYSIS

Faraday's fascination with electricity and his background in chemistry both found a natural expression in electrolysis, in which he was also to perform ground-breaking work. In 1833 he was the first to state the basic laws of electrolysis, namely that: (1) during electrolysis the amount of substance produced at an electrode is proportional to the quantity of electricity used and (2) the quantities of different substances left on the cathode or anode by the same amount of electricity are proportional to their equivalent weights.

CHARLES DARWIN

1809–1881

CHRONOLOGY • **1831–36** Darwin takes the job of unpaid naturalist aboard HMS Beagle • **1859** Publishes *The Origin of Species* • **1871** Publishes *The Descent of Man* • **1881** Dies and is buried at Westminster Abbey

The spark for Darwin's accomplishments was ignited with the 1831 HMS *Beagle* expedition, which was to chart coastlines in the South Americas and other areas of the Pacific. Darwin, supposedly studying religion at that time, had become increasingly absorbed with natural history and persuaded the Professor of Botany, John Henslow, to put him forward for the post of unpaid naturalist on the Beagle's voyage. He thereby abandoned his university studies. His father, and initially the vessel's Captain FitzRoy, resisted, but he eventually persuaded them to let him take part in the five-year expedition.

▸ THE GALAPAGOS

During the journey, Darwin made many geological and biological observations, but it was his time spent around the Galapagos Islands which would end up having the most significant impact on him. The ten islands are relatively isolated, even from each other, and as such act as a series of distinct observatories through which Darwin could draw comparisons. He noted that the islands shared many species of flora and fauna in common, but that each land mass often displayed distinct variations within the same group of organisms. For example, he famously noted fourteen different types of finch across the

'Man, with all his noble qualities, still bears the indelible stamp of his lowly origin'

islands, notably with different shaped beaks. In each instance the particular beak seemed to best suit the capture of that bird's prevalent food source, whether it be seeds, insects or fish.

Over the ensuing years, and upon his return to England, Darwin pondered on the reasons for the variations in the finches and other plants and animals. He soon surmised that the birds had descended from a single parent species, rather than each springing up independently and thus acknowledged the idea of evolution, a concept which had existed for some time but was not widely accepted. Darwin began looking for an explanation for this evolution. One text which had a particular impact on him was Thomas Malthus's 1798 work *An Essay on the Principle of Population* which Darwin read in 1838. Malthus had been concerned with overpopulation resulting in famine, and the possible competition for food which could ensue. Darwin immediately saw that this could also be applied to the animal world too, where only those best adapted to food collection in their environments would survive. Those that could not compete would die out and the characteristics of the successful animals, which may have occurred in the first place by chance, would be passed on to future generations. As environments changed and animals moved about, success criteria would change, gradually resulting in variations within species, as had happened with the finches. Ultimately new species would also be created.

CHALLENGING THE NOTION OF GOD

Unfortunately, such a hypothesis would challenge the commonly held view of man as the lord of the earth, specifically created and placed upon the planet in God's image, as described in the Bible. Darwin was implicitly suggesting that man had evolved by chance over thousands of years. He correctly anticipated uproar and resistance to his ideas, particularly from religious leaders. Consequently, he kept his theories dark for twenty years while he gathered additional evidence to back up his case.

He finally published in 1858. He did this jointly with Alfred Russell Wallace (1823–1913), whose independent ideas were remarkably similar to Darwin's. They agreed to a joint public declaration of their hypotheses by submission of a paper to the Linnean Society. Darwin followed this up with a more detailed account in 1859 containing evidence he had collected over the previous decades called *On The Origin of Species by Means of Natural Selection.*

The predicted outcry ensued and a fierce debate followed, but Darwin already had a number of friends, particularly Thomas Huxley, known as 'Darwin's Bulldog', who would vigorously defend his ideas. This left Darwin free to follow through further implications of his hypothesis in other works, including the 1871 text *The Descent of Man,* which articulated the idea of the evolution of the human race from other creatures.

THE LEGACY OF DARWIN

Darwin's ideas took a long time to become generally accepted (even today they are not embraced by everyone), challenging as they did all previous conceptions of what it meant to be human. As has been the case with so many scientists, he encountered paticularly fierce opposition from the Church, whose members preferred the safety of a sacred text to the uncertainties of observation and experiment.

The idea of evolution through natural selection is, however, at the heart of modern biology. The man who disappointed his father for lack of academic interest had eventually gone on to turn an entire branch of academia on its head.

JAMES JOULE

1818–1920

CHRONOLOGY • **1840** Joule discovers the properties of Joule's Law • **1840s** Determines the principle of the conservation of energy • **1849** *On the Mechanical Equivalent of Heat* published • **1852–59** Together with William Thomson (later Lord Kelvin) he describes the Joule-Thomson effect

Well into the nineteenth century, scientists still did not fully understand the properties of heat. The common belief held that it was some form of transient fluid, retained and released by matter, called 'caloric.' Gradually the idea that it was just another form of energy, expressed as the movement of molecules, began to gain ground. It came to be understood in no small part due to the work of an Englishman, James Joule, who contributed much to the founding of the science of thermodynamics in the process.

▸ JOULE THE BREWER

Joule was actually a brewer by trade and not a full time scientist. Working throughout his life at his wealthy father's brewery, he had not benefited from a formal education, and indeed never attended university nor held an academic post throughout his life, which makes his findings all the more remarkable.

His interest in the phenomenon of heat, however, led to his father building him a laboratory near the brewery. The subject would dominate Joule's studies for the rest of his life.

The discoverer of the First Law of Thermodynamics was a brewer by trade

▸ UNDERSTANDING HEAT

Joule began by examining the relationship between electrical current and resistance, and the heat that they produced. In 1840 this led to his first major achievement, the expression of 'Joule's Law' which mathematically determined the link between current and the resistance in the wire through which it passed in terms of the amount of heat given off. This had added important since it effectively meant that one form of energy was transforming itself into another: electrical energy to heat energy for example, and it undermined the concept of the caloric.

▸ HEAT FROM ENERGY OR WORK

Joule pursued this line of enquiry over the following decade. He proved that heat could be produced from many different types of energy or work, including mechanical energy. He proved it in an experiment in which a paddle was turned by a handle in water and the temperature of the water was seen to rise as a result of this work. Indeed, one of Joule's key skills was his ability to quantify the equivalence of different forms of energy. He used his paddle experiment to deduce the amount of mechanical effort needed to be applied to raise the water temperature by one degree Fahrenheit. From this he formulated a value for the work necessary to produce a unit of heat. Afterwards he summarised his results in an 1849 paper *On the Mechanical Equivalent of Heat,* which brought him acclaim.

▸ HEAT IN GASES

Joule went on to study the role of heat and movement in gases. In 1848 he provided the first estimate of the speed at which gas molecules moved. From 1852 until the end of the decade – together with William Thomson, who later became Lord **Kelvin** (1824–1907) – he continued experiments in this area, and described what became known as the 'Joule-Thomson effect'. This demonstrated how most gases actually lose temperature on expansion due to work taking place to pull apart the molecules. It was a discovery put to wide practical use in the growth of the refrigeration industry later in the century.

▸ JOULE'S LAW

In addition to the First Law of Thermodynamics (see below), Joule also discovered the law that bears his name. It describes the conversion of electrical energy into heat, and states that the heat (Q) produced when an electric current (I) flows through a resistance (R) for a time (t) is given by $Q=I^2Rt$.

The SU unit of energy or work, the joule, is also named after him.

The First Law of Thermodynamics

Joule's most significant discovery was a finding he made during the course of his 1840s experiments. He determined what later became known as the first law of thermodynamics: the principle of the conservation of energy. This was a natural extension of his work on the ability of energy to transform from one type to another. Joule contended that the natural world had a fixed amount of energy within it which was never added to or reduced, but which just changed form.

Julius Robert von Mayer (1814–78) and Hermann Ludwig Ferdinand von Helmholtz (1821–94), independent of Joule and each other, came to similar conclusions at around the same time and are also credited with the discovery of the law.

LOUIS PASTEUR

1822–1895

CHRONOLOGY • **1862** *Mémoire sur les corpuscles organisés qui existent dans l'atmosphère* (*Note on Organized Corpuscles that Exist in the Atmosphere*) published. It puts an end to centuries of debate on the theory of spontaneous generation • **1880** An accident by an assistant leads to Pasteur's development of vaccines • **1885** Pasteur successfully uses his rabies vaccine on a nine-year-old boy, Joseph Meister • **1892** Produces a successful vaccination against anthrax

Louis Pasteur's name is best remembered for his development of the process of 'pasteurisation.' Though Pasteur was a chemist his most significant breakthroughs were in medicine. Indeed, he is considered by many to be the most important figure in nineteenth century medical research. Much of this reputation hinges on his development of a vaccine against rabies. After Edward **Jenner's** (1749–1823) breakthrough of a vaccine against smallpox made at the end of the previous century, little more had been done to take advantage of the potential of this treatment against other diseases. In 1880, however, Pasteur was to recognise and manipulate a chance occurrence that he noticed in his laboratory to finally systemise a scientific approach to the development of vaccines.

▸ BACTERIA IN CHICKENS

Some chicken cholera bacteria had accidentally been left alone for a long period. Pasteur noticed

Pasteur's process contributed greatly to improving the fermentation of wine and beer

that when he injected this into chickens they did not develop, or only suffered a mild form of, the disease normally associated with the bacteria. When he later injected the same chickens with fresh bacteria, they survived, while others which had not received the earlier treatment quickly died. Pasteur drew parallels between this result and the work of Jenner and set about deliberately applying the approach to other diseases.

RABIES AND PASTEURISATION

By 1882 he had successfully produced a vaccination against anthrax, a disease which seldom affected humans, but which could devastate stocks of sheep and cattle. By 1885 he had developed a vaccine, extracted from the spines of infected rabbits, to successfully treat animals for rabies.

Pasteur's apprehension at performing a trial on humans was cast aside when a nine-year old called Joseph Meister was brought to him. The boy had been bitten several times by a rabid dog. Pasteur injected him with the new vaccine and the boy survived. Word of the success spread and the following year over 2500 infected patients received the same treatment. As a result, fatalities dropped to less than 1%. As well as the immediate benefit and fame that Pasteur's development brought him, it also prompted a rush by other scientists to begin searching for new vaccines for other diseases. Several more successes were heralded by the end of the century.

Prior to this, Pasteur had helped limit the spread of tuberculosis and typhoid through the application of his pasteurisation process. This was developed during his studies on the fermentation of milk and alcohol. Through microscopic examination and other experiments he conclusively countered the prevailing argument of the day which held that it was merely a chemical process. Pasteur proved that microorganisms were essential for fermentation to take place. He also found that potentially dangerous microbes existing in milk, such as those which caused tuberculosis and typhoid, could be destroyed by heating the liquid for about thirty minutes at 63°C. This is now known as pasteurisation, still used to treat milk.

During the same period of work, Pasteur also conclusively disproved 'spontaneous generation' theories, which had persisted for centuries. He demonstrated that sterilised fluid not exposed to microbes in the air would remain uncontaminated, whereas the moment the liquid was put into contact with them it became spoiled.

In addition, from 1865 he greatly aided the French silk industry. By analysing diseases which decimated silkworms he eventually provided successful recommendations for their prevention. Pasteur undertook important work early in his career on the discovery of asymmetrical molecules in compounds which did much for the later development of structural chemistry.

Pasteur used a similar process to pasteurisation to improve the success of fermentation in the wine and beer-making industries.

A Memory of Pasteur

By the time of his death, Pasteur was world-famous and tributes poured in. Perhaps the most dramatic gesture of all, however, came almost a half-century later. The nine-year-old boy, Joseph Meister, whom Pasteur had saved from rabies, went on to become caretaker at the Pasteur Institute (founded in 1888) where the scientist was buried. In 1940 the Nazis arrived in Paris and ordered Meister to open Pasteur's tomb in order to examine it. Meister chose to kill himself rather than comply with the violation.

JOHANN GREGOR MENDEL

1822–1884

CHRONOLOGY • **1856** Begins his pea-plant experiments • **1865** Mendel first articulates his findings • **1866** *Experiments with Plant Hybrids* published • **1868** Elected as abbot of his monastery where his duties prevent him furthering his research

The work of the Austrian monk, Johann Gregor Mendel, would be at the heart of the future development of biology, and founded a new branch of science in its own right. During his lifetime and for some time afterwards, however, his efforts were largely ignored. Only when others started making similar discoveries in heredity and began looking for related studies was it realised that Mendel had got there decades before them.

▸ JUST TWELVE YEARS

The later impact of Mendel's findings, which effectively act as the starting point for the modern science of genetics, was even more remarkable given the limited amount of his 'career' actually spent researching them. Up to 1856 his time was spent on religious duties or training at his monastery, or in trying to improve his limited early education sufficiently to allow him to pass his teaching examination. Ironically he never succeeded in attaining the qualification, in part due to his lack of success in biology! From 1868 he became abbot of his monastery, located in the modern day Czech Republic, and had to give up the majority of his scientific research. This meant he completed only twelve years of active experimentation.

Mendel, the father of genetics, had his lowest mark in an examination in biology

▸ THE HUMBLE PEA-PLANT

Even the arena for the breakthrough was an unusual one. Mendel's laboratory was the monastery's garden, his subject the humble pea-plant. The monk had been fascinated by what caused the different characteristics of these plants to occur, such as blossom colour, seed colour and height. He decided to undertake a systematic study of when these features occurred in descendent generations. He set about cross-fertilising plants with different characteristics and recording the results.

▸ UNDERSTANDING HEREDITY

Common assumption at the time was that when two alternate features were combined, an averaging of these features would occur. So, for example, a tall plant and a short one would result in a medium height offspring. The statistical results Mendel collaborated, however, proved something entirely different. Across a series of generations of descendents, plants did not average out to a medium, but instead inherited the original features (for example, either tallness or shortness) in a ratio of 3:1, according to the 'dominant' trait (in the example of height, this was the tall characteristic). He explained this by assuming each parent carried two possibilities for any given trait, for example a tall 'gene' (as we now know it) and a short one for height, or a dark gene and light one for seed colour, or gene 'A' and gene 'B' for 'X' trait. Only one gene from each parent would carry into the offspring (now described as Mendel's law of segregation), however, giving four possible combinations: AA, AB, BA and BB. The 3:1 ratio would be achieved because the 'dominant' gene would feature whenever it were present. So if 'A' were the dominant factor, it would occur three times in four, with the 'B' scenario only occurring when a BB result was obtained. He also noted the different pairs of genes making up the characteristics of the pea plant, such as the two causing height, the two causing seed colour, and so on, when crossed occurred in all possible mathematical combinations, independently of each other. This is now described as Mendel's law of independent assortment, and offered him a simple statistical model for predicting the variety of descendents, backed up by ongoing experimental proof.

▸ DELAYED RECOGNITION

Mendel first articulated his results in 1865, and published them in an article of 1866 entitled *Experiments with Plant Hybrids*. He was frustrated that the conclusions were largely ignored in his lifetime and it was only when three other scientists, Hugo de Vries (1848–1935), Karl Erich Correns (1864–1933) and Erich Tschermak von Seysenegg (1871–1962) independently came across similar experimental evidence in 1900 that Mendel's work was rediscovered. Its importance in explaining principles of heredity across all sorts of life forms (although with refinement in some areas) was soon realised, and it was later used to underpin **Darwin's** argument for natural selection too. The science of what is now known as genetics gradually evolved, and Mendel's position as its, albeit unwitting, founder became cemented in history.

MENDEL'S LEGACY

Although he did not gain any recognition for his work on heredity during his lifetime, he was well respected and liked by his fellow monks and townspeople. Nowadays, Mendel is regarded as the father of the study of genetics

JEAN-JOSEPH LENOIR

1822–1900

CHRONOLOGY • **1807** Francois Isaac de Rivaz builds an early internal combustion engine powered by oxygen and hydrogen • **1859** Lenoir demonstrates his electric-spark internal combustion engine • **1860** Patents his engine • **1863** Uses his engine to power a vehicle • **1876** Nikolaus August Otto introduces a superior four-stroke engine

Unbelievably, the origins of the internal combustion engine, today at the heart of the automobile industry, go back to 1680 and the famous Dutch scientist Christiaan **Huygens** (1629–95). In that year he outlined a design for a primitive form of such an engine which used gunpowder to spark and drive its pistons. The design, however, was never constructed and it would be over a century before anyone would come close to reviving the idea. This time it was a Swiss inventor, Francois Isaac de Rivaz, who in 1807 actually built an early internal combustion engine powered by oxygen and hydrogen. He constructed an accompanying vehicle in which to place it, but the design, being largely impractical, was commercially unsuccessful. It is for this reason, then, that the design patented by the Belgian Jean-Joseph Étienne Lenoir in 1860 is regarded as the first viable internal combustion engine, beginning the revolution in the transport industry which would change the world.

▸ THE ELECTRIC-SPARK INTERNAL COMBUSTION ENGINE

Even as child, Lenoir was fascinated by the

'When I am tall, I will make machines, new machines, machines working all alone'

power of mechanical devices. When only twelve years old he is said to have declared, 'When I am tall, I will make machines, new machines, machines working all alone.' Finding little use for his inventive talents in the small town in which he grew up, Lenoir moved to Paris at the age of sixteen. He initially worked with electricity, making a number of breakthroughs in electro-plating as well as electrical devices for use in the railway industry. But it was engines which absorbed Lenoir and, after many years of design and finally construction, he demonstrated his electric-spark internal combustion engine in 1859, patenting it the following year.

Liquid petroleum fuels were not in use at that time, so Lenoir's engine used a combination of uncompressed air and coal gas to power it in a two-stroke design. Although the method was still primitive by today's standards, it was commercially successful. Lenoir worked on subsequent improvements and had sold some 500 versions in and around Paris by 1865. By that point he had already demonstrated its versatility, using it to power a boat in 1861 and a land-based three-wheeled vehicle in 1863.

▸ LATER DEVELOPMENTS

Lenoir's designs would not endure for long, though. The Frenchman Alphonse Beau de Rochas patented the now much more common four-stroke internal combustion engine design in 1862. He did not, however, actually build the engine, which is why the German Nikolaus August Otto (1832–91) became much more renowned for his commercially successful four-stroke model in 1876. It worked using the 'Otto Cycle,' his explanation for the series of actions which took place on each of the strokes to power the engine. On the first induction stroke, the fuel is drawn into the engine, it is then compressed on the second stroke, ignited on the third using an electric-spark and the residue of gases emitted on the fourth. This process drives a piston up and down which, with the appropriate connecting rods, can be used to power a multitude of devices, most commonly, of course, land vehicles.

Otto's engine still used a combination of gas and air to fuel it and it would not be until the pioneering work of Gottlieb **Daimler** (1834–1900) and others that the petrol internal combustion engine would be invented and following it the widespread manufacture of motorcycles and cars. Indeed, Daimler had worked for Otto and been instrumental in the development of the engine attributed to the latter. It was only when he later set out on his own, however, that the automobile revolution truly began.

His engine typically produced about two horsepower, which was not really big enough to power a vehicle of any size and with its fuel mixture of coal-gas and air was only around four per cent efficient. They were of sturdy construction, however, with some models still running perfectly after twenty years of continuous operation.

FURTHER ACHIEVEMENTS

In addition to his engine, Lenoir also developed a number of other inventions. They include an electric brake for trains (1855), a motorboat using his engine (1886), and a method of tanning leather with ozone.

LORD KELVIN

1824–1907

CHRONOLOGY • **1834** William Thomson enters Glasgow University at the age of ten • **1852–59** Together with James Joule he describes the Joule-Thomson effect • **1858** Patents a telegraph receiver, called a 'mirror galvanometer', for use with submarine cables • **1892** Thomson enters the House of Lords and becomes Lord Kelvin

William Thomson, a Scotsman, was clearly destined for great things from a very young age; he entered Glasgow University at the age of just ten! After studying scientific subjects there he went on to Cambridge, graduating in 1845, and the following year was offered a post as chair of natural philosophy (physics). He accepted and remained in the position for over fifty years, a half-century during which he would exercise significant international authority in his field.

THERMODYNAMICS

Working over such a long period gave Thomson the opportunity to experiment with a wide range of subjects, mostly within the sphere of physics. His key influence, however, was in two areas: thermodynamics and electromagnetism. The former involved a large amount of collaborative work with James Joule, the other dominant British authority on the subject during the nineteenth century, who is among those credited with the first law of thermodynamics. The wider acceptance of this conservation of energy principle, which states energy cannot be created or

The Kelvin scale, used mainly for scientific purposes, defines absolute zero as -273°C

destroyed, was in part due to Thomson's systematic description of it in a paper of 1852. Moreover, the Scotsman went on to independently enunciate and publicise the second law of thermodynamics which described the one way nature of heat flow. Heat can only flow spontaneously from a hotter to a colder body and never from a colder to a hotter one. The German Rudolf Clausius (1822–88) also arrived at the same conclusion during a similar period to Thomson. Together with Joule, Thomson discovered the 'Joule-Thomson effect', whereby most gases actually lose temperature on expansion (from a small nozzle) due to work taking place to pull apart the molecules. This finding was of decisive importance in the growth of the refrigeration industry later in the century.

Thomson also attempted to estimate the ages of the Sun and Earth from theoretical work on the rate at which a hot sphere would cool, but his estimates were significantly below modern calculations due to no knowledge of the heat produced by the phenomenon of radioactivity.

▸ THE KELVIN SCALE

One of the reasons Thomson's peerage name, Kelvin, is so widely known today is because of the work he carried out in defining an absolute temperature scale. He undertook theoretical work to predict that -273.16° Celsius is the point at which the molecules in a substance reached their least energy; nothing could become colder below this point named 'absolute zero'. from this he proposed the Kelvin temperature scale, which has the same increments as the Celsius scale, but defines 0 K (degrees kelvin) as absolute zero. 0°C is 273.16 K and the boiling point of water is 373.16 K. The Kelvin scale is widely used by the scientific community today.

▸ ELECTROMAGNETISM

In electromagnetism, Thomson studied the work of **Faraday** (1791–1867) and sought to add to or reinforce his findings. In particular, he attempted to offer some of the mathematic backup Faraday lacked when expounding his theories, helping to gain acceptance for the idea of electromagnetic fields. He also contributed thoughts for the basis of the electromagnetic theory of light, although in this, as with his mathematical work, he would only be partially successful.

It would require James Clerk **Maxwell** (1831–79) to later pull together the work of Faraday and Thomson into a definitive, mathematically sound hypothesis. Thomson did, however, highlight the correct voltages for underwater telegraph signal cables, and was then instrumental in the project which laid the first transatlantic cable, completed in 1866. Due to these successes Thomson became rich and in 1892 was elevated to the peerage, taking the name by which he is best remembered today, Lord Kelvin.

FURTHER ACHIEVEMENTS

William Thomson risked his life by personally taking part in the hazardous process of laying the first transatlantic cable that was to make him his fortune.

He introduced the concept of an absolute zero temperature, a temperature below which nothing can fall.

A man with many interests, Thomson also concerned himself with geophysical questions about tides, the shape of the Earth, atmospheric electricity, thermal studies of the ground, the Earth's rotation, and geomagnetism. He held the chair of natural philosophy at the University of Glasgow for 53 years.

JAMES CLERK MAXWELL

1831–1879

CHRONOLOGY • **1861** Maxwell produces the first ever colour photograph • **1864** *Dynamical Theory of the Electric Field* published • **1873** *Treatise on Electricity and Magnetism* published • **1888** Heinrich Rudolph Hertz discovers radio waves, confirming Maxwell's theories of undiscovered types of wave

The Scottish physicist James Clerk Maxwell's breakthroughs in electromagnetism came largely in the early 1860s while he was a professor at King's College, London. He examined **Faraday's** idea concerning the link between electricity and magnetism interpreted in terms of fields of force and began to search for an explanation for this relationship. Maxwell soon saw that it was simple: electricity and magnetism were just alternative expressions of the same phenomena, a point he proved by producing intersecting magnetic and electric waves from a straightforward oscillating electric current. Furthermore, Maxwell worked out that the speed of these waves would be similar to the speed of light (186,000 miles per second) and concluded, as Faraday had hinted, that normal visible light was indeed a form of electromagnetic radiation. He argued that infrared and ultraviolet light were the same and predicted the existence of other types of wave – outside of known ranges at that time – which would be similarly explainable. The discovery of radio waves in 1888 by Heinrich Rudolph **Hertz** (1857–94) would later confirm this.

'The most profound and fruitful [ideas] *since the time of Isaac Newton'* *Albert Einstein on Maxwell*

▸ ELECTROMAGNETISM

But Maxwell did not stop there. In 1864, he published his Dynamical Theory of the Electric Field which offered a unifying, mathematical explanation for electromagnetism. The text was based around four equations, now known simply as 'Maxwell's equations,' which outlined the relationship between magnetic and electric fields. He later wrote another piece on this association, published in 1873 under the title Treatise on Electricity and Magnetism.

▸ MAXWELL AND BOLTZMANN

While Maxwell's most outstanding achievements were in explaining electromagnetic radiation, he also undertook important work in thermodynamics and would offer important kinetic explanations for the behaviour of gases. This involved building on the idea of the movement of molecules in a gas. The Scotsman proposed that the speed of these particles varied greatly. Again he used his mathematical skills to produce a statistical model which would reinforce the ideas behind this research, now known as the Maxwell-Boltzmann distribution law (the last part of the name coming from the Austrian Ludwig Eduard Boltzmann who independently discovered the same explanation). Amongst other things, the convincing explanation that heat in a gas is the movement of molecules would finally do away with the caloric fluid theory of heat.

▸ COLOUR PHOTOGRAPHY

Maxwell's other accomplishments involved the deduction that all other colours can be created from a mix of the three primaries. In 1861 he applied this discovery practically in photography, producing one of the first ever colour photographs. Earlier in his career Maxwell had studied Saturn's rings and concluded that they were made up of lots of small bodies and could not be either a liquid or whole solid as had previously been speculated. In 1871, he returned to Cambridge and became the first Professor of Physics at the Cavendish Laboratory, which he helped to establish. The laboratory became world-renowned, dominating the progress of physics for many decades and producing countless leading scientists.

It is highly possible that the Scotsman himself could have led many other breakthroughs had it not been for his tragically early death. He contracted and died from cancer aged just 48.

Although regarded as a slow learner by his first tutors, William Hopkins, one of the country's brightest minds, recognised his great ability at university. Einstein also described the change in the conception of reality in physics that resulted from Maxwell's work as 'the most profound and the most fruitful that physics has experienced since the time of Newton.'

MAXWELL'S SCIENTIFIC LEGACY

He may not be as well known, but James Clerk Maxwell's standing as a scientist is often considered by many to be on a par with Isaac Newton and Albert Einstein.

Like those other great scientists, he offered explanations for physical phenomena which would revolutionise our understanding of them. He forged a path for scientists to follow by taking the experimental discoveries of Faraday (1791–1867) in the field of electromagnetism and providing a unified mathematical explanation for an achievement which had evaded other minds for so long.

ALFRED NOBEL

1833–1896

CHRONOLOGY • **1864** Nobel's nitroglycerine factory explodes, killing his brother • **1866** Invents dynamite • **1876** Invents blasting gelatin • **1886** Invents ballistite • **1896** The Nobel Foundation is set up to comply with the terms of Nobel's will • **1901** The first Nobel prizes are awarded

Alfred Bernhard Nobel is a unique entrant in this book. His research and inventions were important but perhaps not significant enough in themselves to demand inclusion in this compilation. However, the fact that his name is indirectly associated with not just one, but scores of scientists who changed the world through the prize named after him, certainly does.

▸ A MAN OF THE WORLD

Although he was a Swede, born and brought up in Stockholm, most of Nobel's education took place in Russia. His family moved to the country in 1842 to join his father, an engineer, who had taken on a supervisory position in St. Petersburg. During his education there Nobel showed a talent for languages, mastering Russian, French, English, German and Swedish, but it was chemistry which truly captured his interest. In 1850 he went on to study the subject in Paris, before spending a number of years in the United States. He then went back to St. Petersburg until 1859 where he stayed.

When Nobel was not allowed to rebuild his factory, he carried out research on a barge

▸ NITROGLYCERIN

Nobel eventually returned to his native Sweden and began putting his chemical knowledge to practical use. He set up a factory to manufacture the relatively unstable liquid explosive, nitroglycerin, to serve the growing market for it in civil engineering. But in 1864, perhaps predictably, disaster struck. There was a massive explosion, destroying Nobel's factory and, tragically, killing five people, one of whom was his own brother Emil. The accident set Nobel on a determined path to develop a more stable explosive but the government would not allow him to rebuild his factory, so he had to resort to carrying out his chemical research on a barge.

▸ STABLE EXPLOSIVES

In 1866 he made his breakthrough. He found that the liquid explosive became safe to handle if it was absorbed into a substance called kieselguhr and packed into small 'sticks.' Nobel called the invention dynamite and successfully gained patents for it in the United Kingdom and the United States. The relatively safe, yet powerful, nature of the explosive became widely popular and was a commercial success. He went on to improve the effectiveness of his invention, developing a more formidable substance called blasting gelatin in 1876, and another compound ten years later called ballistite. His other inventions included a series of detonating devices. These replaced the need for a live spark to fire his explosives, further improving their safety.

▸ THE NOBEL PRIZES

The success of Nobel's dynamites as well as his interest in oil helped him to obtain vast personal wealth. Ironically, for a man who had spent most of his life developing explosives, Nobel was a pacifist. Although he hoped the devastating potential of his inventions would act as a deterrent to war, he feared how they might be abused in the future. This was one of the reasons he chose to leave much of his fortune to funding the establishment of a series of awards, one of which included an accolade for peace. There was another dedicated to literature, with the remaining three presented for achievements in the sciences. The first Nobel prizes for medicine (or physiology), physics and, of course, chemistry were awarded in 1901 and since then they have become synonymous with excellence in their related fields. They are awarded on an annual basis, according to the terms of Nobel's will, 'to those who, during the preceding year, shall have conferred the greatest benefit on mankind.' By their very definition the prizes are presented to scientists who have changed the world, and encourage others to endeavour to do so, and as a consequence have established Alfred Nobel as a scientist who did the same.

FURTHER ACHIEVEMENTS

The sixth Nobel award, the prize for economics, was added in 1968 by the Bank of Sweden, and was first awarded in 1969.

Alfred Nobel was prolific as an inventor, obtaining over 350 patents in various countries: artificial leather, silk, and the blasting cap were among his many inventions.

His wealth came not only from explosives, but also from holdings in his brothers' oil companies, as well as extensive involvement with the Swedish arms industry, particularly the Bofors Company. Ironically, for someone who made his name with the ivention of dynamite, Nobel was a committed pacifist.

WILHELM GOTTLIEB DAIMLER

1834–1900

CHRONOLOGY • **1885** Daimler patents the first petroleum-injected internal combustion engine • **1885** Invents the motorcycle • **1886** Invents the four-wheeled petrol-driven automobile • **1889** Constructs an improved two-stroke, four-cylinder engine • **1899** Produces the first Mercedes automobile • **1893** Karl Benz produces the first mass-produced automobile • **1926** Merger of Daimler and Benz

The German Gottlieb Daimler spent much of his life working with engines long before he made the breakthrough which would change the way the world travelled. When success did arrive, however, it followed quickly and dramatically. Daimler had become convinced early that steam power was outdated. A temperamental workaholic, he perfected the first petrol internal combustion engine, and produced the first motorcycle and the first four-wheeled petrol driven car, within only a few years of each other. Daimler was a cosmopolitan man and founded companies in England and France as well as Germany.

▸ EARLY WORK

The foundations for Daimler's work had already been laid by the earlier pioneering efforts of Jean Joseph Étienne **Lenoir** (1822–1900), Alphonse Beau de Rochas (1815–93) and Nikolaus August Otto (1832–91) in the creation of two and four-stroke gas-fuelled internal combustion engines. Daimler himself had been involved in these early developments as technical director at Otto's

Daimler, the man who made the automobile revolution possible, never liked driving

factory from the early 1870s, and had been instrumental in the development of the four-stoke engine and the 'Otto cycle' which drove it.

▸ THE BREAKTHROUGH

In 1882 Daimler left Otto's employment to set up in business with Wilhelm Maybach (1846–1929), an engineer who had worked for him in the old company and who was undertaking pioneering work on the use of liquid petroleum as a possible fuel source in the internal combustion engine. Although the fuel had been known about for thousands of years and had been commercially available for decades, it had been of no use in the developing internal combustion engine industry because the liquid could not be compressed in the same manner as gas. Together, though, through their development of a carburettor, Daimler and Maybach made the breakthrough which would allow them to take advantage of the fuel source.

The carburettor converted the liquid petroleum into a thin spray which could be compressed and sparked in a four-stroke engine, just like the earlier gas models. The company's first patents were recorded in 1883 and by 1885 they had fully developed a modern-style lightweight petroleum- injected engine which would become the basis for the burgeoning automobile industry.

▸ THE FIRST VEHICLES

Daimler took immediate advantage of his engine, using it to power his 1885 'Reitwagen', the world's first motorcycle. In 1886 he invented the first four-wheeled petrol-driven automobile by using his engine to power a stagecoach. In between times, however, he had missed out on the claim to the world's first internal combustion engine motorcar. Although it was driven by a less advanced 0.75 horsepower engine, Karl Benz (1844–1929), a name also still famous within the automobile industry, had designed and constructed a three wheel vehicle with a superior electrical ignition.

▸ MERGER WITH BENZ

In 1889 Daimler constructed an improved two-cylinder, four-stroke engine. The following year Maybach improved this to a four-cylinder, four stroke version. Benz, meanwhile, was making advances in automobile manufacture and was rolled out the first mass-produced car in 1893, the Benz Velo. Both companies became leaders in the car industry, a position they still hold onto today, and combined forces through a merger in 1926. Daimler had long since passed away but Benz was still alive and served on the new company's board for his remaining years.

Ironically, Daimler's death had been hastened by a car journey in bad weather which he had insisted on taking against doctor's orders. What is more, the man who had made the automobile revolution possible had apparently never liked driving!

Daimler, Mercedes and Benz

It was the development of the carburettor that began the age of petroleum powered transport Many of the principles he developed are still used in modern vehicles.

The Mercedes name was originally the name of Emil Jellinek's daughter. Jellinek was a businessman who raced Daimler cars under the pseudonym 'Mercedes'. In 1900 there was an agreement to develop a new engine under the name Daimler-Mercedes. As the engine proved to be virtually unbeatable the name stuck and was registered as a trademark.

A Daimler-powered car won the first international car race, from Paris to Rouen, in 1894.

DMITRI MENDELEEV

1834–1907

CHRONOLOGY • 1860 Mendeleev attends a lecture by Stanislao Cannizzaro which has a profound influence on his later work • 1868–70 The Principles of Chemistry published • 1869 *On the Relation of the Properties to the Atomic Weights of Elements* published, containing the first periodic table • 1893 Becomes director of the Bureau of Weights and Measures • 1955 Element 101 discovered and named mendelevium in his honour

Mendeleev was born the last child of a large family in Tobolsk, Siberia, whose blind father could not support it. His mother enabled the family to survive by setting up and running a glass factory. Under such circumstances Mendeleev's early education was limited although he did attend school. He showed enough promise, though, to encourage his mother to leave Siberia in 1848, shortly after his father had died and the glass factory had burnt down, in a quest to enable him to enter university. Mendeleev was denied entry to a number of academic schools before settling to study science at the Pedagogic Institute in St. Petersburg. Here he excelled and qualified as a teacher in 1855 before being given additional opportunities over the following years to study at universities in Russia and overseas.

▸ HIS GREATEST INLUENCE

In 1860 Mendeleev attended an important chemistry conference at Karlsruhe where the Italian, Stanislao Cannizzaro (1826–1910), passionately

Discoverer of the periodic table, Mendeleev was to miss out on a Nobel Prize by one vote

announced and backed his rediscovery of the distinction between molecules and atoms (originally made in 1811) by **Avogadro** (1776–1856). Understanding of the atomic weights of elements had been confused for half a century without this distinction and Cannizzaro's speech was a profound influence on the development of Mendeleev's later work.

During the 1860s Mendeleev returned to St. Petersburg, finally becoming Chair of Chemistry at the university in 1866. He became aware during this period that chemistry lacked a comprehensive teaching textbook, so set about writing his own. This was finally published in 1869 under the title *The Principles of Chemistry*, setting a new standard. It was while he was researching this book that Mendeleev returned his attention to the atomic weights of elements, and introduced his card game into the equation.

▸ THE STRUCTURE OF ELEMENTS

Mendeleev wanted to list the known chemical elements in a structured way. Other scientists had tried to do the same in the past but were unsuccessful in finding a uniform way in which to list them or, indeed, in even deciding criteria by which to arrange them. So Mendeleev decided to write the properties of each element on a single card and began placing them in different formations according to various principles. He quickly discovered that if he positioned the elements according to atomic weight in short rows underneath each other, the resultant columns seemed to share common properties. The British chemist, John Alexander Reina Newlands (1837–98), had independently made a similar observation in 1864 but had had his observations ignored.

▸ THE PERIODIC TABLE

Mendeleev took his work a step further. He drew up a 'periodic table' of the elements according to their atomic weights and the common properties he found in the columns. He realised that for this scheme to work it was necessary to leave spaces for elements which he believed were as yet undiscovered. He could, though, predict their likely properties and weights and was vindicated over the coming years when gallium, scandium and germanium were discovered to slot into the gaps he had left.

Mendeleev also believed that some atomic weights, such as that for gold, had been miscalculated and he re-estimated their details to fit his structure. Again, more accurate measurements would later prove Mendeleev's assumptions to be correct. He first published his table in 1869. The text was not widely accepted at first, but eventually it became the standard method of classifying the chemical elements, restructuring the entire subject of chemistry and greatly aiding scientists of all disciplines in understanding the properties and behaviour of the elements.

Mendelevium, St Petersburg & The Table

Mendeleev predicted three yet-to-be-discovered elements including eke-silicon and eke-boron and his table did not include any of the Noble Gases which were still unknown. Mendeleev also investigated the thermal expansion of liquids, and studied the nature and origin of petroleum. In 1890 he resigned his professorship and in 1893 became Director of the Bureau of Weights and Measures in St. Petersburg.

The element with the atomic number 101 was discovered in 1955 and named mendelevium as a tribute to the great Russian scientist who narrowly missed out on the Nobel Prize for Chemistry in 1906. He lost by one vote.

WILHELM CONRAD RÖNTGEN

1845–1923

CHRONOLOGY • 1868 Röntgen receives his doctorate for his thesis *States of Gases* • 1894 Begins experimenting with cathode rays • **8 November 1895** Inadvertently discovers X-rays • **28 December 1895** Reveals his finding to the world • **1901** Becomes the first person to be awarded the Nobel Prize for Physics • **1912** X-rays finally understood as a result of the work of Max Theodor Felix von Laue

Today the uses of X-rays, particularly in hospitals, are well known to the general public, yet little more than a century ago leading physicists were not even aware of their existence. It would take a chance discovery by a German named Wilhelm Conrad Röntgen to change that and begin a process which would not only result in an understanding of X-rays, but lead on to pioneering work in radioactivity.

'X' RAYS

Röntgen was a successful scientist in his own right long before he stumbled upon X-rays. He was a Professor of Physics at the University of Würzburg in Germany from 1888 and had conducted research in many areas. But he was largely unknown to the wider world until 28 December 1895 when he unfurled the exciting discovery for which he would subsequently be remembered. The story of the rays Röntgen named 'x', because of their mysterious properties on his first finding them, however, had actually

Exposed to Röntgen's X-rays, bones would appear as shadows against a screen

begun a few weeks earlier in Röntgen's laboratory, on 8 November 1895.

He had been undertaking some tests involving little understood cathode rays when he noticed something unusual. He knew that the cathode rays emitted from the device he was using to project them could only travel a few centimetres, yet he suddenly noticed that another item in the darkened room became illuminated during the test. It was a screen covered in a substance called barium platinocyanide and Röntgen realised straight away that the glow could not have been caused by the cathode rays as the object was over a metre away. He thought that perhaps it indicated some unidentified radiation being emitted when the rays hit the glass wall of the projection device. He began excitedly investigating the properties of his accidental discovery.

▸ PICTURES OF BONES

Before announcing his finding to the world he uncovered many of the properties of the rays, including some of the factors which would go on to make them so useful in the future. For example, Röntgen discovered that the rays would pass through many different kinds of matter including metals, wood and, significantly, human limbs. Indeed, bones would appear as shadows against a screen or photographic plate allowing an X-ray image of them to be taken. He also found that the rays travelled in straight lines and were not knocked off course by electric or magnetic fields. But he was unsure exactly what the rays were; they had some characteristics in common with light rays, but did not reflect or refract like light. It was not until 1912 that they were fully understood, when Max Theodor Felix von Laue (1879–1960) showed that they were a form of electromagnetic radiation with a wavelength shorter than visible light.

▸ MEDICAL APPLICATION

The benefits of X-rays in medicine were quickly brought into common use, and as they were better understood they were applied to other areas, such as the study of the structure of molecules and in researching the properties of crystals. Others scientists ran into new phenomena as a by-product of their researches into X-rays, most notably Antoine-Henri **Becquerel** (1852–1908) who began to understand radioactivity as a result of his investigations. By the same token, it took time for the potentially harmful effects of X-ray radiation to be understood, and Röntgen's health was affected by his experiments.

Röntgen did, however, become an early beneficiary of the Nobel Prize. In 1901 he was the first person to receive the award for physics in recognition of his discovery.

Röntgen's Other Achievements

Wilhelm Röntgen was the first person to take X-ray photographs. His pictures included amongst other things images of his wife's hand.

After his discovery it only took him six weeks to determine many of the properties of X-rays. The development was to be instrumental in the later discovery of radioactivity.

Röntgen also worked and researched in other scientific fields: elasticity, capillarity, the specific heat of gases, conduction of heat in crystals, piezoelectricity, absorption of heat by gases, and polarised light.

Sadly, as a result of his experiments Röntgen and his technician were both affected by radiation poisoning.

THOMAS ALVA EDISON

1847–1931

CHRONOLOGY • **1870** Edison's first commercially successful invention, the universal stock ticker • **1875** Sets up his laboratory at Menlo Park • **1877** Patents the carbon button transmitter, still used in telephones today • **1877** Invents the phonograph • **1879** Invents the first commercially incandescent light

Amongst the plethora of universally renowned inventions that Thomas Edison is credited with is the expression, 'Genius is one percent inspiration and ninety-nine percent perspiration.' As a summary of the work ethic that the man applied to his own life, it could not be more appropriate.

▸ TRIAL AND ERROR

That is not to undermine the highly creative and original mind that was at the root of Edison's inventions. The fact that his nature and approach to his science were the complete antithesis to almost every academic and inventor of his time, scorning high-minded theoretical and mathematical methods and relying more often than not on trial and error in practical experiments, was a kind of genius in itself. Commentators have attributed Edison's unique mindset to a host of causes, from his lack of formal education, having left school at twelve, to his increasing loss of hearing from the age of fourteen, ultimately to the point of virtual deafness, which allowed him to focus without distraction on his work. His mother reported that from the earliest age he

'Until man duplicates a blade of grass, Nature can laugh at this "scientific" knowledge'

questioned everything he learnt, but others might just put it down to the 'one percent.'

Edison simply refused to accept that 'impossibilities' could not become facts without relentless experimentation and results which convinced him to the contrary. In Edison's own words, 'I find out what the world needs. Then I go ahead and try to invent it.'

With some 1093 patents singularly or jointly held in his name by the time he stopped inventing at 83, nobody could doubt that Edison meant what he said. He was the most prolific inventor the world had ever known, filing a patent once every two weeks of his working life. Given such a statement, the question is not how he changed the world, but which one of his inventions changed the world the most.

HIS GREATEST WORK

Was it, for example, the phonograph, the first ever sound-recording machine, designed and invented in 1877, surprising even Edison himself when it actually worked? Perhaps it was the first commercially incandescent light bulb, successfully produced in 1879 after more than 6,000 attempts at finding the right filament until landing on a solution in carbonised bamboo fibre. Could it have been the creation of the first commercial electric light, heat and power system, centrally generated to provide power directly into homes and businesses – set up by Edison in Lower Manhattan 1882 – ultimately leading to the creation of the company General Electric? Or was it the development of devices for recording and playing moving pictures, the Kinetograph and the Kinetoscope respectively, available commercially from 1894, leading on to silent movies and the industry which followed? Then there are the other inventions remarkable in themselves: the carbon button transmitter, still used in telephones today, which made Bell's telephone audible enough for practical and commercial exploitation; the dictaphone; the mimeograph; the electronic vote recording machine, his first patented invention; or the universal stock ticker, his first commercially successful invention sold in 1870 for $40,000, enabling him to fund the research which led to his later inventions.

REVOLUTIONARY APPROACH

Edison's revolutionary approach of establishing dedicated research and development centres full of inventors, engineers and scientists, working day and night on testing and building, brought many of his ideas to fruition. This began with his laboratory in Menlo Park, New Jersey, in 1876. Not only did these centres help Edison practically complete his own inventions but they also changed the rest of the business world's approach to research and development.

THE LEGACY OF EDISON

Arguably the best-known American of his generation, Thomas Edison was considered to be retarded at school due to his hearing difficulty, and attended only occasionally for five years.

Despite this inauspicious start Edison unquestionably changed the world, holding as he did 1093 patents singularly or jointly. Yet this most prolific inventor felt he had only scratched the surface of the possible. 'Until man duplicates a blade of grass,' he once said, 'nature can laugh at this so called "scientific" knowledge,' adding, 'We don't know one millionth of one percent of anything.'

ALEXANDER GRAHAM BELL

1847–1922

CHRONOLOGY • **1870** Following the death of Bell's two brothers from tuberculosis, the Bell family emigrates to Canada • **1873** Bell becomes Professor of Vocal Physiology at Boston University • **1875** His multiple telegraph is patented • **1876** Bell patents the telephone

Although Alexander Graham Bell had put into practice the previously fanciful notion of voice communication using a wire before he was even thirty, his path towards the invention of the telephone was, physically at least, a long one. A Scotsman, born and brought up in Edinburgh, Bell was mostly taught at home with some limited education at the University of Edinburgh and University College, London. His development of the telephone, however, took place in Boston, in the United States in 1876. Between times, he had taken up a teaching post in Elgin, Scotland. It was here that he began to study the sound waves which would prove so important in the creation of his revolutionary device.

▸ EMIGRATION

After this worked with his father in London – like his son, Bell senior was a trained speech therapist. Tragedy struck shortly afterwards with the successive deaths of his two brothers from tuberculosis. This acted as the spur for the remaining members of the Bell family to emigrate to Canada in 1870. But by this time Alexander had

The inventor of the telephone, Bell devoted much of his life to working with the deaf

also contracted the disease. He successfully convalesced on arrival in Canada and in 1871 was well enough to move to the US city of Boston which provided the setting, cast and finance for his master creation. He began first, however, by giving a series of lectures on his father's Visible Speech language, a system of phonetic symbols which enabled the deaf to converse. Further work with the deaf continued and in 1873 he took up a professorship in vocal physiology at Boston University.

▸ IDEAS INTO PRACTICE

Bell's studies on sound waves were by now at an advanced stage, but his practical experiments were not. It was perhaps most important at this juncture then that he met the dexterous handy-man, Thomas Watson, who would help Bell's theoretical ideas become physical reality with the building of the Scotsman's designs. Watson also raised money to fund the work from the enthused parents of two of his deaf students, one of whom would eventually become his wife, Mabel Hubbard.

▸ SOUND WAVES

Bell's plans for voice communication were based around a single, simple concept. He believed that sound waves from the mouth could be converted into electrical current if the appropriate device could be created to make the conversion. Once this had been done, sending the current along a wire would be a relatively simple job, before positioning another device at the opposite end to reconvert the current into sound. After several taxing years working with Watson on perfecting the conversion device, he finally succeeded in producing a piece of equipment which would change the world at least as much as any other single creation before or since. Bell's telephone was patented in March 1876 and, although many disputes would follow concerning priority and copyright, the Scotsman took his place in history.

Still young, however, Bell used the money brought in by his invention, associated awards and his company AT & T, to fund the building of laboratories for further research. Just as Thomas Alba **Edison** (1847–1931) later improved the viability of Bell's phone through his carbon transmitter button, so Bell enhanced Edison's phonograph.

In addition, the Scottish inventor also worked on, among other things, sonar recognition, flying machines and the photophone, another sound transmitting device, this time employing light as its medium of transmission. Bell's work with the deaf continued too, including the development and improvement of teaching methods, as well as a period spent educating the now famous deaf and blind student, Helen Keller. Bell was also instrumental in the establishment of the international journal *Science*.

Bell's Legacy

For the man who is best remembered for an invention that allowed people to communicate over long distances using just the human voice, it is perhaps a little ironic that Alexander Graham Bell dedicated much of his life to working with the deaf.

When Bell was not teaching people who couldn't hear, he was inventing or dreaming up ideas for inventions. While the telephone was far and away the most successful of these, his passion meant that he became involved in a wide range of projects, many of them nothing to do with his principle invention, the telephone.

ANTOINE-HENRI BECQUEREL

1852–1908

CHRONOLOGY

- **1875** Becquerel begins research into various aspects of optics
- **1876** Takes up teaching post at the École Polytechnique in Paris
- **1888** Obtains his doctorate from the École
- **1899** Elected to the French Academy of Sciences
- **1896** Discovers radioactivity
- **1903** Awarded the Nobel Prize for Physics jointly with Marie and Pierre Curie

Wilhelm Conrad **Röntgen's** discovery of X-rays towards the end of 1895 would spark a flurry of investigations into the properties of the new phenomena. One of those stimulated into action by Röntgen's finding was the Frenchman Antoine-Henri Becquerel. Whilst attempting to further investigate X-rays in 1896, he chanced upon what is now known as radioactivity, opening up a whole new branch of scientific research.

THE ENGINEER

Of course, by the time of the discovery for which he is remembered, Becquerel was already an established scientist, himself coming from a family noted for their scientific achievements. His grandfather and father were respected physicists in their own right and both had gone on to take up professorships at the French Museum of Natural History in physics, a feat which Antoine-Henri would also later mimic. In fact, his own son, Jean (1878–1953), would continue the family tradition following a similar path.

Investigating X-rays in 1896, Becquerel chanced upon the phenomenon of radioactivity

Like his immediate predecessors, Antoine-Henri was educated at the École Polytechnique in Paris, as well the School of Bridges and Highways, in engineering and the sciences. It was no surprise then that Becquerel eventually became chief engineer of the Department of Bridges and Highways. Nonetheless, he worked in parallel for much of his career in science and held many academic posts. Most notably, by 1895 Becquerel had become chair of physics at the École Polytechnique, and, again following in the footsteps of his father and grandfather, he had been honoured with membership of the prestigious Académie des Sciences in 1889, in recognition of his scientific endeavours. But Becquerel still, however, lacked his 'big break,' the achievement which would mark his place in the history of science. Almost certainly, he would have had little idea his innocuous researches into X-rays would take him there.

▸ AN INSPIRED HYPOTHESIS

The Frenchman's breakthrough began with a single hypothesis. Becquerel believed there was a possibility Röntgen's X-rays might also be responsible for the glowing or 'fluorescence' given off by some substances after being placed in sunlight. If this were the case, he deduced, the rays would make an impression on a covered photographic plate, passing through the protection as Röntgen had proven. Becquerel began to experiment on this basis. It just so happened one of the 'fluorescent' substances in which he had particularly expert knowledge was uranium, having undertaken prior work investigating its compounds. So, it was natural for Becquerel to use such a compound in his experiments and he found that after exposure to sunlight, the material did indeed mark the photographic plate.

It was when Becquerel packed his experimental equipment away, however, that his truly exciting discovery was made. After several days left in the dark, he took out the apparatus again, by now not fluorescent, and was amazed to discover the uranium compound, even after prolonged absence from sunlight, was still giving off sufficient radiation to impress upon the covered photographic plate (also stored next to the compound). He quickly realised this result was not down to X-rays but a new phenomenon for which he had no sound explanation, but which happened independent of any sunlight-induced luminescence. Further investigation would isolate the uranium as the cause of the 'radioactivity,' a name given to the occurrence not by Becquerel but Marie **Curie** (1867–1934), famous for her later researches into the subject. Becquerel was jointly awarded the Nobel Prize for physics with Marie Curie and her husband Pierre Curie (1859-1906) in 1903 for his work on radioactivity. The SI unit of radioactivity, the becquerel, is named after the Frenchman.

Further Achievements

The significance of Becquerel's chance discovery was not immediately realised and it was not until the Curies returned to the subject in 1898 that its potential and impact were first understood. One important observation which Becquerel himself later noted was the possibility of the use of radioactive materials in the medical world after he was burned by some radium in his pocket in 1901. Subsequent development would result in the use of the radiotherapy so common in cancer treatment today.

PAUL EHRLICH

1854–1915

CHRONOLOGY • **1882** Robert Koch discovers the tuberculosis bacillus • **1885** *Das Sauerstoff-Bedürfniss des Organismus* (*The Requirement of the Organism for Oxygen*) published • **1892** Ehrlich shows that mothers pass on antibodies through their breast milk • **1908** Receives the Nobel Prize for Physiology (jointly with Élie Metchnikoff) • **1909** Discovers an arsenic-based compound that combats syphilis

After the work of Edward **Jenner** (1749–1823) and Louis **Pasteur** (1822–95), the role and value of vaccinations were widely realised in the fight against disease. By the start of the twentieth century, however, there remained many untreatable fatal illnesses. Scientists began looking for alternative ways of conquering disease. One who was particularly successful and who, in the process, founded a new approach to the discovery of cures, was the German Paul Ehrlich.

▸ THE PRINCIPLE OF STAINING

Earlier in his career, Ehrlich had been deeply impressed by the development of a new finding involving the 'staining' of cells to highlight them when studied under the microscope. Some of these dyes only stained particular types of microorganism and Ehrlich was instrumental in creating a dye which illuminated the tuberculosis bacillus discovered by Robert Koch (1843–1910) in 1882. This was an important achievement in itself, becoming a technique widely used in the diagnosis of tuberculosis.

Ehrlich's 'magic bullet' became the cure for diseases such as tuberculosis and syphilis

THE MAGIC BULLET

The principles behind staining remained central to most of the other work undertaken by Ehrlich during the rest of his career, and would provide the inspiration for the achievement for which he is most remembered. From around 1905 Ehrlich began to thoroughly research his hypothesis that if a dye could latch solely onto harmful bacteria (as he had proved with his previous work on tuberculosis), then perhaps other chemicals would behave in a similar way. Instead of illuminating the disease-causing microorganisms, however, he hoped they would kill them. The chemical which would become the basis for the proof of Ehrlich's theory was arsenic. This was an element potentially fatal to humans, but which in certain compounds he found could be used effectively to kill bacteria without a harmful number of side effects. Ehrlich at last completed the successful trial of the 'magic bullet' as a treatment for disease in 1909. An arsenic-based compound that he had been testing hunted out and killed the organism which caused syphilis. The following year he launched his treatment under the name Salvarsan, and it was hugely popular in combatting the disease, a widespread and unpleasant affliction often resulting in insanity and death. Moreover, the technique Ehrlich had employed was regarded as the foundation of chemotherapy, the treatment of disease by the use of synthetic compounds to locate and destroy the organisms causing an illness. It was an approach which would go on to have vital importance in combatting so many other diseases, most noticeably cancer-causing cells.

A NOBEL PRIZE

In between his research on staining techniques and his cure for syphilis Ehrlich had jointly received a Nobel Prize for Physiology (in 1908) for a different discovery. From around 1889 to the turn of the century he was deeply involved with immunology and it was for this he received his award. He is often considered to be the founder of the modern approach to this area of science for his systematic and quantitative methods in attempting to understand it. He put forward theories on how the immune system worked and the role of antibodies. He also undertook a number of experiments designed to measure the increasing strength of the immune system in animals after repeated exposure to different types of disease-causing bacteria. This led to breakthroughs in the preparation of treatments for diphtheria and the development of techniques for measuring their effectiveness. Indeed, it was the later recognition of the limitation of these types of cures which would lead directly to Ehrlich's new approach to chemotherapy.

EHRLICH'S LEGACY

In our modern world with ready access to penicillin and other antibiotics, it is easy to forget the dreadful impact that diseases like smallpox and tuberculosis had on previous societies. Diseases which are now to all intents and purposes eradicated could spell a miserable death, even as little as fifty years ago. This is certainly the case with tuberculosis. In his obituary the London Times *paid tribute to Ehrlich's achievement in opening new doors to the unknown, acknowledging that, 'The whole world is in his debt.'*

NIKOLA TESLA

1856–1943

CHRONOLOGY • **1883** Tesla invents an induction motor • **1884** Arrives penniless in the United States • **1885** Westinghouse Electric buys the rights to Tesla's alternating current inventions • **1891** Invents the 'Tesla coil' • **c. 1899** Discovers terrestrial stationary waves • **1917** Tesla receives the Edison Medal

Few contemporaries of Thomas **Edison** (1847–1931) took him on and won, but a man who could make such a claim was one of the great American inventor's own former employees. Nikola Tesla, an eccentric electrical engineer, born in modern day Croatia but an emigrant to the U.S in 1884, was given work by Edison when he first landed in his new country. Contrasting personalities and conflicting ideas about electricity made the relationship a short one, sparking a bitter feud which would ultimately change the way the world received much of its power.

A WAY TO TRANSPORT ELECTRICITY

The story begins earlier, though, back in Europe. The brilliant, eccentric and often troubled mind of Tesla was apparent from a young age. Although he did not come from an academic family background, there was a history of inventors in his ancestry and his father worked hard on developing Tesla's mental abilities. Despite interruptions to his childhood education due to frequent sickness and the severe trauma caused by the death of his older brother, Dane, Tesla progressed into higher education, taking up a place at the university of Graz in Austria.

The idea of the transmission of electricity without wires became a later interest for Tesla

Whilst at the university, Tesla was exposed to demonstrations of existing generators and electric motors and began to ponder better ways of creating and transporting electricity. He later came up with an idea involving a rotating magnetic field in an induction motor which would generate an 'alternating current' (now known as a.c.). Most electricity being created at the time for use in homes, offices and factories involved a direct current (d.c.) which had its limitations, particularly the cost of generating it, its difficulty in being transported over long distances and its need for a commutator. By contrast, Tesla would later prove his alternating current could travel safely, efficiently and cheaply over long distances. His invention of an induction motor, in line with his earlier ideas in 1883 was the first big step on that road. His next move was to sell it.

▸ AC OR DC?

Telsa decided to emigrate to America, arriving penniless but soon finding work by making use of his electrical engineering skills. Edison employed him, Edison fell out with him and Edison got rid of him within a year. But Edison's rival, George Westinghouse, wooed Tesla. In 1885 his company, Westinghouse Electric, bought the rights to Tesla's alternating current inventions and a war of electricity began. Edison and others believed in and, probably more importantly, had a financial interest in, direct current and wanted to make it a success. It was already the standard way of generating and supplying electricity. Westinghouse and Tesla believed their method was ultimately more adapted to the job and fought hard to promote it. In spite of Edison's attempts at damaging the reputation of alternating current by claiming it was unsafe (which Tesla would later refute by grand demonstrations lighting lamps using only his body, by allowing his a.c. current to flow through it), the greater benefits of alternating current were soon realised. With the subsequent invention of better transformers in its transportation, alternating current became the standard, with d.c. increasingly confined to only specialist applications. The trend continues to this day.

▸ THE TESLA COIL

In 1891, Tesla built on his knowledge to invent the 'Tesla coil' which was even more efficient at producing high frequency alternating current. It had many applications and still today is widely used in radio, television and electrical machinery. Using this and his turn of the century discovery of 'terrestrial stationary waves,' which basically meant the planet earth could be employed as an electrical conductor, he produced some spectacular demonstrations. Tesla generated self-made 'lightning-strikes' over a hundred feet long, and he once lit 200 lamps, unconnected by wires, stretched over 25 miles. Indeed, the idea of the widespread transmission of electricity without wires became a particular area of interest for Tesla in his latter years. The 'tesla', the SI unit of magnetic flux density, is named in honour of him.

FURTHER ACHIEVEMENTS

Tesla was also a prolific inventor. His inventions include: the telephone repeater, the rotating magnetic field principle, the polyphase alternating-current system, the induction motor, alternating-current power transmission, the Tesla coil transformer, wireless communication, radio, fluorescent lights, and more than 700 other patents.

SIR JOHN JOSEPH THOMSON

1856–1940

CHRONOLOGY • **1860** Thomson enters Manchester University aged 14 • **April 1897** Announces his discovery of electrons • **1906** Receives the Nobel Prize for Physics • **1908** Thomson is knighted

Towards the turn of the twentieth century, a time when many physicists believed most of the important discoveries in their subject had already been made, the Englishman John Joseph Thomson arrived and blew any such notions away.

Along with other scientific issues the nineteenth century had cleared up much of the confusion in atomic theory. Scientists believed, for example, that they now largely understood the properties and sizes of the atoms contained within elements; without question hydrogen was the smallest of all. So when 'J.J.' Thomson announced the discovery of a particle one thousandth of the mass of the hydrogen atom, he rocked the scientific world.

▸ THE CATHODE RAY DEBATE

An early starter, Thomson attended classes in theoretical physics – a new subject at the time and one not on offer at all universities – at the University of Manchester when he was only fourteen. His most important discovery came while he held the chair of the now famous Cavendish Laboratory at Cambridge, a post he had taken over in 1884 and held until 1919. He had decided to investigate the properties of cath-

Instead of 'corpuscle', the tiny negatively charged particles were renamed 'electrons'

ode rays, now known to be a simple stream of electrons, but at the time the cause of widespread debate among scientists. The rays were visible, like normal light, but were quite clearly not normal light. Were they perhaps some form of X-ray? Most thought almost certainly not. To clear up the argument Thomson devised a series of experiments which would apply measurements to these cathode rays and clarify their nature.

▸ MEASURING THE MASS OF PARTICLES

The rays were created by passing an electric charge through an airless or gasless tube. By improving the vacuum in the tube Thomson quickly demonstrated that the rays could be deflected by electric and magnetic fields, a result which had not been observed before.

From this he concluded that the rays were made up of particles, not waves. Thomson then saw that the properties of the rays were negative in charge and didn't seem to be specific to any element; indeed, they were the same regardless of the gas used to transport the electric discharge, or the metal used at the cathode. Thomson devised a way of measuring the mass of the particles and found them to be consistently about a thousandth of the weight of the hydrogen atom. From these findings he concluded that cathode rays were simply made up of a jet of 'corpuscles,' and, more importantly, that these corpuscles were present in all elements. He announced his discovery of the sub-atomic particle in April 1897 and in the process opened up a whole new branch of scientific research.

Thomson's conclusions were soon widely accepted but his terminology was not. Instead of the word 'corpuscle,' the tiny negatively charged particles were renamed 'electrons' and have been a fundamental part of the understanding of atomic science ever since.

▸ THE CAVENDISH LABORATORY

Thomson's position within the Cavendish Laboratory meant he also became involved in a range of other important physics projects, most notably involving the discovery of certain isotopes and aiding the development of the mass spectrograph. He was a superb teacher and leader, playing a vital part in the development of the reputation the Laboratory would gain as the world's leading authority in physics. Seven of his pupils went on to gain Nobel Prizes and, indeed, Thomson himself received the award for physics in 1906, as well as a knighthood in 1908: all from a man who had originally intended to go into engineering! Thomson had studied the sciences instead because he could not afford the fee to become an engineer's apprentice – his father had died in 1872. It was a strange quirk of fate for which physics would always be grateful.

FURTHER ACHIEVEMENTS

Although there are many claimants to the title of Father of modern physics, John Joseph Thomson's is probably as justified as anybody's. It was Thomson's discovery of the electron in 1897 which opened up a whole new way of looking at the world. Not only was matter composed of particles not even visible with a modern electron microscope (as scientists from Democritus to Dalton had predicted) but it also appeared that those particles were composed of even smaller components themselves. Following Thomson, the discovery of these particles raised questions about the structure of matter that remain unanswered today.

SIGMUND FREUD

1856–1939

CHRONOLOGY • **1886** Freud sets up his private clinic in Vienna • **1895** *Studies in Hysteria* published • **1896** Coins the phrase 'psychoanalysis' • **1899** *The Interpretation of Dreams* published • **1905** *Three Essays on the Theory of Sexuality* published • **1923** *The Ego and the ID* published

Sigmund Freud's popular impact remains profound even today. Yet for a scientist who changed the world, some critics would argue that his methods were at best unscientific and at worst downright reckless. Indeed, later thinkers in the fields of psychology and psychiatry have long since discredited many of his conclusions but still the Austrian's influence pervades. Whatever the rights or wrongs of his 'scientific' deductions, Sigmund Freud remains the benchmark by which others working in the same field must compare themselves and compete against.

▸ MEDICAL BEGINNINGS

Freud's entry into science was far less controversial. He began by studying medicine at the University of Vienna in 1873 and went on to take up a position at a hospital in the same city from 1882. It was time spent working with the French neurologist Jean-Martin Charcot (1825–93) in Paris from 1885, however, which set him on the path of his future career. Here he worked with patients suffering from hysteria and began to analyse the causes of their behaviour. Additional research with Josef Breuer back in Vienna during the early 1890s helped develop

'The interpretation of dreams is the royal road to the unconscious activities of the mind'

the basis for all of his future work, culminating in the publication of *Studies in Hysteria* in 1895.

▸ THE IDEA OF 'FREE ASSOCIATION'

In common with views generally held at the time, at the heart of Freud's conclusions was a belief that mental illness was normally a psychological rather than a physical brain disease. Once one accepted this premise then Freud's introduction of the idea of 'psychoanalysis' for diagnosing the causes of mental disorder (and indeed ultimately to explain all mental behaviour) was a logical one. One of the innovative tools he developed to aid in this was the idea of 'free association.' Rather than hypnotise people as was traditional, Freud advocated this method whereby patients enunciated thoughts or ideas which came into their consciousness without prior contemplation or analysis.

▸ DREAM THEORY

From this Freud believed he could make an insight into the 'unconscious' of a patient and, in particular, the 'repressed' thoughts and emotions (often related to past negative experiences) which their 'conscious' prevented from being articulated or enacted upon. For Freud, having a patient understand and acknowledge their repressed desires was a route to therapy and ultimately the treatment of a mental disorder. He also believed that dreams offered a major insight into repressed thoughts held in the unconscious mind. Indeed, his most prominent work which fully established his revolutionary approach – was entitled *The Interpretation of Dreams*, published in 1899.

While many critics could stomach, if not necessarily agree with, Freud's interpretations up until this point, he caused an outcry with his 1905 *Three Essays on the Theory of Sexuality*. His conclusions included the explanation that most repressed behaviour was in essence suppression of sexual impulses and, most shockingly, this activity began in infancy. It was here that he also introduced the now notorious concept of the Oedipus complex, a phrase used by Freud to describe feelings of sexual attraction of a child for its parent of the same sex, and hostility to the parent of the other sex. This phase , Freud claimed, speculatively at best, was one that all children passed through.

Gradually, however, Freud's analyses would gain credibility, if not necessarily with everyone, and certainly by the 1920s they had entered the popular consciousness on a global scale. He wrote many other texts including the 1923 *The Ego and the ID*. Freud effectively redefined the 'unconscious' as the 'ID,' an intangible collection of base impulses such as instincts and emotions present in the mind from birth. With experience, living and structure, aspects of the ID would gradually help formulate a person's 'ego.'

FREUD BY NAME, FREUDIAN BY NATURE

Freud's legacy remains as much in the tools of language that he has bestowed on the modern world as anything else. Terms he introduced or of which he altered the meaning to give them our now common understanding include: psychoanalysis, free association, the ID, the ego, neuroses, repression, the Oedipus complex and, of course, the Freudian slip. The structured, systematic approach he brought to analysing an inherently difficult-to-quantify subject also pervaded the work of his successors in the field.

HEINRICH RUDOLF HERTZ

1857–1894

CHRONOLOGY • **1878** Hertz begins his PhD at the University of Berlin • **1880** receives his PhD • **1885** Appointed Professor of Physics at Karlsruhe Technical College • **1888** Discovers radio waves

Hertz came from a wealthy background and undertook his higher education initially at the University of Munich. In 1878, he began a PhD at the University of Berlin which he completed in 1880, still only twenty-three. By 1885, he was professor of physics at Karlsruhe Technical College, taking up a similar post at the University of Bonn in 1889. He had already completed his most memorable work by this point and within just five years he had died from blood poisoning.

TESTING MAXWELL'S PREDICTIONS

The experiments for which Hertz became famous were undertaken in 1888. He had been developing them for about three years but had been considering them, at least theoretically, for a lot longer. His tutor whilst studying his PhD had suggested in 1879 experimental investigative work in the subject Hertz eventually examined, but it took the German several years to obtain the necessary equipment and facilities in order to conduct the tests. At their basis was an interest in James Clerk's Maxwell's vital prediction that there were almost certainly other forms of electromagnetic radiation similar in behaviour to

The name 'hertz' is familiar today to anyone who has ever switched on a radio

infrared, ultraviolet and normal visible light which lay out of known ranges at the time and would subsequently be discovered. Hertz hypothesised if this were true he could experimentally search for these waves through creating apparatus to detect certain electromagnetic radiation. He devised a machine which contained a circuit of electricity but with a gap in it which a spark would leap across when he chose to close the circuit. He reasoned if Maxwell's theory were true, appropriately sensitive equipment should pick up electromagnetic waves distributed by the spark and so he constructed the equivalent of an antenna. He placed this device across the room from the spark-creating circuit and, sure enough, the antenna detected waves. He called the waves 'Hertzian waves' but what he had in fact discovered, and they later became commonly known as, were radio waves.

▸ THE SPEED OF RADIO WAVES

Further experimental investigation showed these radio waves to have exactly the properties Maxwell had predicted. First and foremost, like other forms of electromagnetic radiation, they travelled at the speed of light. They could be reflected and refracted and made to vibrate in the manner of other waves. Indeed, as well as being significant for their importance as a newly found phenomena in their own right, Hertz's discovery of radio waves and their properties was just as essential in that they conclusively and experimentally proved Maxwell had been correct when suggesting light, and additionally heat, waves were all forms of electromagnetic radiation.

▸ A USELESS DISCOVERY?

The German did not immediately see himself, however, the true significance of his experimental results beyond merely proving Maxwell's theories had been correct. When asked what physical application could be made of his discovery, Hertz answered, 'It's of no use whatsoever. This is just an experiment that proves Maestro Maxwell was right. We just have these mysterious electromagnetic waves that we cannot see with the naked eye. But they are there.' Others did not accept such a conclusion as easily, though, and when Hertz published the methods and results of his experiments they began looking at ways of exploiting these radio waves.

▸ MARCONI'S DEVELOPMENT

Unfortunately, Hertz did not live long enough to see the practical use one of those inspired by his essay, Guglielmo **Marconi** (1874-1937), would make of his discovery. The Irish-American transimitted radio signals over increasing lengths towards the end of the century, and succeeded in sending a signal across the Atlantic by 1901.

THE LEGACY OF HERTZ

His surname is one with which the world is familiar even today. In honour of Heinrich Rudolf Hertz's achievements, the SI unit of frequency, the hertz, was named after him. Indeed, the fact that people most commonly encounter his name when tuning in their radio is indicative of his accomplishment in itself. What is perhaps less well known is that the German physicist made his breakthrough at a very young age. By the time he was just thirty-six, he was dead.

MAX PLANCK

1858–1947

CHRONOLOGY • **1892** Planck appointed Professor of Theoretical Physics at the University of Berlin • **1 January 1900** First public enunciation of quantum theory • **1900** 'On the Theory of the Law of Energy Distribution in the Continuous Spectrum' published • **1918** Planck awarded the Nobel Prize for Physics

When did the 'modern' scientific era actually begin? Throughout the nineteenth century, there had been many advances in all aspects of science which could arguably be seen to have launched a new foundation. But for physics at least, the answer is simple: 1 January 1900. That was the day the German Max Planck gave the first public enunciation, albeit to his son, of quantum theory. It was a notion which completely abandoned assumptions made in classical physics and it founded a whole new age.

QUANTUM THEORY

To a degree, Planck stumbled across the concept that would change the scientific world through an element of chance. He had been undertaking theoretical work in thermodynamics and it was his search for a hypothetical answer to an inexplicable problem in physics at that time which led to a solution reflected in reality. The German, like many other scientists before him, had been considering formulae for the radiation released by a body at high temperature. He knew it should be expressible as a combination of wavelength frequency and temperature, but the 'irregular'

Planck stumbled across the concept that would change the scientific world by chance

behaviour of hot bodies made a consistent prediction difficult. For a theoretically perfect form of such a matter known as a 'black body', therefore, physicists could not predict the radiation it would emit in a neat scientific formula. Earlier scientists had found expressions which were in line with the behaviour of hot bodies at high frequencies, and others found an entirely different equation to show their nature at low frequencies. But none could be found which fitted all frequencies and obeyed the laws of classical physics simultaneously.

▸ PLANCK'S CONSTANT

Not that such a conundrum bothered Max Planck. Instead, he resolved to find a theoretical formula which would work mathematically, even if it did not reflect known physical laws. The answer he soon found was a relatively simple one: the energy emitted could be expressed as a straightforward multiplication of frequency by a constant which became known as 'Planck's constant' (6.6256×10^{-34} Js). But this only worked with whole number multiples (e.g. 1, 2, 3, etc.) which meant that for the formula to have any practical use at all, one had to accept the radical assumption that energy was only released in distinct non-divisible 'chunks', known as quanta, or for a single chunk of energy, a quantum. Up until that point it had been assumed that energy was emitted in a continuous stream, so the idea it could only be released in quanta seemed ridiculous. It completely contradicted classical physics. But Planck's explanation fitted the behaviour of radiation being released from hot bodies. Moreover, the individual quanta of energy were so small that when emitted at the everyday, large levels observed in nature, it seemed logical energy could appear to be flowing in a continuous stream.

▸ BIRTH OF A THEORY

In this way classical physics was cast into doubt and quantum theory was born. Planck announced his results to the wider public in his 1900 paper 'On the Theory of the Law of Energy Distribution in the Continuous Spectrum'. It naturally caused a stir, but when Albert **Einstein** was able to explain the 'photoelectric' effect in 1905 by applying Planck's theory, and likewise Niels Bohr in his explanation of atomic structure in 1913, the notion suddenly did not seem so ridiculous. The abstract idea really could explain the behaviour of physical phenomena and consequently Planck was quickly elevated to the status of Germany's most prominent scientist. He was awarded the Nobel Prize for Physics for his breakthrough in 1918.

FROM CLASSICAL PHYSICS TO QUANTUM MECHANICS

The classical physics of Newton and Galileo provides us with laws capable of explaining the ordinary, everyday world around us. However, experiments conducted early in the twentieth century began to produce results that could not be explained by classical physics. One example was the discovery that if the electrons of an atom orbited the nucleus in the way classical physics predicted, they would spiral down into the nucleus within a very short time, and the atom would cease to exist. As this clearly was not the case, it became clear that another way of dealing with atomic and subatomic particles was required. This discovery, allied to Planck's quanta theory of energy, led to the development of quantum mechanics.

LEO BAEKELAND

1863–1944

CHRONOLOGY • **1863** Baekeland born in Ghent, Holland • **1887** Appointed Professor of Physics and Chemistry at Bruges • **1888** Returns to Ghent as Assistant Professor of Chemistry. • **1889** Frustrated by academic life, Baekeland settles in America while on honeymoon • **1899** Baekeland's first company, a manufacturer of photographic paper, is bought by the Kodak Company for $1,000,000 • **1909** Sets up General Bakelite Company (GBC) • **1939** GBC becomes a subsidiary of the Union Carbide and Carbon Corporation

Baekeland worked at first as a photographic chemist and in 1891 he opened his own consulting laboratory. In 1893 he began to manufacture a photographic paper, which he called Velox, and six years later his company was bought out by the Kodak Corporation for one million dollars. Now financially independent, Baekeland returned to Europe to study at the Technical Institute at Charlottenburg.

Throughout the nineteenth and into the twentieth century, the United States of America became increasingly influential in many arenas. Its impact on the world of science was no exception. The country was also renowned for its encouragement of entrepreneurship – anybody who could combine a talent for scientific originality with a flair for business not only stood a good chance of discovering something which could change the world, but could also become incredibly wealthy off the back of it! The Belgian born

Baekeland chased the American dream and was rewarded very handsomely indeed

immigrant Leo Hendrick Baekeland was one such chemist. He chased the American Dream and was rewarded very handsomely indeed.

▸ MAKING PLASTIC

The creation for which Baekeland is best remembered and certainly the one which would have the most impact on the modern world was his development of the first widely useful synthetic plastic, a product which would find many future applications. His journey of discovery began in 1905. Baekeland began work on new chemical experiments after returning to the United States from a spell of study in Europe at the Charlottenburg Technical Institute. He had decided to try and produce a synthetic version of shellac: thin plates created by melting naturally occurring crimson-coloured lac. For this purpose he chose to work with a product of formaldehyde and phenol discovered over thirty years earlier but never commercially developed. The scientist quickly found his search for synthetic shellac to be a lost cause. Nevertheless, he became interested in the properties of the materials he was working with. After further experimentation he found that if he brought the formaldehyde and phenol together under a combination of high temperature and pressure, they produced a stiff, hardwearing resin. He had, in fact, produced the first plastic by thermosetting and was quick to realise the potential for its wide-ranging commercial use. In 1909 he launched his product under the name Bakelite and it became hugely popular both industrially and domestically. The family of plastics quickly grew and the world was never quite the same again.

▸ PHOTOGRAPHIC PAPER

Although the launch of Bakelite was commercially very successful, Baekeland had made his fortune long before this through another extremely useful invention which took advantage of his prior chemical knowledge. The Belgian had moved to the United States in 1889 after visiting during his honeymoon. Prior to this time, however, he had held academic posts in his home country, in both physics and chemistry. On arrival in America he decided to abandon formal study and took a job in a photographic laboratory. Bringing his knowledge of science to the fore, he invented a special type of photographic paper which could be developed under artificial light. Baekeland then left regular employment to set up his own company. In 1893 he launched his new creation under the brand name Velox. It was the first photographic paper in the world to be both successful and widely used, and became instrumental in the growth of the photographic industry. This was reflected in the one million US dollars paid for Baekeland's firm just six years later by the Kodak Corporation, a huge sum in Baekeland's time and not a small amount today.

The Influenece of Baekeland

Baekeland made not just one important contribution to the modern world through his chemistry, but two. The United States had served the fulfilment of his dreams well and he had served his adopted country likewise. This was acknowledged not just in the money Baekeland made but also in the many awards and honours which were bestowed upon him for his efforts, including the presidency of the American Chemical Society in 1924.

The synthetic bakelite was in effect the first plastic, one of the most common materials in everyday use in the modern world.

THOMAS HUNT MORGAN

1866–1945

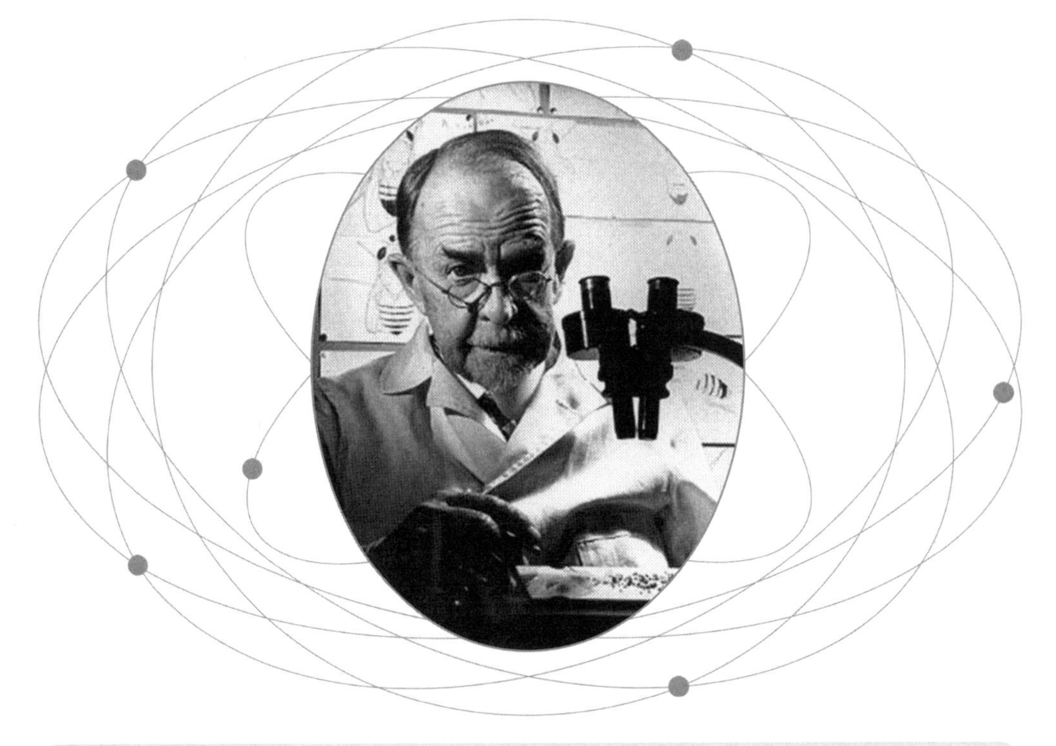

CHRONOLOGY • 1908 Morgan begins his breeding experiments with fruit flies • 1911 Produces the first chromosome map • 1915 Publishes *The Mechanism of Mendelian Heredity* • 1926 Publishes *The Theory of the Gene* • 1933 Awarded Nobel Prize for Physiology

The 'rediscovery' in 1900 of the laws of inheritance first observed by Johann Gregor **Mendel** (1822–84) excited many biologists who at last believed they had found an explanation for hereditary traits, and quite possibly the mechanism to underpin Darwin's theories. One scientist who initially remained unmoved and unconvinced by them, however, was the American Thomas Hunt Morgan.

HEREDITY AND THE CELL

After earlier work in embryology, Morgan dedicated most of the period between 1904–28, while he was Professor of Zoology at Columbia University, to clarifying how hereditary processes worked. He started out with Mendel's laws of segregation and independent assortment and began to critically assess them. It was not so much that he doubted the outcome of hereditary traits predicted by the laws; experimental evidence often seemed to back up the mathematical forecasts for characteristics present in descendants that Mendel had suggested. Morgan felt it to be more that they could not accurately reflect the process of arriving at the end result, in

Morgan began breeding the fruit fly in 1908, work which was to make him famous

particular the law of independent assortment. The reason the American felt this way was because it had been separately established that chromosomes – long thread-like matter present in the nucleus of a cell which grew and divided during cell splitting – clearly played an important part in inheritance. Yet there were far fewer chromosomes in living things than there were 'units of heredity' (renamed 'genes' in 1909 by the Dane Wilhelm Johannsen). To Morgan, this meant that groups of genes had to be present on a single chromosome. This would implicitly invalidate Mendel's law of independent assortment (which dictated that hereditary traits caused by genes would occur in all possible mathematical combinations in a series of descendents, independent of each other).

▸ BREEDING THE FRUIT FLY

From 1908 Hunt began to investigate, breeding the fruit fly, which has just four pairs of chromosomes. It is for this work that he eventually became famous. Early into his studies, he observed a mutant white-eyed male fly which he extracted for breeding with ordinary red-eyed females. Over subsequent generations of interbred offspring, the white-eyed trait returned in some descendants, all of which again turned out to be males. It was exactly the link Morgan had been looking for. Clearly, certain genetic traits were not occurring independently of each other but were actually being passed on in groups. At this point, though, Morgan realised that rather than invalidate Mendel's law of independent assortment, a simple adjustment was all that was required to unite it with his belief in the importance of chromosomes and produce an all- encompassing, proven thesis. He suggested that the principle of independent assortment did apply, but only to genes found on different chromosomes. For those on the same chromosome, linked traits would be passed on, usually a sex-related factor with other specific features (for example, the male sex and the white-eyed characteristic in the fruit fly). Otherwise, Morgan now accepted Mendel's laws.

▸ THE CHROMOSOME MAP

The results of his work had convinced Morgan that genes were arranged on chromosomes in a linear manner and could actually be 'mapped'. Further testing showed that the linked traits that Morgan had previously observed could occasionally be broken during the exchange of genes that occurred between pairs of chromosomes during the process of cell division. The American suggested that the nearer on the chromosome the genes were located to each other, the less likely the linkages were to be broken. Thus, by measuring the occurrence of breakages he could work out the position of the genes along the chromosome. Consequently, in 1911, he produced his first chromosome map showing the position of five genes which were linked to gender characteristics. Just over a decade later Morgan, together with other scientists, had mapped two thousand genes on the chromosomes of the fruit fly.

Further Achievements

Of Morgan's many books two in particular deserve special attention: The Mechanism of Mendelian Heredity *(1915) and* The Theory of the Gene *(1926). They laid the basis for understanding Mendel's observations and, along with work by later geneticists, helped to provide the microscopic science required to reinforce Charles Darwin's conclusions. In 1933 Morgan received the Nobel Prize for Physiology.*

MARIE CURIE

1867–1934

CHRONOLOGY • **1893** Curie graduates in physics from the Sorbonne. She is top of her class • **1898** Discovers the elements polonium and radium • **1903** Awarded the Nobel Prize for Physics (jointly with her husband Pierre Curie and Henry Becquerel) • **1910** *Treatise on Radioactivity* published • **1911** Awarded the Nobel Prize for Chemistry

Quite aside from her practical achievements, Marie Curie's is also important in the history of science for the pioneering role she played in opening up the subject to other women. She was arguably the first globally renowned and accepted female scientist and as such forged a path for all those of her sex who followed her. Her scientific discoveries in themselves were vital to understanding the new phenomenon of radioactivity. This was reflected in the fact that she was awarded not one, but two Nobel Prizes.

The majority of Curie's scientific work would take place in France, where she spent most of her life from 1891. Her country of birth, however, was Poland where she was born under the name Marya Sklodowska. Despite both her parents' status as teachers she grew up in relative poverty there. This was further accentuated when she was forced to move to Paris in order to obtain a higher education in physics, a level of study women were unable to undertake in her home country at the time. She graduated and shortly afterwards met her future husband, Pierre Curie (1859–1906), at the Sorbonne, where she studied and he worked.

Even today, Marie Curie's notebooks of her studies remain too radioactive to handle

He was a respected physicist in his own right and it is no surprise that the two began working together in 1895, not long after they married.

▸ IN THE FOOTSTEPS OF BECQUEREL

The stimulus for the couple's later achievements would come initially from Marie's hunt for an area of research to undertake for her postgraduate studies. Encouraged by Pierre, she decided to further investigate the exciting, new discovery of radioactivity made by Henry **Becquerel** (1852–1908) in 1896. Curie's investigations into better understanding the properties of the phenomenon soon yielded results. Becquerel had proved that uranium was radioactive. Curie, wanting to find out which other elements were, quickly discovered that thorium shared similar traits. She went on to conclusively prove that radioactivity was an intrinsic atomic property of the element in question – uranium for example – and not a condition caused by other outside factors.

Curie's next achievement was to actually discover two new elements in 1898 through her researches, which she called polonium and radium, both highly radioactive, especially the latter. She had tracked down these elements after realising uranium ore had a greater level of radiation than pure uranium, thereby correctly deducing that the ore must have contained other more radioactive, hidden elements. After these discoveries, Curie sought to obtain large enough quantities of the new substances to further understand their properties. Unfortunately, because of the minute amounts in which radium in particular was present in uranium ore, this meant she, along with her husband, had to wade through tonnes of the stuff for several years just to obtain a tenth of a gram by 1902. This, at least, allowed the calculation of the atomic weight of the new element to be made, as well as other work on its properties.

▸ A QUESTION UNANSWERED

There was one question the Curies never fully got to the bottom of, however. What exactly was the radiation which came from these elements? Ernest **Rutherford** (1871–1937) would take the credit for the answer to this question with his explanation of 'alpha,' 'beta,' and later 'gamma,' rays, but Marie did observe that the radiation was made up of at least two types of rays with distinct individual properties.

Sadly, Marie Curie would eventually die from leukaemia, which is thought to have been caused by her long exposure to radiation. At the time of her work with radioactive elements, the risks associated with radiation were not known and so no precautions were taken. Even to this day, her notebooks from her period of radioactive study remain too dangerous to examine.

MARIE CURIE'S LEGACY

Marie Curie would go on to be elected to her husband's former post of Professor of Physics at the Sorbonne, and become the institution's first female professor in the process. Her achievements in this position included the establishment of a research laboratory for radioactivity in 1912. The laboratory would go on to become world-renowned for its contribution to physics. This was due in no small part to a gift bestowed on Curie by the United States in 1921 which greatly facilitated the centre's work: a gram of the rare radium.

She received her second Nobel Prize in 1911 (her first was awarded with Pierre jointly in 1903), this time in chemistry, in recognition of her discovery of polonium and radium.

ERNEST RUTHERFORD

1871–1937

CHRONOLOGY • **1902** Rutherford establishes new branch of physics with Frederick Soddy: radioactivity • **1908** Awarded Nobel Prize for Chemistry • **1911** Establishes nuclear theory of the atom • **1914** Becomes Sir Ernest Rutherford • **1919** Develops proton accelerators (atom smashers)

After the discovery by Antoine-Henri **Becquerel** (1852–1908) of radioactivity in 1896, a number of scientists became responsible for a deeper understanding of the new phenomenon, including of course Marie **Curie** (1867–1934). However, the person who perhaps did most to bring a full understanding of radioactivity to the world, and greatly develop nuclear physics in general, was Ernest Rutherford.

‣ THE CAVENDISH LABORATORY

The New Zealand-born scientist won a scholarship to the Cavendish Laboratory in Cambridge in 1895, working under the eminent J. J. **Thomson** (1856–1940). He went on to become a professor at the McGill University in Montreal in 1898. Here he put forward his observation that radioactive elements gave off at least two types of ray with distinct properties. These he named 'alpha' and 'beta' rays.

‣ GAMMA RAY THEORY

In 1900 he confirmed the existence of a third type of ray, the 'gamma' ray, which was distinct in that it remained unaffected by a magnetic force, while

The discovery that atoms could simply decay away from an element was remarkable

alpha and beta rays were both deflected in different directions by such an influence.

▸ DISCOVERING THE HALF-LIFE

It was also in Montreal that Rutherford met the British chemist Frederick **Soddy** (1877–1956). Between 1901 and 1903, the two collaborated on a series of experiments related to radioactivity and came to some startling conclusions. They showed how, over a period of time, half of the atoms of a radioactive substance could disintegrate through 'emanation' of a radioactive gas, leaving 'half-life' matter behind. The notion that atoms could simply decay away from an element was quite remarkable. Moreover, during the process the substance spontaneously transmuted into other elements – a revolutionary finding.

After the collaboration with Soddy ended, Rutherford went on to examine alpha rays more closely. He soon proved through experimental results that they were simply helium atoms missing two electrons (beta rays were later shown to be made up of electrons and gamma rays, actually short X-rays). During this period he moved back to England, to take up a post at Manchester University. Here he worked with Hans Wilhelm Geiger (1882–1945) to develop the Geiger counter in 1908. This device measured radiation and was used in Rutherford's work on identifying the make-up of alpha rays. He went on to use it even more significantly in his next major advance.

▸ BOUNCING ALPHA PARTICLES

In 1910 Rutherford had proposed that Geiger and another assistant should undertake work to examine the results of directing a stream of alpha particles at a piece of platinum foil. While most passed through and were only slightly deflected, about one in eight thousand bounced back virtually from where they had come! Rutherford was astonished, describing it later as 'quite the most incredible event that has ever happened to me in my life. It was almost...as if you had fired a fifteen-inch shell at a piece of tissue paper and it came back and hit you.'

He didn't let it fox him. In 1911 he put forward the correct conclusion: the reason for the rate of deflection was because atoms contained a minute nucleus which bore most of the weight, while the rest of the atom was largely 'empty space' in which electrons orbited the nucleus much as planets did the Sun. The reason that the one in eight-thousand alpha particles bounced back was because they were striking the positively charged nucleus of an atom, whereas the rest simply passed through the spacious part. It was a vital discovery on the path to understanding the construction of the atom and would greatly aid Niels **Bohr** in his related revelations of 1913.

Further Achievements

During the First World War Rutherford served in the British Admiralty. Afterwards, in 1919, he was appointed to the chair of the Cavendish Laboratory at Cambridge. In the same year he made his final major discovery. Working in collaboration with other scientists, he found a method by which he could artificially disintegrate an atom by inducing a collision with an alpha particle. Essentially, what we now know as protons could be forced out of the nucleus by this smash. In the process the atomic make up of the substance changed, thereby transforming it from one element to another.

In this first instance he transmuted nitrogen into oxygen (and hydrogen), but went on to repeat the process with other elements.

THE WRIGHT BROTHERS

WILBUR 1867–1912 ‣ ORVILLE 1871–1948

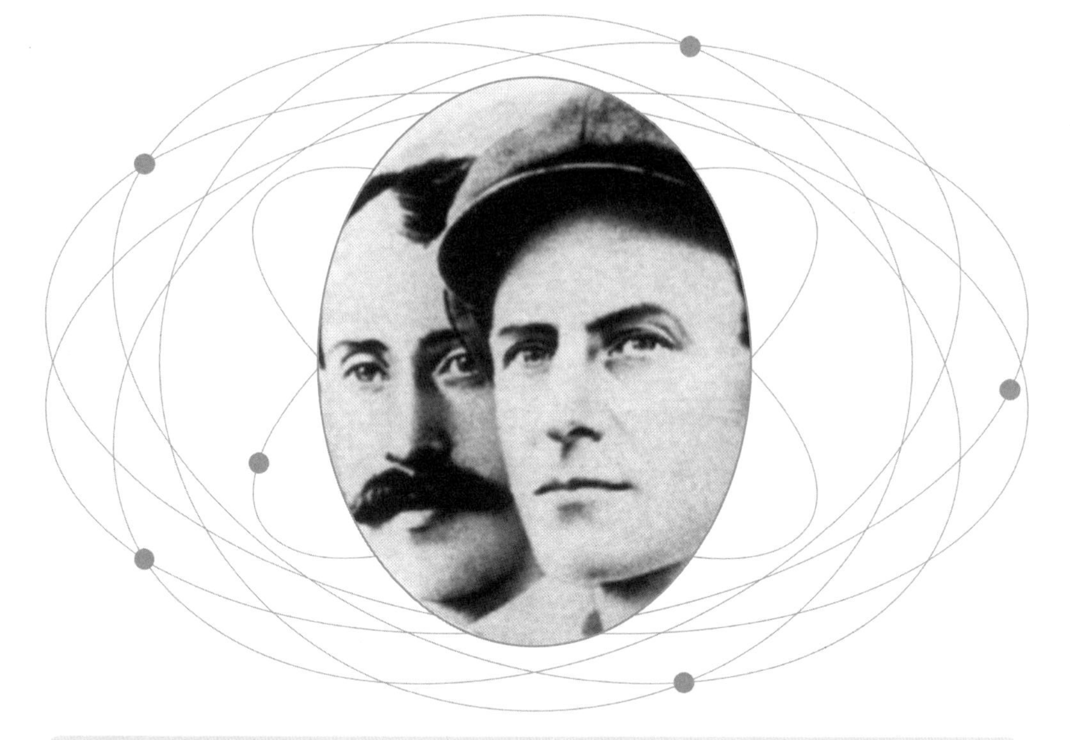

CHRONOLOGY • **4 June 1783** First public demonstration of the Montgolfiers' hot air balloon • **1896** German Otto Lilienthal dies in a glider accident. This marks the beginning of the Wrights' serious interest in flight • **17 December 1903** Orville makes the first successful powered, sustained and controlled flight in heavier-than-air machine 'The Flyer' • **1908** The Wrights reveal their improved machines to the public • **8 August 1908** First public flight in Europe at Le Mans in France

The Wright brothers' interest had not always been in flying. Their first joint project, after leaving school, was starting and publishing their own newspaper. They later moved into bicycles, opening their own shop building and selling them. So it was only from 1896 their curiosities turned to aircraft. Others had already built gliders and it was the death in a crash of one of the early pioneers of these machines, the German Otto Lilienthal (1848–96) which initially drew their attention to the subject. Lilienthal had made a number of advances in understanding aerodynamics, so the brothers began by studying his and other inventors' progress to date.

‣ SYSTEMS OF CONTROL

The Wrights quickly realised whilst much of the focus in the pioneering of gliders had been on making the craft stable, this had been at the expense of any real kind of control. So they systematically began experimentation on control

The Wrights quickly realised that gliders, though stable, lacked control

mechanisms which would eventually include ground-breaking mechanisms for the twisting of their aeroplane's wings, inspired after observing the flight of birds when searching for aerodynamic clues from nature.

▸ MASTERING THE AERODYNAMICS OF THE GLIDER

The Americans' research also involved the building and testing of unmanned gliders which they refined after each experiment. By 1900 they had built their own manned glider. They further improved their testing with the construction of a wind tunnel to aid their research in 1901 and by the following year they had made enough findings to build the most successful glider anywhere in the world up to that time. It was almost inevitable, therefore, that once they had mastered the aerodynamics of aircraft construction, it would only be a matter of time before they took the next logical step of adding motors to their invention. Before they could do this, however, they needed a method of converting the electrical power of the motor into forward thrust. They found this method in the propellor.

The Wrights tested numerous prototypes of propellor, which they conceived as moveable wings rotating around fixed axes, until they were finally satisfied. They then built their own 12 horsepower engine which powered two propellors and fitted it to their biplane, The Flyer.

The first ever successful manned flight in a powered aircraft took place, like all of their other tests, at Kitty Hawk, North Carolina on 17 December 1903. Wilbur had made an unsuccessful launch a few days earlier, but it was Orville who became the first to go airborne in a controlled manner on this date, initially covering just 120ft. By the end of the day, they had both made successful, longer flights.

▸ AIRCRAFT ON DISPLAY

Over the next few years the brothers worked on improved versions of their early aircraft, waiting until they had much more reliable models in place before demonstrating their advances to the public. Indeed, it was not until 1908 that they revealed their machines: Wilbur gave a display in France and Orville in Virginia, USA, within a few days of each other. They went down well, for within a year the brothers had received the backing to build their planes commercially and with great success in both the USA and Europe. Sadly, Wilbur caught typhoid fever in 1912 and died, leaving Orville to inherit the business and a sizeable fortune when he sold it in 1915.

FURTHER ACHIEVEMENTS

Although the human race had been airborne for over a hundred years before Orville Wright and his elder brother Wilbur (1867–1912) came along, it had struggled to make the next logical step in the progress of flight. For thousands of years mankind had dreamed of flying until the Montgolfiers had fulfilled that fantasy with their hot air balloon in 1783. This had taken advantage of the fact that some gases were 'lighter-than-air' and as such would propel a suitable device, such as a balloon, upwards. For the century afterwards, however, inventors had fantasised about achieving another seemingly impossible dream; powered flight with 'heavier-than-air' machines. Many tried to create such devices during the nineteenth century, and many failed. Then the Wright brothers arrived with a completely different approach and by 1903 had invented and tested the world's first motor-powered aeroplane.

GUGLIELMO MARCONI

1874–1937

CHRONOLOGY • **1896** Marconi files first patent for radio equipment • **1897** Sends radio signal almost 9 miles across the Bristol Channel • **1898** Sets up the Wireless Telegraph and Signal Company, Ltd • **1909** Awarded the Nobel Prize for Physics jointly with Ferdinand Braun • **1918** Sends first radio message from England to Australia

Some people make remarkable discoveries and others take advantage of them. Guglielmo Marconi was not in essence a scientist but instead a brilliant collaborator and manipulator of other scientists' findings. The most important of these was Heinrich Rudolf **Hertz's** (1857–94) breakthrough (the discovery of radio waves in 1888), with ramifications even he could not have realised. One person who did quickly realise their significance, however, was the Irish-Italian Marconi. And it was he, not Hertz, who ended up with a Nobel Prize for Physics in 1909.

FIRST EXPERIMENTS

As well as being more interested in the scientific implications of his findings than the practical or commercial use of them, the main reason Hertz had not been able to take much advantage of his discovery was because he had died shortly afterwards in 1894. At exactly that time, Marconi was beginning to undertake his first experiments with radio waves after having his interest stimulated when learning about the work of Hertz and others during his studies in Bologna and Livorno. Both his father and mother had been wealthy even before they had married, so Marconi had

The 2,000-mile radio transmission across the Atlantic in 1901 convinced the last doubters

the advantage of a large family estate and limited financial worries in which to begin his experiments.

▸ WORK IN LONDON

Indeed, the size of the estate became important as he made his first developments. It was a mile and a half long so he had the benefit of a large private testing ground across which to send his radio waves. By 1895 he had made sophisticated enough equipment to emit and receive waves the entire length of his grounds. Whilst Marconi was convinced of the importance of his test results, few others in Italy were, so he set off for London, England instead in an attempt to stimulate backing for his work there.

He was far more successful. Government, military and postal departments quickly became interested in the potential uses of the new technology and within a few more years, enthusiasm was widespread. This came largely after his first overseas radio transmission between Britain and France in 1899, after which he was brought to the attention of the public. During the intervening period, he had also been involved in the setting up of a company to promote his developments, renamed in acknowledgement of his centrality to it as the Marconi Wireless Company Limited in 1900. By that time he had also held successful radio transmission trials on floating vessels on behalf of the British navy, after which naturally followed further converts to his cause.

▸ ACROSS THE ATLANTIC

But the event which made Marconi world famous, and the one which silenced many of the doubters about the practical and scientific uses of his equipment who still persisted, was the two thousand mile transmission of Morse code across the Atlantic in 1901. Many had thought this an impossible task, not least because the curvature of the earth would prevent it, but Marconi had been convinced it was possible and was duly rewarded for his perseverance.

Both before and after this time, the Irish-Italian took out many patents for the equipment he had successfully developed and added to. Indeed, his life remained absorbed by improvements to radio technology virtually up until his death, perhaps most notably making advances in short-wave technology to set up an international network for radio broadcasts by 1927.

Marconi and the Titanic

'Those who had been saved, had been saved through one man, Mr. Marconi.... and his marvellous invention' (The Right Hon. Herbert Samuel, the Postmaster General April 18, 1912).

Although there was initially a great deal of scepticism regarding the usefulness of Marconi's invention, once he had demonstrated its potential, it quickly became something of a craze. One particular area it was warmly welcomed was shipping; before long, every ocean-going liner had its 'Marconi man'. One reason as many survivors were picked up from the Titanic when she went down was her Mrconi men, Jack Phillips and Harold Bride, who stayed on the ship transmitting a distress call until the power failed, and the radio room was awash. Harold Bride survived the ordeal, and later joined the Royal Navy; Jack Phillips went down with his ship.

FREDERICK SODDY

1877–1956

CHRONOLOGY • **1898** Graduates from Merton College, Oxford with first class honours in chemistry • **1901–03** With Ernest Rutherford, discovers the phenomenon of atomic disintegration • **1904–14** Lecturer in Physical Chemistry at Glasgow University • **1919** Appointed Dr. Lees Professor of Chemistry at Oxford University

While perhaps not as pre-eminent as his contemporary Ernest Rutherford (1871-1937), Frederick Soddy was still a significant influence in understanding the behaviour and implications of radioactivity, particularly of radioactive decay. The English chemist's first breakthrough came after working with Rutherford between 1901–03 while at the McGill University in Montreal. Moreover, the later solitary work which this led to for Soddy would help scientists understand an important aspect of the atomic design of elements.

DECAYING ATOMS

The work Soddy undertook with Rutherford came shortly after he took up his position as demonstrator of chemical experiments in Montreal. Their collaboration yielded radical results, most notably the revolutionary idea that elements could disintegrate into a 'half-life' by 'emanating' atoms spontaneously. In the process of this decay, matter could transmute into various other elements, an astonishing conclusion.

These findings in themselves established Soddy's reputation, but his later results in related work were just as important. He turned his atten-

Soddy's discovery of isotopes cleared away a good deal of chemical confusion

tion to the array of apparent 'new' elements which had been discovered during the transmutation caused by radioactive decay. The problem scientists were facing was that there had been so many recent discoveries there were clearly were not enough spaces in which to fit them in the periodic table. Yet each element definitely had a distinct atomic weight and specific time period ascribed to it before it reached its 'half-life' state. Meanwhile, other scientists were trying to produce these new elements artificially by breaking-down closely related elements but were inexplicably failing.

▸ ISOTOPES

Soddy's 1913 solution was another simple, yet incredible, one. He maintained that even though the atomic weights and half-lives of the 'new' elements were different, they otherwise shared identical chemical properties with known existing elements and were therefore variations of the same element. So, for example, the 'new' elements thorium C and radium D, which had different atomic masses and half-lives from each other and from the element lead, were all chemically the same and were therefore all simply lead. This also explained why scientists had not been able to break down artificially the closely related elements, as expected, because the substances they were starting with and seeking to produce were chemically identical! Soddy named the variations 'isotopes' and in a stroke cleared up the confusion which had previously surrounded the 'new' elements. It would take the later discovery by James **Chadwick** (1891-1974) of the neutral, but weight bearing, neutron to fully explain how the differences in atomic mass were possible whilst still retaining the same atomic number, but Soddy's explanation offered enough for science to make some sense of the new order in the meantime.

▸ RADIOACTIVE DISPLACEMENT

In the same year as his isotopic explanation, Soddy also expressed the radioactive displacement law. This stated that when an alpha particle is emitted from a decaying substance, its atomic number is reduced by two, and its atomic weight by four (explicable by the fact that an alpha particle is simply a helium nucleus with corresponding atomic figures). Likewise when a beta particle (a negatively-charged electron) is emitted, the atomic number is increased by one.

Soddy was awarded the Nobel Prize for chemistry in 1921 for his work on isotopes, after which his interest moved into other academic disciplines and he ceased active involvement in chemical research.

SODDY'S LEGACY

It was during his time at Glasgow, between 1904 to 1914 that Soddy did much of his most important work in chemistry, including his formulation of the 'Displacement Law', which stated that emission of an alpha particle from an element causes that element to move back two places in the Periodic Table. He also formulated the concept of isotopes during thie period, the realisation that elements can exist in two different states, with different atomic weights, while remaining chemically identical. Soddy devoted much of his later time to fields other than chemistry, evolving theories which were never widely accepted.

ALBERT EINSTEIN

1879–1955

CHRONOLOGY • **1902** Einstein begins work at Swiss patent office • **1905** Publishes three seminal papers on theoretical physics, including the 'special' theory of relativity • **1916** Proposes general theory of relativity; is proved correct three years later • **1922** Wins Nobel Prize in Physics • **1933** Emigrates to Princeton, N.J. • **1939** Urges Franklin D. Roosevelt to develop the atom bomb • **18 April 1955** Dies in his sleep

Of the essays written by Einstein in 1905, arguably the most influential was his enunciation of a 'special' theory of relativity, which advanced the idea that the laws of physics are actually identical to different spectators, regardless of their position, as long as they are moving at a constant speed in relation to each other. Above all, the speed of light is constant. It is simply that the classical laws of mechanics appear to be obeyed in our normal lives because the speeds involved are insignificant.

▸ THE SPEED OF LIGHT

But the implications of this principle if the observers are moving at very different speeds are bizarre and normal indicators of velocity such as distance and time become warped. Indeed, absolute space and time do not exist. Therefore if a person were theoretically to travel in a vehicle in space close to the speed of light, everything would look normal to them but another person standing on earth waiting for them to return would notice something very unusual. The space ship would appear to be getting shorter in the

'Science without religion is lame; religion without science is blind'

direction of travel. Moreover, whilst time would continue as 'normal' on earth, a watch telling the time in the ship would be going slower from the earth's perspective even though it would seem correct to the traveller (because the faster an object is moving the slower time moves). This difference would only become apparent when the vessel returned to earth and clocks were compared. If the observer on earth were able to measure the mass of the ship as it moved, he would also notice it getting heavier too. Ultimately nothing could move faster than or equal to the speed of light because at that point it would have infinite mass, no length, and time would stand still!

▸ A GENERAL THEORY OF RELATIVITY

From 1907 to 1915, Einstein developed his special theory into a 'general' theory of relativity which included equating accelerating forces and gravitational forces. Implications of this extension of his special theory suggested light rays would be bent by gravitational attraction and electromagnetic radiation wavelengths would be increased under gravity. Moreover mass, and the resultant gravity, warps space and time, which would otherwise be 'flat', into curved paths which other masses (eg, the moons of planets) caught within the field of the distortion follow.

Amazingly, Einstein's predictions for special and general relativity were gradually proven by experimental evidence. The most celebrated of these was the measurement taken during a solar eclipse in 1919 which proved the sun's gravitational field really did bend the light emitted from stars behind it on its way to earth. It was the verification which led to Einstein's world fame and wide acceptance of his new definition of physics.

Einstein spent much of the rest of his life trying to create a unified theory of electromagnetic, gravitational and nuclear fields but failed. It was at least in keeping with his own remark of 1921 that 'discovery in the grand manner is for young people and hence for me is a thing of the past.'

▸ $E=MC^2$

Fortunately, then, he had completed three other papers in his youth (in 1905) in addition to his one on the special theory of relativity! One of these included the now famous deduction which equated energy to mass in the formula $E=mc^2$ (where E=energy, m=mass and c=the speed of light). This understanding was vital in the development of nuclear energy and weapons, where only a small amount of atomic mass (when released to multiply by a factor of the speed of light squared under appropriate conditions) could unleash huge amounts of energy. The third paper described Brownian motion, and the final paper made use of Planck's quantum theory in explaining the phenomenon of the 'photoelectric' effect, helping to confirm quantum theory in the process.

FURTHER ACHIEVEMENTS

Almost inevitably, Einstein was also drawn into the atomic bomb race. He was asked by fellow scientists in 1939 to warn the US President of the danger of Germany creating an atomic bomb. Einstein himself had been a German citizen, but had renounced his citizenship in favour of Switzerland, and ultimately America, having moved there in 1933 following the elevation of Hitler to power in his home country. Roosevelt's response to Einstein's warning was to initiate the Manhattan project to create an American bomb first.

After the war Einstein spent time trying to encourage nuclear disarmament.

ALEXANDER FLEMING

1881–1955

CHRONOLOGY • **1929** Fleming publishes first report on antibacterial properties of penicillin • **1939** Indirectly provides penicillin to Howard Florey and Ernst Chain • **1944** Becomes Sir Alexander Fleming • **1945** Awarded the Nobel Prize for Medicine jointly with Florey and Chain • **1955** Dies of a heart attack in London

Alexander Fleming had had a quite unremarkable life up until the chance discovery of a mould in his laboratory in September 1928. Even the decade after he made the find which would go on to save millions of lives, little changed. It took until 1940 for the 'penicillin' to be produced from the fungi in practically useful quantities. Although it was in fact another team of people altogether who facilitated the latter development, it was Fleming who became revered as a hero.

The son of a Scottish farmer, Fleming came from humble background, and began his working life at the age of sixteen as a shipping clerk in London, England. After inheriting a small amount of money, and following suitable encouragement from this brother who was a doctor, Fleming decided to study medicine. In 1902, he joined St. Mary's Hospital Medical School in London, where he remained for the rest of his career barring a period from 1914-18 putting his medical skills to good use for the war effort.

▸ AN INTEREST IN BACTERIOLOGY

Fleming became increasingly interested in bacte-

The discovery of penicillin was due as much to luck as to scientific study

riology. Indeed, it was his wartime experiences which made him realise there was a need for a non-toxic drug to combat the millions killed by the bacteria which infected wounds. After he rejoined St Mary's therefore, he searched for a naturally occurring bacteria-killer and focused initially on what he believed were the body's own sources: tears, saliva and mucus from the nose. In 1922, he had his first success, producing lysozyme, an enzyme produced by the body. It killed certain bacteria naturally, but Fleming could not produce it in sufficiently concentrated quantities to be of medical use.

▸ A STROKE OF LUCK

The search continued, although even scientists sometimes have to take a holiday, and ironically it was a two-week break which led to Fleming's ultimately world-changing discovery. Before leaving for his vacation in 1928, however, the bacteriologist had been examining some dishes containing staphylococcus bacteria, which turned out to be the first in a sequence of rather fortunate events. He accidentally left one of the dishes exposed to the air before he departed and it became infected with Penicillium notatum. The form of infection in itself was lucky as it was only because it was being studied elsewhere in the hospital that it was present to contaminate Fleming's sample at all, and a cold spell of weather in Fleming's absence allowed the fungi which developed to grow.

Although Fleming had luck on his side in the first instance, the fact he was a skilled bacteriologist was also vital. On returning from his holiday, he noticed a mould had grown in the infected dish and, rather than simply wash it out, was sufficiently interested to examine it further. He noticed clear patches around the edges of the contamination and correctly deduced that there was something in the Penicillium notatum which was killing the staphylococcus bacteria. On further testing he found it was a useful killer of many forms of bacteria, but again it occurred in sufficient quantities to be of much further use.

▸ THE DEVELOPMENT OF PENICILLIN

So it was left to the spur brought about by World War Two over a decade later and a new team of scientists before the quest for a non-toxic antibiotic was revived and 'penicillin' as Fleming had named his finding was revisited. The Scot supplied the team led by Howard Walter Florey and which included a chemist called Ernst Boris Chain with a sample of his mould. By 1940 the team had proved penicillin's potency in fighting infections in mammals and soon afterwards made the breakthroughs necessary for it to be produced on an industrial scale.

The importance of Florey and Chain to the story was at least acknowledged when, along with Fleming, they were jointly awarded the Nobel Prize for Physiology in 1945.

FURTHER ACHIEVEMENTS

Natural and semi-synthetic versions of penicillin would go on to be mass-produced, saving millions of lives during the war and even more afterwards as it was used to combat a whole series of bacteria-causing diseases. Fleming would be hailed as a saviour by a public in need of heroes and was knighted in 1944, although the team of later scientists had arguably done most to make penicillin useful. Fleming himself said of his role, 'My only merit is that I did not neglect the observation and that I pursued the subject as a bacteriologist.'

ROBERT GODDARD

1882–1945

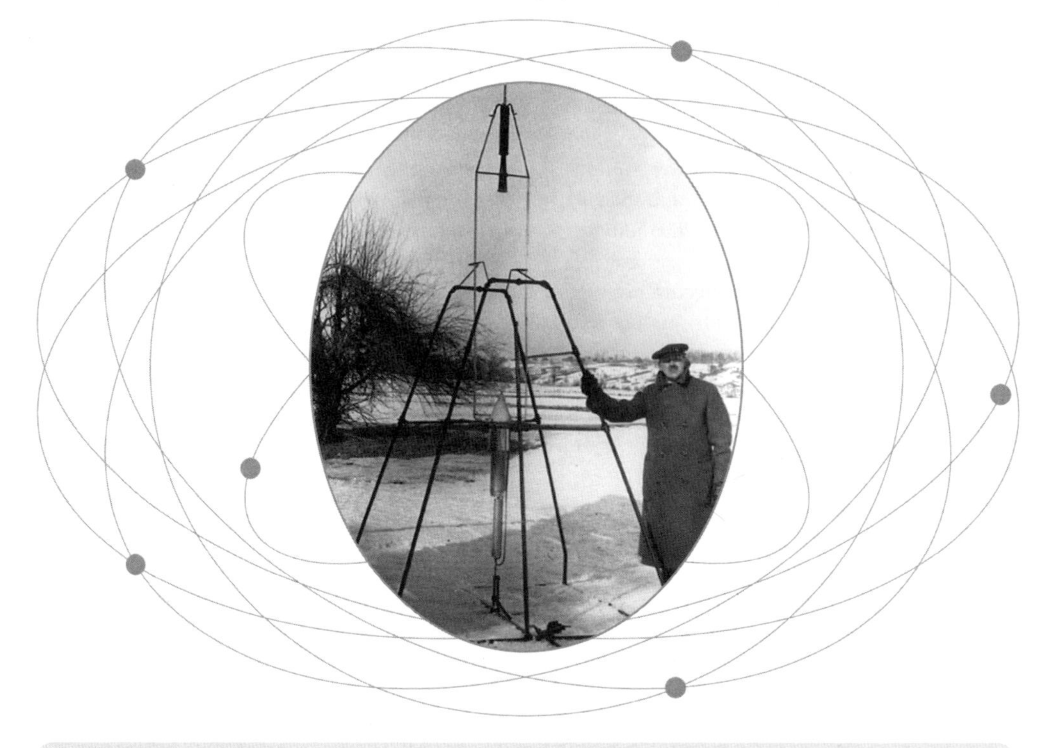

CHRONOLOGY • **1908** Goddard studies Physics at Clark University • **1915** Demonstrates rocket engines can produce thrust in a vacuum • **1926** Launches first liquid-fueled rocket, which reaches an altitude of almost 15 metres • **1930** Starts working in Roswell, New Mexico, where he develops supersonic and multi-stage rockets and fin-guided steering

'It has often proved true that the dream of yesterday is the hope of today and the reality of tomorrow.' (Robert Hutchings Goddard, 1904). This is probably not the measured reflection of the average high school leaver, but then the pioneer of modern rocketry was never going to be an average student. From the age of seventeen, the American Robert Goddard had known exactly what he was going to do with his life. One October day in 1899, he had had an all-consuming vision while chopping branches off a cherry tree in his garden of '...how wonderful it would be to make some device which had even the possibility of ascending to Mars. I was a different boy when I descended the tree from when I ascended, for existence at last seemed very purposive.'

▸ OF SCEPTICS AND IGNORANTS

A few years later when the *New York Times* found out about Goddard's vision, the scientist was upset to find that the newspaper not only failed to share it but also mocked him for it. By this stage, Hutchings had already written a paper entitled *A*

'Every vision is a joke until the first man accomplishes it'

Method of Reaching Extreme Altitudes which he had published in 1919, outlining his advances in rocketry to date and his hopes of a future landing on the moon. The 13 January 1920 editorial in the *New York Times* scoffed at the doctor of physics (as he was by then), teaching at Clark University in Worcester, Massachusetts. It rebuked him for lacking 'the knowledge ladled out daily in high schools' that it would be impossible for a rocket to move forward outside of the earth's atmosphere because there was no atmosphere for it to push against in order to gain propulsion.

If the column writer had bothered to delve a little deeper he would have found that Goddard had already gone a long way towards refuting exactly this objection. As early as 1907 he had completed mathematical calculations to show a rocket could thrust in a vacuum, and had backed this up with a physical experiment showing just such a concept in 1915. Internationally, others too were already envisaging space travel and were beginning to undertake work to that end, most notably a Russian called Konstantin E. Tsiolkovsky (1857–1935) and the German Hermann Oberth (1894–1989).

▸ GODDARD'S RESPONSE

The criticism directed against Goddard by the editorial merely spurred him on further. Responding to the article with the determined statement, 'Every vision is a joke until the first man accomplishes it,' he soon afterwards started making strides towards the rocket which would lay the foundations for space travel. He began to work with liquid fuels, rather than gunpowder, realising this was the most likely method by which he could power his dreams. By 1925 he had developed a prototype rocket fuelled by gasoline and liquid oxygen which succeeded in lifting its own weight for the first time in a controlled test. Just three months later, on 16 March 1926, the world's first full launch of a liquid-fuel propelled rocket took place. At his 'aunt' Effie's farm in Auburn, Massachusetts, Goddard sent a ten foot rocket just 41ft into the air and 184ft in distance over two and a half seconds. But it had worked.

Over the next decade, and thanks to better funding from the Guggenheim family for his pioneering work after 1930, Goddard improved and successfully launched over thirty more rockets, gradually increasing their altitude and reliability. He filed patents for better control, better guidance and better fuel pump mechanisms. By 1935 he had launched a rocket which travelled faster than the speed of sound. His efforts culminated in a launch on 26 March 1937 with a rocket which reached 1.7 miles in altitude, then a record.

Yet in spite of his successes, the US government largely ignored Goddard's work until the space race gathered full momentum in the 1940s and 1950s. It turned to Goddard's developments as a base from which to begin. Indeed, the government was eventually forced to pay one million dollars for patent infringement to Goddard's widow in acknowledgement of the use they had made of his designs.

Goddard Vindicated

By the time man landed on the moon in 1969, Goddard had long since passed away. But the 'Correction' to the 1920 editorial by the New York Times *three days before Neil Armstrong's historic first lunar steps vindicated the man who had played a large part in ultimately putting him there. 'It is now definitely established,' the newspaper wrote, 'that a rocket can function in a vacuum as well as in an atmosphere. The* Times *regrets the error.'*

NIELS BOHR

1885–1962

CHRONOLOGY

- **1911** Bohr receives PhD from the University of Copenhagen
- **1913** *On the Constitution of Atoms and Molecules* published
- **1914** Begins two years of work with Ernest Rutherford in Manchester
- **1916** Returns to Denmark to head the newly created Institute of Theoretical Physics
- **1922** Receives Nobel Prize for Physics
- **1943** Bohr, who has a Jewish mother, leaves occupied Denmark and journeys to Los Alamos, USA, where he works on the atomic bomb project
- **1955** Bohr organises the first Atoms for Peace conference in Geneva

Few twentieth century theoretical physicists are referred to in the same breath as Albert **Einstein** (1879–1955) but the Dane Niels Bohr is one of them. He made major contributions to validating the concept of quantum physics set out by Max **Planck** (1858–1947) in 1900, solved issues concerning the behaviour of electrons in Ernest **Rutherford's** atomic structure and was involved in the development of the first atomic bomb.

▸ THE COLLAPSING ATOM

Bohr gained his PhD at the University of Copenhagen in 1911 then moved briefly to the Cavendish Laboratory at Cambridge before settling in Manchester to work with Ernest Rutherford. The New Zealand physicist had just established his 'planetary' model of the atom: a tiny central nucleus bore most of the weight, around which electrons spiralled in a series of orbits. But there was a problem with this model.

Electrons only existed in 'fixed' orbits where they did not radiate energy

Classical physics insisted that if the electrons moved around the atom in this way, the energy they radiated would ultimately expire and the electrons would collapse into the nucleus. In 1913 Bohr resolved this issue and simultaneously validated Rutherford's model by applying Planck's quantum theory to it. He argued, from the perspective of quantum theory, that electrons only existed in 'fixed' orbits where they did not radiate energy. Quanta of radiation would only ever be emitted as an atom made the transition between states and absorbed or released energy. Only at this point in time would electrons 'move,' hopping from a lower to a higher-energy orbit as the atom took on energy, or jumping down an orbit as it emitted it (producing light in the process).

Bohr calculated the amount of radiation emitted during these transitions using Planck's constant. It fitted physical observations. Also, when he applied this to hydrogen atoms and the wavelengths of light that they should have released under this principle, he again found his calculations matched real world examples. It was a bizarre concept to grasp, as had been Planck's initial enunciation of quantum theory, but here was another practical example which validated it.

▸ CORRESPONDENCE AND COMPLIMENTARITY

Bohr made another important addition to quantum theory and one to the school of quantum mechanics which succeeded it. In the former in 1916 he enunciated the 'correspondence principle': in spite of the huge apparent differences between the two, the laws which govern quantum theory at the microscopic level should still correspond with our understanding of classical physics as observed on the larger 'real world' level.

Later in 1927, in quantum mechanics, Bohr added the 'complementarity principle'. This argued that debates over whether light, as well as other atomic objects, behaved in a wave-like or particle-like fashion were futile because the equipment used in experiments to try to prove the case one way or the other greatly influenced the outcome of the results. Instead all results only gave a partial glimpse of the answer to any atomic test and therefore had to be interpreted side-by-side with all other results to give a broader 'sum-of-the-parts' understanding. This idea sat neatly beside theories offered by the likes of Louis de **Broglie** (1892–1987), Werner **Heisenberg** (1901–76) and Max Born (1882–1970).

THE LEGACY OF BOHR

Bohr's theoretical and practical involvement in the physics which led to the creation of the first atomic bomb was dramatic. In 1939 he developed a theory of nuclear fission (splitting a heavy atom's nucleus to release huge amounts of energy suitable for an atomic bomb) with John Archibald Wheeler (b.1911) from Bohr's 'liquid-drop' description (1936) of the way protons and neutrons bonded in the nucleus. He realised, ominously, that the uranium-235 isotope would be more susceptible to fission than the more commonly used uranium-238. Ultimately his findings would make their way to the US atomic bomb project, especially after Bohr escaped to America to flee occupied Denmark and acted as a consultant to the team.

Bohr was, however, uncomfortable with the implications of the new technology and dedicated much of the rest of his life to encouraging the control and limitation of nuclear weapons, founding the Atoms for Peace Movement for physicists with similar opinions to his own.

Bohr was awarded Nobel Prize for Physics in 1922.

ERWIN SCHRÖDINGER

1887–1961

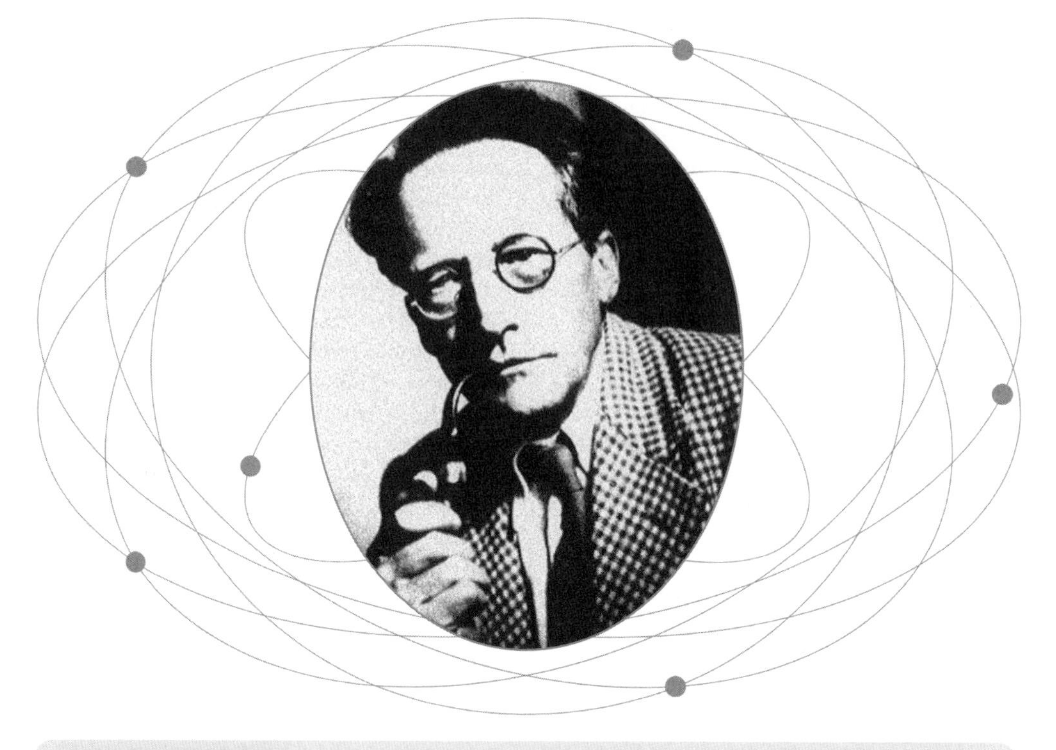

CHRONOLOGY • **1908** Schrödinger enters University of Vienna: studies mathematics and theoretical physics under Mertens and Wirtinger • **1926** Publishes his paper in which he outlines the elements of quantum wave mechanics • **1927** Joins the faculty of the University of Berlin, along with Albert Einstein • **1944** Publishes *What is Life?*

The mid-1920s was open season in the field of quantum theory and one of the many physicists who waded in with a new influential direction was Erwin Schrödinger. Born in Vienna, to a prosperous merchant family, Schrödinger had a grandmother who was half Austrian, and half English, the English side of the family originating from Leamington Spa, and he grew up speaking both English and German in the home. Schrödinger was taught at home by a private tutor until he was ten. The Austrian scientist developed what became known as 'wave mechanics,' although like others, including Einstein, he later became uncomfortable with the direction quantum theory took after doing so much in the first place to validate it.

▸ NOT PARTICLES BUT WAVES

Schrödinger's own development was built largely on the back of the 1924 proposal by Louis **de Broglie** (1892-1987) that particles could, in quantum theory, behave like waves. Whilst the Austrian Schrödinger was attracted to this explanation, he was troubled by certain implications of it. Essentially, he felt de Broglie's equations

Rejecting the idea of particles, Schrödinger argued everything was a form of wave

were too simplistic and did not offer a detailed enough analysis of the behaviour of matter, particularly at the subatomic level. So he took things a stage further and removed the idea of the particle completely! In its place, he argued everything was a form of wave.

▸ A WAVE EQUATION

Amazingly, between 1925 and 1926 he was able to calculate a 'wave equation' which mathematically underpinned this argument and the science of quantum wave mechanics was born. Further proof came when the theory was applied against known values for the hydrogen atom, and correct answers were obtained, for example, in calculating the level of energy in an electron. It clearly overcame some of the more woolly elements of the earlier quantum theory developed by Niels **Bohr** (1885-1962) and addressed the weaknesses in de Broglie's thesis.

▸ ADAPTING WAVE THEORY

Indeed, the theory behind wave mechanics was now applied to all sorts of other situations with great effect. Unfortunately, it too had some fundamental weaknesses and Schrödinger was not blind to these. The overriding one was, having done away with particles, it was difficult to offer a physical explanation for the properties and nature of matter. The Austrian came up with the concept of 'wave packets' which would give the impression of the particle as seen in classical physics, but would actually be a wave. The justifications he offered, though, were found not to add up.

▸ THE PROBABILISTIC INTERPRETATION

This left Schrödinger's work susceptible to being superseded by that of others, just as his had improved on those whose ideas came before him. Shortly afterwards, the probabilistic interpretation of quantum theory based on the ideas of **Heisenberg** (1901-76) and Born (1882-1970) took hold. This effectively proposed matter did not exist in any particular place at all, being everywhere at the same time, until one attempted to measure it. At that point the equations they put forward offered the best 'probability' of finding the matter in a given location. Whilst this is still widely accepted as the most adequate explanation today, Schrödinger joined **Einstein** and others in condemning such a loose, probabilistic view of physics where nothing was explainable for certain and essentially cause and effect did not exist.

Ironically, Paul Adrien Maurice Dirac(1902-84), another important influence in quantum mechanics, went on to prove Schrödinger's wave thesis and the alternative probabilistic interpretation he abhorred were mathematically, at least, the equivalent of each other. Schrödinger shared a Nobel Prize for Physics in 1933 with Dirac.

Schrödinger's Cat

This famous 'animal' is in fact part of a thought experiment designed by Erwin Schrödinger in the 1930s, to try and explain the problem that, contrary to all logic, atoms can exist in two states simultaneously, decayed and undecayed. Schrödinger uses the analogy of a cat locked in a box with a vial containing deadly poison. The vial's lid contains a radioactive atom. If the atom decays, the released particle opens the vial and the cat dies. This is an example of a quantum system, in which it seems the cat exists in an indeterminate state, because the atom is both decayed and undecayed, implying that the unobserved cat is neither dead nor alive, which is patently absurd.

HENRY MOSELEY

1887–1915

CHRONOLOGY • 1910 Appointed Lecturer in Physics at Manchester University • 1913 Publishes first paper on X-rays, containing the elements of what will become known as Moseley's Law • 1914 Publishes paper asserting existence of three elements between aluminium and gold • 1915 Moseley is killed during the Gallipoli landings in Turkey during the First World War

There were a number of future Nobel Prize winners working with Ernest **Rutherford** at Manchester University in the few years before the outbreak of the First World War. One who could well have gone on to win such an award but was robbed of a promising scientific career by the war itself was the Englishman Henry Gwyn Jeffreys Moseley. In spite of only a few years given to him to make any kind of academic progress at all, though, he still had time to achieve enough to place his name amongst the scientific greats.

A SCIENTIFIC BACKGROUND

Moseley was born into an academic family, his father a leading anthropologist and zoologist and his grandfather, as well as being a priest, noted for his efforts in physics, mathematics and astronomy. So, after growing up in Weymouth, Dorset, it was little surprise Moseley proved capable enough to go on to Oxford University to study natural science. He graduated from there in 1910 and was keen to follow in the family's scholarly tradition by immediately joining Rutherford's team in Manchester.

Killed at Gallipoli aged just 28, Moseley did enough to ensure his place among the greats

Initially, like so many of the others there he worked on trying to further understand radiation, particularly of radium. Moseley soon became interested by x-rays, however, and learning new techniques in measuring their frequencies. A method had been developed for placing metals in an x-ray tube and using crystals to diffract the emitted radiation, which had a wavelength specific to the element being experimented upon.

▸ EXAMINING X-RAY SPECTRA

In 1913, Moseley examined the X-ray spectra of more than thirty metallic elements and recorded the frequencies of the lines produced. He noted the lines moved according to the elements' atomic weight. Moreover, he soon deduced the frequencies of the radiation produced were related to the squares of certain incremental whole numbers. These integers were in themselves indicative of the 'atomic number' of the elements, and therefore their position in the periodic table. In addition, Moseley noted this atomic number was the same as the positive charge of the nucleus of an atom (and by implication also the number of electrons with corresponding negative charge in an atom).

By uniting the charge in the nucleus with the atomic number, and therefore an element's position in the periodic table, Moseley had found a vital link between the physical atomic make-up of an element and its chemical properties (as indicated by its position in the periodic table). Indeed, the step change meant the properties of an element were now considered much more in terms of its atomic number than its atomic weight, as had previously been the case.

▸ RE-EXAMING MENDELEEV

By realigning the periodic table according to this atomic number rather than the atomic weight, certain inconsistencies in the **Mendeleev** version could be ironed out in a logical fashion. Most notably, 'Moseley's law' (the principle he expressed outlining the link between the X-ray frequency of an element and its atomic number), predicted that there were several missing elements in the periodic table as he improved and restructured it according to his findings. Naturally, he was able to forecast the undiscovered elements' atomic numbers, weights and other properties from their expected position in the table. Over the ensuing years, the absentees were found and they slotted into their designated places.

A LIFE CUT SHORT

Sadly, Moseley did not live long enough to see his predictions come true. On the outbreak of the First World War he signed up with the Royal Engineers. He was killed the following year by a sniper's bullet through the head at the Battle of Suvla Bay in Gallipoli. At the age of just twenty-seven science had lost one of its brightest young sparks, already widely acknowledged for his completed work, and with the potential for further proving his genius cruelly snatched away in a single shot.

EDWIN HUBBLE

1889–1953

CHRONOLOGY • **1919** Hubble joins the staff of the Mount Wilson Observatory • **1923** Proves that the universe extends beyond the edges of our home galaxy, the Milky Way • **1925** Creates the first useful scheme for classifying galaxies • **1929** Demonstrates that the universe is expanding • **1936** Publishes *The Realm of the Nebulae*,the most popular science book of the year • **1990** The giant orbiting Hubble telescope is named in his honour

The man who completely changed our view of the bubble in which we exist was almost lost to astronomy, first to boxing, then to law. The young Hubble was such a fine fighter during the days of his astronomy and mathematics degree at the University of Chicago boxing promoters tried to persuade him to turn professional. He refused the offer. He did not turn down the chance to go to Oxford University in the United Kingdom on a Rhodes Scholarship to study law in 1910, though. He duly gained a BA in 1912 and contemplated a career in law on returning to the United States. In comparison to astronomy, he found the subject boring, however, so instead returned to Chicago to gain his PhD in the field of study he loved. After serving and being injured in the First World War, he finally had the chance to observe the stars professionally, taking up a post in the Mount Wilson Observatory in California in 1919, where he would spend the rest of his career.

The astronomer was lucky in that shortly after

As Hubble measured the distances of galaxies from the Earth, he found they were receding

he arrived, the observatory built a brand new 100-inch telescope, which was the most powerful in the world at that time. It allowed Hubble to view the skies in a level of previously unseen detail. He quickly took full advantage of this privilege. The American was particularly interested in the many 'nebulae' in the skies, all of which were thought to be clouds of dust within our own Milky Way galaxy. Indeed, it was thought at the time there was only this one galaxy in all, which according to the measurements of **Halley's** contemporary and rival, Harlow Shapley (1885-1972), was approximately 300,000 light years across (this was subsequently revised to 100,000 light years).

Focusing on the Andromeda nebula, Hubble used a technique developed by Shapley himself to ascertain that this 'cloud' was some 900,000 light years away from earth and therefore clearly outside the Milky Way. Moreover, Hubble soon came to realise these spiral-shaped nebulae were in fact other galaxies, much like our own. There were literally millions of them in the sky, containing billions of other stars. The results were breathtaking, completely changing our perception of the size of the universe, and brought Hubble fame overnight.

THE GALAXIES RECEDE

Moreover, during the next few years, Hubble continued measuring the distances of the galaxies from earth and found they seemed to be moving away from it, or 'receding.' In addition, the greater the distance between the earth and the galaxy, the faster the latter seemed to be receding. By 1927, Hubble came to the only logical conclusion: the universe, which most astronomers had believed was static, was in fact expanding. Other scientists had for the first time hinted at this possibility a few years earlier but now Hubble had provided conclusive evidence. Indeed, **Einstein** himself had developed an earlier theory which required the universe to be moving either inwards or outwards for it to work, but had changed it because astronomers had told him the universe was definitely static. He later referred to this alteration, on hearing the universe was actually expanding, as 'the greatest blunder of my life.'

HUBBLE'S CONSTANT

By 1929, Hubble had measured the distances of enough galaxies to announce his formulation of 'Hubble's constant.' He had worked out the speed at which the galaxies were recessing to be distance multiplied by his constant. Although Hubble overestimated the size of the constant, his formula was valid. Corrections since have allowed astronomers to estimate the radius of the universe to be a maximum of 18 billion light years and its age to be between 10 to 20 billion years old. Hubble went on to provide a system of classifying galaxies which is still largely in use.

HUBBLE'S LEGACY

Edwin Hubble, most widely known today because of the space telescope named after him, revolutionised our understanding of the cosmos. In the same way that that telescope hoped to improve our perception of the universe after its launch in 1990, so the American provided the most incredible 'picture' of space that humans had ever known, some sixty-five years earlier.

The notion of a universe which was expanding allowed later scientists to, amongst other things, find consensus on the origin of space and settle on the big-bang theory. Indeed, the principle of an expanding cosmos has been at the heart of astronomical theory ever since.

SIR JAMES CHADWICK

1891–1974

CHRONOLOGY • **1911–1913** Chadwick graduates from Manchester Universtity, and spends the next two years working for Ernest Rutherford • **1913** Goes to Berlin to study under the renowned Hans Geiger • **1920** Rejoins Rutherford at the Cavendish Laboratory in Cambridge • **1932** Discovers the neutrino • **1935** Chadwick awarded the Nobel Prize for Physics

The Englishman James Chadwick had a distinguished career in physics, primarily as an assistant to Ernest **Rutherford** (1871–1937), even before he made the breakthrough which secured his entry into this book. His solving of one of the last remaining mysteries concerning the basic structure of the atom through his discovery of the neutron, however, saw his elevation from little known researcher to celebrated physicist.

In 1910, Chadwick rejoined Rutherford, having previously worked for him in Manchester, when the latter took on his role as head of the Cavendish Laboratory at Cambridge. They worked successfully there together until 1935, when Chadwick left to become professor of physics at Liverpool University. Chadwick's earlier work at Cambridge largely involved the showering of elements with alpha particles to see the transmutation and other effects this would have. An important spin off of this was the deduction that the nucleus of the hydrogen atom, the positively charged proton with an atomic weight of one,

Chadwick's discovery of the neutron made possible the development of the atom bomb

was actually present in the nucleus of every other atom, just in larger quantities.

▸ THE WEIGHT OF AN ATOM

This work still left difficulties, though, in explaining the atomic weight of atoms. Chief amongst them was the fact the mass of known components of an atom simply did not add up. Protons seemed to account for around half of the weight and were matched in number by an equal amount of negatively charged electrons to counter their positive charge. But the weight of an electron was one-thousandth that of a proton, so still approximately half of the atomic weight of elements was unaccounted for. One leading theory suggested the missing mass was also that of protons whose charges, and therefore they themselves, were 'hidden' by additional electrons bedded inside the nucleus. The problem with this idea, though, was when nuclei fell apart, there was no evidence for these additional electrons.

▸ A SHOWER OF ALPHA PARTICLES

Finally, Chadwick solved the conundrum in 1932 after reinterpreting the results of an experiment carried out by Irène and Frédéric Juliot-Curie (Irène was Pierre and Marie **Curie's** daughter). The couple had found in 1932 that when the element beryllium was showered with alpha particles, the resultant radiation could force protons out of substances containing hydrogen. They had concluded the emission causing this reaction was made up of gamma rays, but Chadwick soon proved there was no way the rays could do this. Rather, it was far more likely neutrally charged subatomic units, which he named neutrons, with the same weight as protons could force this reaction and therefore were what made up the radiation. Rutherford had hinted at the existence of a similar particle back in 1920 but now there was strong evidence for it. The explanation was widely accepted and at last the riddle of atomic weight had been solved: a similar number of neutrons to protons in the atom of an element would make up the remaining fifty per cent of the previously 'missing' mass.

A knighthood followed for Chadwick in 1945, partly for this discovery but largely for his scientific service to Britain during the Second World War. Indeed, Chadwick's career was greatly affected by both World Wars but in vastly different ways. He had been robbed of four years of his scientific development, spending the First World War imprisoned in a racing stable after having had the misfortune of being in Germany to undertake work with Hans Geiger (1882–1945) at the outbreak of hostilities. The next time round, however, he spent it mostly in the USA as an effective head of the British delegation working on the development of the atomic bomb.

Chadwick received the Nobel Prize for physics in 1935 for his discovery of the neutron.

Further Achievements

Chadwick's discovery of neutrons – elementary particles devoid of any electrical charge – was crucial to the development of nuclear physics. In contrast with the helium nuclei (alpha rays) which are charged, and therefore repelled by the considerable electrical forces present in the nuclei of heavy atoms, the neutron is capable of penetrating and splitting the nuclei of even the heaviest elements, creating the possibility of the fission of uranium-235. This made possible the atomic bomb. For this epoch-making discovery Chadwick was awarded the Nobel Prize for Physics in 1935.

FREDERICK BANTING

1891–1941

CHRONOLOGY • **1916** Banting graduates MD from Victoria College, Toronto, and joins the Canadian Army Medical Corps • **1918** Awarded the Military Cross for gallantry in action and invalided out of the army • **1921** With Charles Best, begins study of the role of the pancreas in diabetes • **1923** The pair produce and patent insulin; pharmaceutical firm Eli Lilley begin industrial production of insulin; they are awarded the Nobel Prize for Physiology • **1939** Joins Canadian army medical unit at outbreak of Second World War • **1941** Killed when his plane crashes en route to Britain from Newfoundland

Up until the 1920s diabetes had been a ruthless killer. In 1921, after only a few months experimentation, the Canadian Frederick Grant Banting led a breakthrough in its treatment and virtually overnight offered the possibility of saving millions of lives.

Banting had graduated in medicine in 1916 from Victoria College in Toronto. In 1920, after returning from the First World War decorated with the Military Cross for bravery, Banting founded a practice in London, Ontario. At the same time he undertook research work at the local medical school, focusing on studies related to the pancreas.

▸ INITIAL RESEARCH

Earlier research had shown there was almost certainly some link between the pancreas and diabetes, but at the time it was not understood

In the age of AIDS and Ebola, one can forget that diabetes too was once a killer disease

what this was. We now know a hormone within the pancreas controls the flow of sugar into the blood stream. Diabetics are lacking this function and, untreated, are gradually killed by uncontrolled glucose input into the body's systems. Banting did not have the confirmation of this knowledge but he suspected the cause might be something to this effect.

His specialisation within pancreatic studies was in areas called the islets of Langerhans. Banting believed that the islets might be the most likely to produce the kind of hormone, if it existed, which controlled the glucose levels in the body. He figured that if this hormone could be extracted it might be viable as an injected treatment to diabetes sufferers.

▸ BANTING AND BEST

In 1921, Banting began a series of tests along with Charles Herbert Best (1899–1978) who was a research assistant at the University of Toronto, after being put in touch with him by Professor John James Rickard MacLeod (1876–1936) who also worked at the university. The two investigators were assigned a laboratory at the university by MacLeod and some dogs upon which to experiment. They extracted matter from the islets of Langerhans in the dogs' pancreas after preventing other pancreatic fluids from entering, in an effort to extract as pure samples as possible. The scientists then removed the pancreas from some dogs to hopefully induce diabetes. This soon happened, so their next step was to try to treat the dogs with their extract. It worked. The symptoms of the disease were soon under control.

▸ PRODUCTION OF INSULIN

Following this success, Banting and Best, at the suggestion of MacLeod, decided to further purify their extracted treatment before testing it on humans. This task was assigned to James Bertram Collip (1892–1965), a biochemist, and the solution he produced was named insulin. Human trials took place in 1923 with immediate impact. Dying patients were restored to health and suddenly diabetes could be managed within the realms of a normal lifestyle. The same year, industrial production of insulin began from pigs' pancreas and patients around the world soon received the life-saving benefits.

Banting was awarded the 1923 Nobel Prize for Physiology. It was shared, rather unfairly, with MacLeod, who had only contributed a limited amount to the discovery, but not with Best, who had been actively involved. To redress the balance a little, Banting shared his portion of the prize money with Best and MacLeod with Collip. The heroics Banting displayed during the First World War were called upon again in the Second as he undertook dangerous research into the effects of poisonous gases. Sadly, this time he succumbed, not to the gases, but in an air crash while flying from Canada to the United Kingdom to share his research with the British.

A Single Discovery Saves Millions of Lives

While many of the scientists in this book are known for a number of discoveries or inventions, others are equally celebrated for just one. This is the case with Sir Frederick Banting. When one thinks of killer diseases, one tends to think of AIDS, or Ebola. It is difficult to think of diabetes in the same category: and yet, prior to the discovery of insulin, diabetes routinely meant a death sentence to the millions unlucky enough to have it. Thanks to Sir Frederick Banting, this is no longer the case.

LOUIS DE BROGLIE

1892–1987

CHRONOLOGY • **1913** De Broglie graduates • **1914** Conscripted into the French army, where he remains until the end of the war in 1918 • **1924** At the Faculty of Sciences at Paris University delivers a thesis *Recherches sur la Théorie des Quanta* (Researches on the quantum theory) • **1927** Demonstrates the wavelike properties of electrons and other subatomic particles • **1929** Awarded Nobel Prize for Physics for his work on subatomic particles • **1952** Awarded the first Kalinga Prize by UNESCO for his efforts to explain modern physics to the layman • **1987** Dies in Paris, aged 95

Louis de Broglie probably had one of the more unusual family histories in quantum physics, with a name to match. The 'Prince' element of his title reflected his an honour bestowed upon his ancestors for service to the Austrians during an earlier war. But de Broglie was primarily a Frenchman of aristocratic background, which also meant he later inherited the title of 'Duc' upon becoming the head of his family on the death of his father.

▸ FROM HISTORY TO SCIENCE

His initial studies in history had been followed by a period working at the radio station based at the Eiffel Tower during the First World War, meaning it was a circuitous route to physical acclaim. His job at the famous landmark was the stimulus for his interest in science, and led to de Broglie signing up for study in physics at the Sorbonne after the war. This was to have a major impact.

If waves could behave like particles, why could not particles behave like waves?

▸ OF WAVES AND PARTICLES

De Broglie may have taken some time to turn to science, but he made a swift impression. Indeed, it was his 1924 doctoral thesis which formed the basis of his fame. The key theme concerned a natural extension of the quantum theory which had already been put in place. **Einstein** (1879-1955) had suggested in one of his 1905 papers the mysterious 'photoelectric' effect could be explained by an interpretation which included electromagnetic waves behaving like particles; indeed the waves were in fact constructed from a stream of particles (called 'quanta' or 'photons'). What de Broglie did was simply turn this on its head: if waves could behave like particles, why should particles not behave like waves?

▸ ELECTRON BEHAVIOUR

His conclusion was indeed they could and he formulated a theoretical proof of his idea involving the behaviour of electrons. In classical physics, these were unquestionably considered to be particles, distinct pieces of physical matter. By applying quantum theory, however, de Broglie was able to show an electron could also act as if it were a wave with its wavelength calculated by simply dividing 'Planck's constant' by the electron's momentum at any given instant. Although the proposal sounded extremely theoretical it was, remarkably, found to be plausible by experimental evidence shortly afterwards.

▸ WAVE–PARTICLE DUALITY

As a result it sparked a fierce debate concerning the 'wave-particle duality' of matter and questions about which interpretation was correct. **Schrödinger** (1887-1961), **Heisenberg** (1901-76) and Born (1882-1970), amongst others, would soon offer compelling arguments. To an extent, Niels **Bohr** (1885-1962) provided some context around the issue in 1927 by stressing the futility of the debate, pointing out the equipment used in experiments to try to prove the case one way or the other greatly influenced the outcome of the results. A principle of 'complementarity' therefore had to be applied which suggested all of the experimental proof one way or the other to be a series of partially correct answers which had to be interpreted side-by-side for the most compete picture. Eventually, though, the 'probabilistic' theories of Heisenberg and Born largely won out. At this juncture, the point at which cause and effect had logically been removed from atomic physics, De Broglie, like Einstein and Schrödinger, began to question the direction quantum theory was taking, and rejected many of its findings.

Further Achievements

There were countless physicists wading into the debate surrounding quantum theory in the mid-1920s. Most of the key contributors were established professors, doctors or specialists in physics which makes Louis de Broglie's background as a historian all the more remarkable. His unusual route to scientific prominence did not make him any less effective when he finally got there, however, proposing a quantum theory which would once again open a whole new branch of investigation.

ENRICO FERMI

1901–1954

CHRONOLOGY • **1923** Fermi studies under Max Born at Göttingen, Germany • **1934** Discovers slow neutrons • **1938** Awarded Nobel Prize for Physics • **1939** Escapes Europe and moves to the US • **1942** Achieves man-made nuclear chain reaction • **1949** Argues against development of the H-bomb

Enrico Fermi, arguably Italy's most talented scientist of the twentieth century and quite possibly since **Galileo**, could have had no idea of the eventual outcome of the experimental work he undertook in Rome in the mid-1930s. He was systematically working his way through the elements to study the effects on them of a neutron-bombarding technique he had discovered. Most yielded predictable, or certainly not extraordinary, results. When he arrived at uranium, the heaviest naturally occurring element, however, something very odd happened, which was to have enormous impact on physics and beyond.

A few years later, in Chicago, Fermi would experience first hand the potential of his discovery. Fermi and his Jewish wife had fled to America following the rise of anti-Semitism in Italy.

▸ THE URANIUM NUCLEUS

Shortly afterwards, he had received reports of a reinterpretation of his uranium bombardment experiment. Fermi himself had been unsure of what had happened, suspecting the possibility that perhaps the uranium had transmuted into new, heavier elements. Now, however, an alternative

'The Italian navigator,' said one commentator, 'has landed in the new world'

explanation was offered by the German scientists Otto Hahn, Fritz Strassmann and Lise Meitner that the uranium nucleus had in fact been broken down into a number of smaller elements. Moreover, this nuclear fission had seen some of the uranium mass converted into potentially huge amounts of energy under the rules of **Einstein's** formula $E=mc^2$. This reinterpretation was leaked out of Germany by Meitner and her nephew Otto Frisch when they escaped the Nazi state.

▸ A NEW WORLD

Fermi immediately saw the impact of the analysis and set to work on reproducing the experiment with Niels **Bohr** on arrival in the US. They confirmed their best and worst fears: using the uranium isotope-235 a nuclear chain reaction could almost certainly be created as the basis of an atomic bomb. Fermi was recruited to the Manhattan Project to ensure the US created a fission bomb ahead of the Germans. Fermi led a team in Chicago seeking to generate a self-sustaining, contained nuclear reaction. By 2 December 1942 his team had created an 'atomic pile' of graphite blocks, drilled with uranium which went on to produce a self-sustaining chain reaction for nearly half an hour. 'The Italian navigator,' as one commentator reported back to the project committee, 'has just landed in the new world.' Less than three years later, the technology would be used in the first atomic bombs with devastating effect.

The innocent discovery back in Italy in the 1930s, which had led to such incredible consequences, had been Fermi's conception of neutron bombardment in the artificial transmutation of elements. The Juliot-Curies had announced in 1934 their discovery that radioactive isotopes could be generated artificially by showering certain elements with alpha particles.

▸ ON NEUTRONS

Fermi had quickly realised that the newly-discovered neutrons would be even more suited to this purpose because their neutral charge would be more likely to allow them to slip into elements' nuclei without resistance. By chance he also found the phenomenon of 'slow neutrons' by placing a piece of solid paraffin in front of his target element during bombardment. This had the effect of slowing the neutrons down before they reached the element, meaning they were exposed to its nuclei for longer and thereby had a much greater chance of being drawn in to create new isotopes. As Fermi now worked through the elements applying these discoveries, he created lots of new radioactive isotopes, which was considered achievement enough for him to be awarded the 1938 Nobel Prize for physics. It was only after he had collected his award that the much more significant consequences of this work when applied to uranium were realised.

FURTHER ACHIEVEMENTS

Earlier in his career, Fermi had established his reputation with important work in theoretical physics. His most notable achievement in this area was his concept of radioactive beta decay. This concerned the theory that a proton could be created from a neutron via the shedding of an electron (a beta particle) and something known as an antineutrino.

It was for his achievements in experimental physics for which the Italian would be remembered, however, leaving behind a world, following his early death from cancer, very different from the one he had entered just over half a century earlier.

WERNER HEISENBERG

1901–1954

CHRONOLOGY • **1922** Heisenberg studies at Göttingen, Germany under Born • **1925** Heisenberg develops his radical approach to quantum theory • **1927** Formulates his famous 'Uncertainty Principle' • **1932** Awarded the Nobel Prize for Physics

Heisenberg's development of matrix mechanics in 1925 sparked a controversy in the rarified world of quantum theory. Like many other physicists, Heisenberg too had been contemplating the debate over whether electrons and other atomic phenomena behaved in a wave or particle-like fashion. Heisenberg found a simple solution. He ignored both arguments altogether! Instead Heisenberg proposed that the only important factor was being able to mathematically predict the occurrence of atomic features which could be measured or observed such as frequency and light emissions. So, he applied algebra to the problem and developed a mathematically based solution which came to be known as matrix mechanics. The predictive and quantifiable powers of this new scheme were excellent and Heisenberg received the Nobel Prize for this development in 1932.

▸ THE UNCERTAINTY PRINCIPLE

They also had a logical extension and it was to this part **Einstein** objected most. Heisenberg expressed it as his 'uncertainty' principle in 1927. In seeking to underline his basis for ignoring the visual idea of the atom and only considering it mathematically, Heisenberg

W9-AIH-892

KNOCK ON WOOD
& Other Superstitions

KNOCK ON WOOD
& Other Superstitions

Carole Potter

BONANZA BOOKS
NEW YORK

For my father who always believed that being the seventh son of the seventh son made him special; and for my mother who believes in *all* superstitions, just in case.

This 1984 edition is published by Bonanza Books,
distributed by Crown Publishers, Inc. by arrangement with
Beaufort Books, Inc.

Manufactured in The United States of America

Library of Congress Cataloging in Publication Data

Potter, Carole.
Knock on wood.

1. Superstition—Dictionaries. I. Title.
BF1775.P67 1984 001.9'6'0321 84-16739

ISBN: 0-517-459442

h g f e d c b a

INTRODUCTION

A SUPERSTITION

—A superstition is a belief or a notion which attempts to fathom the unfathomable, based on neither reason nor fact; sometimes it disregards evidence.

—A superstition is a way to get through a tough situation; it's a set of rules to follow in a game without rules.

—A superstition is a blindly accepted belief which probably dates back thousands of years and has outlived its original meaning and use.

—A superstition takes the fear out of the unknown.

—A superstition is something we think will change our luck.

Some superstitions were designed to help us see into the future, and who doesn't want to know what's ahead for us? And some evolved out of the fear of retribution. If, thought ancient man, we are basically unworthy of the gifts of the gods, then a little bit of luck might come in very handy. So, we'll thank the gods, we'll not tempt them by calling their attention to something wonderful, and we'll knock on wood to keep the evil spirits away and to thank the good spirits for answering our prayers.

If you believe in not walking under a ladder, never inviting thirteen to a dinner party, tying a red ribbon to a baby carriage and never going back for something you've forgotten; if you always throw salt over your left shoulder after having spilled some on the table, or say *gesundheit!* when a friend sneezes; if you are terrified of black cats and cross your fingers for luck; if you wear a charm around your neck or read your horoscope each morning; if you do all of these things, or even some of these things, you're in good company. Most people believe, as Goethe did, that superstitions are "the poetry of life."

Above all, a superstition is something never to be questioned because you will find, as I did, that all superstitions are flawed, albeit highly romantic, contentions. A superstition then, is an unswerving belief that helps make life a tiny bit easier. And what's so terrible about that?

Alas! you know the cause too well;
The salt is spilt, to me it fell;
Then to contribute to my loss,
The knife and fork were laid across:
On Friday, too! the day I dread!
Would I were safe at home in bed!
Last night (I vow to Heav'n 'tis true)
Bounce from the fire a coffin flew.
Next post some fatal news shall tell:
God send my Cornish friends be well!

(John Gay, 1700's
Fables: The Farmer's Wife and the Raven)

KNOCK ON WOOD
& Other Superstitions

ABRACADABRA

See also AMULETS

This word, which we've all said dozens of times in jest, was taken very seriously during the Middle Ages. People believed that the word itself was a cure for fever. A sufferer would wear a parchment amulet around the neck with the word written in an inverted pyramid (in itself a magical symbol, *see* PYRAMID):

ABRACADABRA
ABRACADABR
ABRACADAB
ABRACADA
ABRACAD
ABRACA
ABRAC
ABRA
ABR
AB
A

In some parts of the world, it was believed that a person with fever simply had to write the word several times, each time dropping one letter as shown in the pyramid above. As the letters disappeared so would the fever. The charm was also believed to be helpful against toothaches and infection.

Belief in this magical word probably dates back to the centuries before Christ, but the exact name of the demon or evil spirit it comes from has been lost to history. By the second century A.D. the cabalists, an extreme sect within the Hebrew religion, were using a form of Abracadabra as a charm against evil spirits: *A*, meaning Father; *Ben*, Son; and *Ruach Acadach*, the Holy Spirit.

There was a belief common in ancient times that the mere name of a supernatural being held magical powers. People would evoke a name for protection—as we do today—and they felt that knowing the true name of the proper deity would be a help. Abracadabra became a term in the art of magic as this trend became more popular and the magicians began to use fantastic, incomprehensible words in place of the names of gods. The more obscure the words, thought the magicians, the more magical the powers.

ACHILLES' HEEL

If an arrow struck your heel, it would no doubt be very painful—but would it be fatal? A legend, which has been handed down to us by the Greek poet Homer, tells of a warrior known for his excessive bravery, strength, and most of all, invulnerability. In Homer's tale, Achilles' mother dreamed that her son would be killed in battle. To protect him, she bathed him in the River Styx, which coated him with a protective shield everywhere except on the heel by which she held him. It is said that during the Trojan Wars Paris shot Achilles with a poison arrow, guided by the hand of the god Apollo, which struck his heel and immediately killed the otherwise invulnerable warrior.

Medicine has honored Homer by naming the muscle connecting the calf to the heel the tendon of Achilles.

In today's idiom, any vulnerable spot in a person (or a system) is called an "Achilles' heel."

ACORN

See also OAK TREE

The little acorn, from which the famous great oak grows, has been worn as an amulet since the days of the druids. Druids held particular store in the oak tree and hence the acorn, which represented good luck and long life.

In Asia Minor, the acorn was sacred to the goddess of nature. The Scandinavians believed that the great god Thor, god of thunder, protected the oak tree. They believed that placing an acorn near a window would prevent lightning from striking the dwelling.

Today, some people still strongly believe that an acorn placed on a windowsill will protect them. A dried acorn is also often used as a windowshade pull.

Folklore gives additional significance to the acorn. Since it takes a very long time for an oak to grow from an acorn, the acorn is a symbol for a difficult achievement that has taken a lot of hard work to accomplish.

ALBATROSS

Consider the jinxed albatross. Imagine the horror of Coleridge's ancient Mariner:

> *'God save thee, ancient Mariner!*
> *From the fiends, that plague thee thus!—*
> *Why look'st thou so?'—with my cross-bow*
> *I shot the albatross.*
> *(The Rime of the Ancient Mariner, I)*

It is believed, mostly by sailors, that to kill an albatross is to bring bad luck to the ship and to everyone on it, and death to the person who killed it. An albatross that follows a ship is supposed to carry the soul of a drowned sailor in order to be near his mates. It is also believed that it follows to aid rescuers in case of a shipwreck.

Some think that an albatross sleeps while it flies because its flight seems motionless. Others believe that if the bird flies around a ship bad weather will continue for many days. Looking at a storm at sea, any captain might say, as the ancient mariner did:

Instead of the cross, the Albatross
About my neck was hung.

(II)

ALL SAINTS' DAY (ALL HALLOWS)

See HALLOWE'EN

ALMOND

To the Victorians, the almond was a symbol of stupidity, indiscretion, thoughtlessness, and the impetuousness of youth. They reasoned that since the almond tree blossomed so early in the spring, it left itself wide open to frosts and other damage due to the fickleness of the weather. If the "stupid" almond tree had just waited until later in the season, it could have borne many more blossoms.

The significance of the almond goes back much further in history. The Greeks had a legend about it (indeed, the Greeks had a legend for almost everything). In this story, young Demophon, returning from the Trojan Wars, is shipwrecked and meets a Thracian princess, Phyllis. They fall in love and arrange to marry, but before the ceremony can take place, Demophon learns that his father has died in Athens and he must return for the funeral. He promises to return on a certain date but miscalculates how long it will take him, returning three months later than planned. By this time Phyllis, certain that he will never return, has hanged herself.

The gods, moved by her love, transform her into an almond tree. Demophon is distraught and offers a sacrifice to the almond tree, declaring his undying love. In response, the almond tree blossoms. Impetuous youth and undying love were thus symbolized by the almond prior to the Victorian Age.

Pliny advised in *Natural History*, in 77 A.D. that eating five almonds would prevent drunkenness. He also asserted that if foxes ate almonds with water, they would die.

In the sixteenth century pills containing ground almonds were made for travelers to take, in case they found themselves without food or water.

The Moslems mixed the paste of the almond with mother's milk as a cure for trachoma.

Today, the branch of an almond tree is often used as a divining rod. (*See* DIVINING ROD)

AMETHYSTS

Cleopatra wore them and so did Catherine the Great. Saint Valentine is said to have had one with Cupid's face carved upon it. Egyptian soldiers wore them into battle for a tranquilizing effect in dangerous situations, to insure victories. Perhaps the most wonderful thing about amethysts is the ancient belief that they prevented drunkenness! Greeks in early times believed that the amethyst, which is a wine-colored stone, prevented the wearer from becoming drunk, and so they named it *amethystos*, meaning "not to be drunk." They also carved drinking cups from amethyst to prevent the drinker from becoming intoxicated.

There is a Greek legend about Bacchus, god of wine and revelry, who was angry with the goddess Diana and swore vengeance upon the first mortal he saw. That happened to be a young maiden named Amethyst who came to worship at Diana's altar. When she saw Bacchus's tigers about to pounce on her she called to Diana for help. Diana instantly changed Amethyst into a statue to protect her. Bacchus, as a gesture of contrition, poured wine over the statue, making Amethyst a beautiful grape color and granting her immunity against the narcotic effects of the beverage.

In the Middle Ages, the gem became known as the Bishop's Stone because it was used in the ring a bishop wore on the third finger of his right hand to indicate that he was married to the Church.

For the Hebrews, the amethyst was the stone of Dan, one of the twelve tribes of Israel and stood for judgment, courage, and justice.

Amethyst is the stone of those born under Aquarius and is thought to encourage moderation and self-discipline in all Aquarians who wear it.

The February born will find
Sincerity and peace of mind;
Freedom from passion and from care
If they the amethyst will wear.
(ANON)

Some people believe that an engraved amethyst has extra powers. One engraved with the moon or sun could provide protection against storms. Amethysts with other engravings are thought to protect against drunkenness and thieves, insure pleasant dreams, or keep a wandering husband faithful.

If all this isn't enough, the amethyst is supposed to help against headaches, toothaches, and gout!

AMEN

Why say "amen" before and after prayers? It is believed that a loud, forceful amen, said in unison by worshippers, will keep open the doors of Paradise.

Amen is a sign of agreement, of affirmation, of approval, as the Book of Deuteronomy says,

All the people shall answer
and say Amen.

Macbeth, who probably needed the gates of Paradise to open before him more than most people in history, said:

I had most need of blessing, and 'Amen'
Stuck in my throat.
(Shakespeare, *Macbeth*, II:2)

AMULETS

See also ABRACADABRA
FOUR-LEAF CLOVER
HORSESHOES
RABBIT'S FOOT
TALISMAN

An amulet is an object used mostly to protect a person against evil, or to bring good luck. It can be worn, nailed to a front door, or fastened onto an automobile or baby carriage. An amulet must be a natural item, not man-made, and it may be anything the owner believes has special powers.

An amulet can be a ring, a stone, a book, a horseshoe, or a part of an animal. Probably the most popular amulet in history has been the rabbit's foot. (*See* RABBIT'S FOOT) Parts of animals are very popular amulets and are considered to be especially powerful in bringing good luck. A lion's tooth might be worn for courage; a tiger skin or a bear claw represents strength. In the fifties kids hitched foxtails to their bikes to insure speed.

Bells used as amulets are supposed to frighten away evil spirits and are often seen attached to oxen or horses for their protection. They were once considered very important amulets for travelers. Who knew what dangers lurked in a foreign land?

Golden objects are supposed to bring good luck because gold was believed to be made from the tears of the sun god. Heart-shaped amulets could provide the wearer with lasting love, as well as protection from heart disease.

One of the most popular amulets in Italy is the red pepper. A natural, dried red vegetable is still believed to keep the Devil away, because red is a color that frightens the Devil. (*See* CHARTS, COLORS) When a red pepper dries it resembles the Devil's horn. It is believed that it will keep you safe from his grasp. The protection of this particular amulet, however, has been greatly diluted in recent years because of the commercial manufacture of red plastic peppers. Can you imagine the Devil being afraid of a piece of plastic?

Other popular amulets include coral, mandrake root, four-leaf clover, human hair, teeth, representations of the human hand, medals of saints, birthstone rings, and abracadabra.

The difference between an amulet and a talisman is that an amulet is a passive thing—all you do is wear it. A talisman requires some involvement from the person seeking good luck. You have to kiss, touch, or wave a talisman in order for it to be effective. (*See* TALISMAN)

ANGELS

See STARS

ANKH

The ankh (pronounced like "tank") is a cross-like symbol that comes from ancient Egyptian hieroglyphics and means life. It was held in the hands of the pharoahs and by the various deities and is often translated as their "key to life."

The ankh is not thought to have any special mystical powers today. It regained some of its popularity, however, when author Jacqueline Susann used it in her best-selling novel *The Love Machine*. Consequently it took on explicit sexual connotations.

Another interpretation of the ankh, found in early Greek writings, is the Greek letter *tau* (T) which represents life; topped by a circle, which represents eternity, it implies eternal life.

In any case, the ankh has become an attractive talisman (*See* TALISMAN) which people wear on chains or as a ring.

APHRODISIACS

See also MANDRAKE ROOT
OYSTERS
TOMATOES

By and large, natural aphrodisiacs are a question of mind over matter. Shakespeare, in *Henry IV*, Part I; II:2, wrote: "I am bewitched with the rogue's company. If the rascal have not given me medicines to make me love him, I'll be hanged."

In ancient Rome, chicken soup was considered an aphrodisiac; in ancient Egypt lettuce was supposed to do the trick. Tomatoes (and potatoes) were once considered such strong aphrodisiacs that they were illegal. (*See* TOMATOES)

The word aphrodisiac comes from the name of the goddess Aphrodite, mother of Eros, from whose name we get the word erotic.

The ancients had a strong and abiding faith in the principle of like makes like. In other words, they felt that foods that resembled sexual organs would increase sexual pleasures. For example, asparagus and celery resembled the male organ; oysters (*See* OYSTERS) and clams, the female organs. In the same vein of like makes like, eggs and caviar are thought to be stimulating. (Unfortunately, there is some evidence that this power-of-suggestion school of sexual therapy works in reverse for as many people as it helps. There are many who won't eat these foods for the same reason that others gobble them!)

The botanist Nicholas Culpeper and other herbalists considered the artichoke to be under the influence of Venus. In 1649 he referred to artichokes and said, "Therefore, it is no Marvel if they provoke Lust." (*The Complete Herbal and English Physician Enlarged*)

APPLE

An apple a day keeps the doctor away. (Old English proverb)
To eat an apple without rubbing it first is to challenge the Devil.
(Old English proverb)
A bad woman can't make good applesauce (it gets mushy).
(Old English proverb)

The apple is a favorite means for finding future husbands: Give one twist of the stem for each letter of the alphabet and wherever it breaks is the first initial of your true love.

Apple peel, apple peel, twist then rest,
Show me the one that I'll love best.
Apple peel over my shoulder fly,
Show me the one I'll love 'til I die.

There is a wonderful Scandinavian legend that tells the secret of the gods' eternal youth. They munched on magic golden apples supplied by Idhunn, child of the supreme goddess, Frigga.

The Bible is strangely quiet on the subject of the apple. The fruit with which Eve tempted Adam may not actually have been an apple after all. The Book of Genesis just refers to the fruit of the tree, and the poor apple has taken a bum rap ever since. The apple tree was thought to be the tree of life and knowledge, and the apple remains a symbol of temptation.

The Greeks thought the apple stood for immortality. The Arabs believed it had curative powers. Almost all the English-speaking peoples believed in its magical qualities; and of course, it was used to prove Newton's Law of Gravity.

Perhaps the most famous legend regarding the apple is the American folktale of Johnny Appleseed. (*See* JOHNNY APPLESEED)

Folklore uses for apples abound, such as this sixteenth century proverb:

He that will not a wife wed,
Must eat a cold apple when he goeth to bed.
(Thomas Cogan, *The Haven of Health*)

APPLE A DAY KEEPS THE DOCTOR AWAY, AN

See APPLE

APPLESEED, JOHNNY

The legend of Johnny Appleseed is based on fact. It tells of John Chapman, an eccentric and deeply religious man who roamed the American frontier planting apple trees and nurseries wherever he went. He encouraged the use of the name Johnny Appleseed and even referred to himself that way. He is often called the "American Saint Francis."

Chapman was born in 1774 and turned up in Pennsylvania around 1797. For the next forty-eight years he moved around the frontier, a bearded, hermit-like figure, barefoot and dressed in rags, with a sack of seeds slung over his shoulder.

Let all unselfish spirits heed
The story of Johnny Appleseed.
He had another and prouder name
In far New England, whence he came,
But by this title, and this alone,
Was the kindly wanderer loved and known.
(Elizabeth Akers Allen)

APRIL FOOLS' DAY (ALL FOOLS' DAY)

April prepares her green traffic light
and the world thinks Go
(Christopher Morley, *John Mistletoe*)

But is all as it seems? The world may indeed burst forth in all its greenery—but is it only "contriving a lie for tomorrow," as Jonathan Swift tells us?

In earlier times in France, people were easily tricked in April into thinking that summer had arrived. People who were fooled were thought to be as green as the new grass after a long winter's hibernation. The most popular custom during the month was to send people on a "fool's errand."

The Roman myth concerning Proserpina, who was snatched by

Pluto, the king of the Underworld, to be his bride, provides another popular explanation of April Fools' Day. Ceres, goddess of agriculture and the mother of Proserpina, went running in vain after her daughter when she heard Proserpina screaming during the abduction. This, say some, began the custom of sending gullible people on a "fool's errand." It might also have been starting with Noah sending the dove from the Ark on a fruitless mission.

Most likely the modern customs of April Fools' Day started around 1752, when the Gregorian calendar was adopted in England and the New Year's celebration was changed from the week of March 25 to April 1 to January 1. Practical jokers continued to make calls and bring gifts on the old holiday, April 1.

The first day of April,
You may send a fool wither you will.
(Fuller, *Gnomologia* #6135, 1732)

AQUARIUS

See also ZODIAC

Aquarius is the sign of the Water Bearer. The Egyptians believed that the Nile overflowed every time the Water Bearer dipped his bucket into the river, producing fertile crops. Aquarius became a symbol of the thirst for knowledge and the desire to impart that knowledge to others.

This is the sign of people born between January 20 and February 18. Aquarians are ruled by Uranus and Saturn, and are supposed to be innovative, unconventional people who love human nature. They tend to be undependable, interesting, courageous to the point of irresponsibility, enthusiastic, inventive, and though lovable, often erratic in love.

Wednesdays are good days for Aquarians. The numbers eight and four often work for them, and pastel blues and greens are their best colors.

Aquarians like to write letters, find aviation interesting, and should be careful to eat properly. The amethyst is the birthstone of the Aquarian. (See AMETHYST)

Some famous Aquarians are Franklin Delano Roosevelt, Ronald Reagan, Adlai Stevenson, Marian Anderson, and Paul Newman.

> *'How is your trade, Aquarius*
> *This frosty night?'*
> *'Complaints is many and various,*
> *And my feet are cold!'*
> *said Aquarius.*
> (Robert Graves, *Star Talk*)

ARIES

See also ZODIAC

Aries, the sign of the Ram, is the first constellation of the zodiac because of a very ancient myth that the world was created when the sun entered the sign of the Ram. In mythology, the Ram is often pictured lying on its side, gazing admiringly at its golden fleece.

Aries is the sign of people born between March 21 and April 20. These people, influenced by the Ram, are supposed to possess violent tempers and often die violent deaths.

Tuesdays are good days for Aries people, seven and six are their lucky numbers, and red is their best color. The birthstone for Aries is the diamond. They are ruled by Mars and so are assumed to possess courage. Since theirs is a fire sign, they are enthusiastic enough to inspire other people. They are reputed to be independent people, to be delighted by challenge, and to be good leaders with an ability to plan and develop effective programs.

Aries people are always taking risks. They are extravagant and have strong willpower, good constitutions, and a positive attitude. They are supposed to be difficult to love. Some famous Aries are Robert Frost, Marlon Brando, Nikita Khruschev, Bette Davis, and Tennessee Williams.

ASTRONOMERS' SIGNS FOR THE SEASONS

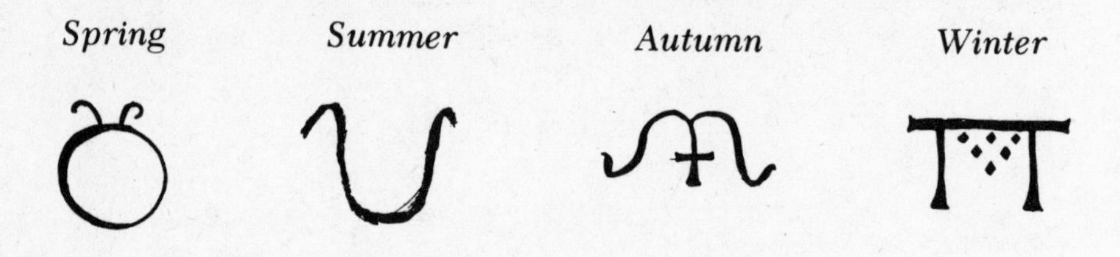

Some Other Astrological Symbols:

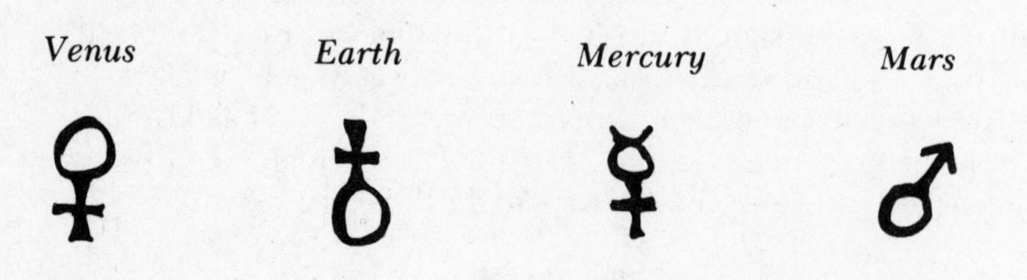

BABIES

First a son, then a daughter,
You've begun just as you oughta.

(Unknown)

If you rock a cradle while empty,
Babies you'll have a plenty.

(Unknown)

Monday's child is fair of face
Tuesday's child is full of grace
Wednesday's child is full of woe
Thursday's child has far to go.
Friday's child is loving and giving;
Saturday's child works hard for a living
But the child that is born on the Sabbath Day
is fair and wise, and good and gay.

(Bray, *Traditions of Devon*)

When you have one,
You can get up and run,
When you have two,
You can go, too.
When you have three,
Better stay where you be
When you have four,
You go no more.

(Unknown)

Ever step over a crawling infant only to be yelled at and told you'll stunt the baby's growth, or been advised to carry a baby to the top of the house before bringing it downstairs for the first time? (It is said that if you have no stairs, you'll get the same good luck by standing on a piece of furniture with the child, or even on top of a box.)

Superstitions about what to do before the birth of a baby are probably as important as those about after it is born. Some, for example, are:

Babies born at midnight can see ghosts.
If your child is born while the moon is on the rise, it will be a girl.
A child carried high is sure to be a girl, and one carried low, to be a boy.
To make certain it's a girl, put a frying pan under your mattress; to insure that it's a boy, place a knife under the mattress.

More examples:
To cut the pain of childbearing, put an ax under the bed.
Mothers-to-be should not witness ugly events because children are marked by what their mothers see, hear, and do during pregnancy.
The sins of the fathers are visited on the sons.
A pregnant woman should always eat the foods she craves or the child will be born with a birthmark in the shape of that food. (How about a chocolate chip cookie?)

Once your child is born, be sure to:
brush the baby with a rabbit's foot; (*See* RABBIT'S FOOT);
put a coral necklace around its neck; (*See* CORAL);
present the child with a fresh egg, a pinch of salt, and a loaf of bread;
make certain the baby sneezes (*See* SNEEZING) to rid itself of evil spirits.

Don't:
let the baby look in a mirror (*See* MIRROR) before it is a year old;
cut the baby's fingernails (*See* FINGERNAILS) before a year—the mother must bite off the growing nails, or the child will grow up to be a thief.
let a cat into the room. (*See* CAT)

Whichever of these superstitions you choose to heed or ignore, the result will still be a splendid child. At least that's what Charles Dickens promised in *Nicholas Nickleby*: "Every baby born into the world is a finer one than the last."

BACHELOR'S BUTTON

The lovely little blue flower known as a bachelor's button has always been associated with love and marriage. During Victorian times it was the flower most favored by young men in love.

The Orientals believe that the flower could foretell a happy marriage. They say that the flower should be picked early in the morning, then placed in the young man's pocket and not looked at for twenty-four hours. If it is still fresh and "true blue" the next day, the couple will have a good marriage. If, on the other hand, it has withered (which is highly likely after twenty-four hours in someone's pocket), then the marriage plans must be terminated. That seems to agree with the Italian proverb that says, "Praise all wives, but remain a bachelor." After all, this way of choosing a bride does seem to have all the odds stacked against it.

Throughout history the word bachelor has implied a novice as well as a man who has never been married. Originally a bachelor was a kind of cowhand. The word might have come from the Latin *baccalarius*, one who cares for cows. The meaning changed during the Middle Ages. Bachelor then referred to a young knight who was serving as a vassal to a member of the gentry.

During modern times it has meant "a man who hasn't been caught yet." If the man lets the bachelor's button make his decision for him, he'll "never get caught."

BAKER'S DOZEN

See also THIRTEEN

A baker's dozen has always meant thirteen items instead of twelve. The phrase goes back as early as the 1500s, and Rabelais included it in his writings. (*Book V*, Chapt. 22)

There are two schools of thought about its origin. One is that when retailers bought their dozen loaves of bread to resell to the public, they got an extra loaf free as a commission. But it is more likely that in earlier times a baker included a thirteenth loaf with every twelve to avoid being accused of short-weighing the bread.

BANANA

The banana is a food some people use to wish upon. Make your wish, then eat your fruit—or cut a slice from the stalk end of the fruit while making a wish. In either case, if you find a Y-shaped mark at the end your wish will come true. (I haven't the faintest idea why.)

BANSHEE

Wailing like a banshee is to be making an unearthly sound to foretell a death in the family. In old Gaelic tradition a banshee was a household spirit, usually a beautiful woman, who took particular interest in the well-being of the family. When death was near, she let out an unholy wail which could be heard throughout the county. Today the woman is usually a hag. In either guise, the banshee's most valuable asset is her ability to foretell death.

BARBER'S POLE

The wonderful candy-striped pole in front of the local barber's shop dates back to fifth-century Rome, when barbers were also surgeons. They treated wounds, pulled teeth, and bled patients.

The white in the pole represents bandages; the red, the blood. Attached to many poles in early England was a brass basin, which represented the bowls used to catch the blood during letting. Barbers performed bloodletting chores right through to the sixteenth century.

Ultimately surgeons took over their rightful practice and the brass bowl began to disappear from the barber's pole. The red and white stripes remain.

BAR MITZVAH

In the Jewish religion, the Bar Mitzvah is a ceremony during which a thirteen-year-old boy assumes the responsibilities of manhood and the responsibilities of his religion. The custom did not begin until the fourteenth century, although writings imply that some sort of symbolic celebration was held when a male reached puberty in ancient times.

The boy, in a loud, clear (and usually high-pitched) voice proclaims, "Today I am a man!"

BATS

There is a popular belief that the Devil, when he isn't doing something to bother people, takes the shape of a bat. Some people believe that the bat is a familiar (a familiar is the servant of a witch). Some believe that ghosts take the form of bats and therefore houses with bats in them are haunted.

To the Chinese and the Poles, bats are a good omen. The Chinese say that bats fly with their bodies down because their brains are so heavy and also believe that the bat is the symbol of long life and happiness.

In most other countries, bats are thought of as evil things. To the Irish, the bat is a symbol of death. There is a common idea that if bats move into a house, the occupants will shortly move out (sounds like common sense to me).

Did you know that the heart of a bat—dried, powdered and

carried in a front pocket—will turn a bullet or stop a man from bleeding to death; and if you wash your face in the blood of a bat, you will be able to see in the dark?

Another superstition that many believe is that if a bat lands on your head, it will not leave until it hears the thunder of an approaching rainstorm.

Incidentally, expressions such as blind as a bat and bats in the belfry indicate that we once knew very little about bats.

BATS IN THE BELFRY

See BATS

BEARDS

See also HAIR

Beware of that man, be he friend or brother,
Whose hair is one color and beard another.
(Unknown)

The properness of a man lives altogether in the fashion of his beard.
(Anonymous, *Humor Out of Breath*, 1608)

A red-bearded man was never any good.
(German Proverb)

Beware of women with beards,
And men without them.
(Basque Proverb)

In 350 B.C., Alexander the Great ordered his troops to shave so that the enemy would not be able to grab them by the beard in order to more easily decapitate them.

Beards have always been an especially important symbol. A beard was a mark of wisdom and was associated with a man's soul. The beard became a symbol of masculinity early in history because young boys and eunuchs had neither body hair nor virility.

The beard also became an almost holy object. People reasoned that since the beard continued to exist even after having been shaved off, it was intended by the gods to always be there.

To the ancients hair, like fingernails, eyelashes, and other nonpermanent parts of the body (*See* FINGERNAILS) were considered pieces of the person never to be discarded. If they fell into the wrong hands the person's soul was in deep trouble. It was unlucky to cut your beard for fear the Devil would get hold of the clippings. Whenever hair was clipped it was immediately burned as a precaution.

The beard is even today so important to the Moslems that they take their oaths by the beard of the Prophet Muhammad; and Orthodox Jews consider it sacred. In ancient Greece, men with beards were commonly considered philosophers. In fact, the ancient Greek word *pogonotrophos* meant "man with a beard" which was synonymous with philosopher.

The beard has been a royal status symbol for centuries. In ancient times, beards of iron were tied with ribbons or straps to the chins of kings and even queens, as depicted in hieroglyphics found on the tombs of Egyptian rulers. The boy king Tutankhamen, for instance, usually had a glorious black beard attached to his chin.

Long ago Egyptians and Babylonians used curling irons and dyes on their whiskers, and at festival times they indulged by using perfumes and powders on their beards. The kings of Persia were known to intertwine their beards with gold.

Women with facial hair have never been popular. As late as the fifteenth century women with hairy warts or chin hairs were condemned as witches and often burned at the stake.

Cleanshaven faces became the rage in Elizabethan England because the queen decided to tax all Englishmen who wore beards!

In *Adagia* (1523), Erasmus wrote, "Since the beard is not completely formed until the age of manhood, it has always been considered an emblem of wisdom." In recent decades beards have become more a symbol of rebellion than of maturity and wisdom.

BEES

Bee venom is an ancient cure for rheumatism, arthritis, and other joint diseases that is even more popular today. The venom, in the form of a cream, can be bought in health food stores and pharmacies throughout the world. In ancient times the treatment was more complicated than simply rubbing in a cream. It involved submitting to two bee stings on the first visit and one at each subsequent visit.

One ancient idea is that bees were messengers to the gods and kept them informed of what was happening among mortals—the local gossip column, so to speak. Later, bees were seen as gods' spies. "Wisdom of the bees" and "ask the wild bee for what the druids knew" are two expressions that probably come from this idea.

The Irish talk to bees all the time, and many people believe that if they don't tell the bees when someone dies, the bees will stop making honey.

A bee flying into your house is supposed to mean that a stranger is coming.

It's good luck if a bee flies into your house, then leaves, but it's bad luck if it dies in your house. It's also bad luck if a swarm of bees comes at you of its own accord, even if this happens only in a dream.

BEETS

In 1649 an herbalist named Culpeper wrote in *The Complete Herbal and English Physician Enlarged* (1649) that beets were good for headaches and colds and helped clean out the liver and the spleen. Some believe, even today, that eating a raw beet each day will keep cancer away.

The Greeks themselves left no record about their use of beets, but we do know that they served them on a silver platter to the great god Apollo in his temple at Delphi. Dare we assume that what was good enough for Apollo was good enough for his subjects?

BEGINNER'S LUCK

See GAMBLING
LUCK

BELL, BOOK, AND CANDLE

The expression bell, book, and candle is actually from a solemn form of excommunication in the Roman Catholic Church. It is more symbolic than religious. When the sentence of excommunication is pronounced—

A bell is rung
The book is closed
A candle is extinguished

—this indicates the spiritual darkness to which the person has just been condemned.

The phrase is also used to indicate that someone is a witch, since being a member of the Church and being a witch are incompatible.

BELLS

See AMULETS
BELL, BOOK, AND CANDLE

BEST FOOT FORWARD

See also RIGHT SIDE OF THE BED

Put your best foot forward is an expression that dates at least to the time of Shakespeare, who had King John declare (in *The Life and Death of King John*, IV:2): "Nay, but make haste the better foot before."

BEST MAN

See WEDDING CUSTOMS

BIRDS

> *A bird in the Hand is worth two in the Bush*
> (Plutarch, *Of Garrulity*, third century B.C.)
>
> *A little bird told me.*
> (Ecclesiastes 10:20
> and Jonathan Swift, *Letter to Stella*, 1711)

The relationship between humans and birds is special, ancient, and in some places revered. It was believed that birds could talk and that certain people could understand them. When a wise man learned to imitate birds he could inform a gullible audience that "a little bird told me" whenever his wisdom was questioned.

Birds were thought to be messengers of departed souls. When a bird tapped on a window or flew into a house, it was assumed that a spirit was in search of another to join it; in other words, it was a messenger of death. From this belief came at least two important superstitions:

If a bird flies into the house, an important message will follow.
It's unlucky to have wallpaper with birds on it in your home.

White birds especially are, to some, a sign of death. Blue jays are also special. They are thought to spend every Friday with the Devil, telling him about the bad things we've done all week. Whenever a blue jay is seen flying overhead carrying twigs he is thought to be bringing fuel to the fires of Hell.

The Mohammedans believe that no bird can be trusted because birds opened the gates of Paradise and let the Devil enter.

You must listen very carefully to tell where a birdcall is coming from, because:

From the north, it's for tragedy
From the south, good for crops.
From the west, good luck will be
From the east, good love.

BIRTH

See BABIES

BIRTHDAY CAKE

Thanks to the Greek goddess Artemis, the goddess of the moon, we have such things as birthday cakes. Artemis's birthday was celebrated with moon-shaped honey cakes with candles on top.

Holy books from many religions say that if a phrase is written upon a piece of food, and the piece is eaten, the person will gain the power in the phrase. Thus we say, "Happy Birthday!"

The superstition connected with birthday cakes actually has to do with the candles: Blow them all out on your first puff and you'll get your wish.

BIRTHDAYS

See also BIRTHDAY CAKE

Celebrating birthdays dates to ancient times, when man first began to understand the stars. It became important to know the exact moment, day, month, and year, of one's birth. With these all-important facts, a horoscope could be drawn, and in those days a horoscope was essential.

The Egyptian pharaohs celebrated their birthdays in a big way. Back then the amount of celebrating a person did generally had to do with the extent of his or her money and power. Rich Romans held circuses.

We know that Cleopatra and Nero enjoyed their birthdays, as did most Romans and Greeks. Augustus held a birthday celebration for himself every month!

Good wishes are offered to the birthday person in an attempt to protect him or her from evil spirits. The chances of spirits causing mischief on a birthday are considered great, because spirits of all kinds are said to be attracted by celebrations and by times of change, and this dangerous combination comes together at birthday time.

The birthday celebration dates to recorded time. The modern party probably started in early Germany, where children were given gifts and had a candle-rimmed cake. (*See* BIRTHDAY CAKE)

Birthday customs changed a lot after Christ was born. Children were named after saints and celebrated the birthday of the saints after whom they were named instead of their real birthdays. Today many children celebrate both days.

The idea of someone "worth his weight in gold" comes from a birthday tradition in the East. For example the Aga Khan III on his sixtieth birthday topped the scales at 243 pounds and received that amount in gold and diamonds.

One of the best birthday ideas comes from China. There people once believed that when a person reached the age of sixty, he or she got to start all over again.

BIRTHMARKS

See also CHARTS, Language of the Mole

A mole on the back of the neck is a sign that the person will be hanged.
(Old English Proverb)

The birthmark, like so many other omens and symbols, has contradictory meanings. It is called the mark of God; it is also called the mark of the Devil.

In fact, birthmarks are unexplained changes in skin pigmentation. But those who believe they are bad luck think they are evil enough to cause cancer, and those who believe they are good luck charms go so far as to throw black pepper on an expectant mother so that her child will be born with birthmarks. Other beliefs are:

More moles on the left side than the right is bad luck.

Oblong moles are bad luck.

If you've got a mole above your chin
You'll never beholden to any of your kin.

The Greeks thought birthmarks were good; and centuries ago, they developed the Language of the Mole as an expression of their love toward those women born with these dark marks. During the decades moles have gone in and out of fashion, and at times a woman was considered inelegant without one.

During the late nineteenth century, women often wore patches like moles called beauty spots, in various shapes and sizes. Depending upon where she put it, a woman could say something special to the men who admired her. For example, a very proper young lady might have worn a beauty patch to indicate that she wanted to be left alone!

A mole on your arm
Can do you no harm.
A mole on your lip
You're witty and flip.
A mole on your neck
Brings money by the peck.
A mole on your back
Brings money by the sack.
A mole on your ear
Brings money year by year.
(Old English rhyme)

BIRTHSTONES

See also AMULETS
CHARTS, Birthstones

The concept of birthstones probably dates back six thousand years. The Egyptians believed that such stones protected them from disease, poisoning, and other calamities. The tomb of the High Priest of Memphis, (from 4,000 B.C.) was found to contain a breastplate with twelve different stones, which were probably amulets worn for protection against the evil spirits in the afterworld.

The idea of birthstones has evolved from that breastplate to the charts now often written on greeting cards. The Egyptians passed their beliefs on to the Hebrews, who assigned a stone for each of the twelve tribes of Israel. These were passed on for the twelve angels of Paradise. Later the stones became the twelve foundations of the Apostles and finally became symbolic of the twelve months of the year. Something that took so long to evolve shouldn't be taken too lightly! Check your birthstone in the CHARTS section, and get yourself a birthstone amulet, quickly.

BITE YOUR TONGUE

See also CHARMS
COUNTERMAGIC

When someone tells you to bite your tongue they are suggesting a countercharm. They don't want what you've just said to come true and they hope the evil spirits haven't heard it.

If you've been lying, or exaggerating outrageously, someone might well say to you "bite your tongue," so that the lie or the exaggeration doesn't come true.

BLACK

See also BLACK CAT
BLACK FOR MOURNING
CHARTS, Colors

Though I am black,
I am not the Devil.
(George Peele,
Old Wives Tale)

Black was the color of death; the color of all things dark, unknown, and often deadly; therefore, it was evil personified.

Some good things are associated with black: It is the color of fertile earth, which ancient civilizations worshipped. It is the color of dignity, elegance, and sophistication. The Egyptians believed that black cats (*see* BLACK CAT) had divine powers.

To the Japanese, black was the color of the soul as it left the body. Both the early Christians and the Cherokee Indians believed it represented death. For the Hindus, it is the color of Siva, the Great Destroyer.

Early African legend tells how people became black by eating the liver of the first ox killed.

Perhaps the first blacklist was Henry VIII's roster of monasteries whose lands he greatly desired to confiscate.

Both the Jolly Roger (the flag of pirates) and the flag of Hitler's SS troops, was black, with a skull and crossbones.

BLACK CAT

See also CAT

It's hard to know which superstition about black cats came first. There is the Norse legend that tells of the chariot of the goddess Freya, which was pulled by black cats. (When the Norse people were converted to Christianity, Freya became a witch and the black cats became black horses—which were uncommonly swift and undoubtedly possessed by the Devil.) The legend goes on to say that after seven years of service as horses, the cats were rewarded by being turned into witches—disguised as black cats.

In ancient Egypt, where cats were revered above all other animals (*See* CAT), the black cat was a lucky symbol. But by the Middle Ages, when notions of witchcraft started running rampant, black cats began to take the rap for everything bad. The black cat was believed to be the mascot or the familiar of witches; and as in the Norse legend, after seven years' service, to become a witch itself. If a black cat crossed your path it was a sign that Satan had been taking notice of you.

Because of its long association with witches, the black cat is the symbol of Hallowe'en. (*See* HALLOWE'EN) It is said that these possessed creatures perch on sleeping babies and old people and suck the breath out of them.

Sailors like black cats, and sailors' wives keep them to insure their husbands' safe return.

If the cat in your house is black,
Of lovers you will have no lack.
(Anonymous)

Good luck will come your way if a strange black cat comes to visit. It turns to bad luck when the cat decides to stay.

In America a black cat crossing your path is bad luck. In England a black cat walking toward you is good luck. In Japan a black cat crossing your path is good luck. Take your pick!

BLACK FOR MOURNING

See also WIDOW'S PEAK

In ancient times, it was thought that death was contagious. It was also believed that some deaths were caused by neglect in some way, and so evil spirits could easily get into the dead body. Mourner wanting to be as inconspicuous as possible wore black, in the belie that it was less likely to attract the attention of death.

This belief was fairly widespread. People were afraid of ghosts anc tried to hide light skin under the black cloth of mourning. In som primitive societies people even painted themselves black; in Afric people often painted themselves white for funeral ceremonies.

The custom of wearing black for mourning is not universal:

In China and Japan they wear white.
In Egypt and Burma they wear yellow.
In South Africa they wear red.
In Ethiopia they wear light brown.
In Syria and Armenia they wear violet.
In Iran they wear light blue.

In America it used to be customary to wear black for six months when mourning. The length of time has gradually dwindled to a few weeks or even days.

BLACK FRIDAY

See FRIDAY
WITCHES AND WARLOCKS

BLARNEY

There are two perfectly wonderful tales about Cormac MacCarthy, the lord who built Blarney Castle, near County Cork, in Ireland. Take your pick.

In the fifteenth century, MacCarthy was embroiled in a lawsuit, which he was sure to lose. He appealed to the druid princess Cliodna, who told him to kiss a stone of the castle and that "words will pour out of you." He did, and they did, and he won his lawsuit. He then moved the stone to an inaccessible spot so that everyone could not get at it and turn Ireland into a country of con artists.

The other story has MacCarthy Mor, lord of the castle, losing a battle to an Englishman named Sir George Carew. MacCarthy agreed to surrender to the British, but each day he put them off with another excuse. Finally Queen Elizabeth I herself tried to get him to surrender, but all she got in return was a long-winded, evasive letter to which she is said to have replied, "This is more of the same Blarney!"

The Blarney stone itself is a limestone triangle, which is supposed to be so difficult to reach that anyone who can hang from his or her heels to kiss the stone will be eternally charming and persuasive.

BLIND AS A BAT

See BATS

BLONDES

See HAIR COLOR

BLUE

See also CHARTS, Colors

Children say:

Touch Blue
And your wish
Will come true.

Brides look for something blue to wear for their wedding ceremonies. (*See* WEDDING CUSTOMS) The pharaohs wore blue for protection. A saying goes: "Wear beads of blue, keep danger far from you."

Blue was the color that protected people against witches simply because witches don't like blue: It's the color of Heaven.

If your love for me is true,
Send me quick a bow of blue.
If you ever of me think
Send me quick a bow of pink.
If you have another fellow
Let me have a bow of yellow
If your love for me is dead,
I'll know it if your bow is red.
(Unknown)

The name Bluebeard evokes an image of a murderous husband, and the blues referred to the "po' man's heart disease."

BLUE MOON

A blue moon isn't impossible, it's just rare. Once in a blue moon doesn't mean never, just not very often.

A blue moon is a result of certain atmospheric conditions that cause the moon to appear blue. This is an astrological oddity, but it happens.

The Egyptians liked the color blue and felt it was a lucky color. They were especially partial to what they called the "thirteenth moon," which their astrologers usually made blue because it completed the astrological cycle with a lucky sign.

BREAD

There are so many superstitions connected with bread that it's a wonder it ever gets eaten. An early idea was that bread was a gift from the gods, never to be eaten without a word of praise. The saying of grace before eating (which comes from the Jewish prayer said over bread) is a symbolic holdover of this idea.

That bread must be broken, not sliced, comes from the same concept: It would be an insult—indeed harmful—to cut the bread since it was a gift of God.

The ancients believed that there were four elements to life:

Water—stream of life
Grain—the fields
Earth—sustenance
Sun—goodness

Each of these elements are found in bread, "the staff of life." To pass the bread among friends was to wish them a long and healthy life.

Remember:

It's bad luck to cut bread from both ends.
Never pass bread around on a knife or a fork.
Never leave a knife stuck in a loaf of bread.
If you toast on a knife you'll be poor all your life.
It's good luck to dream of bread.
When two people reach for the same slice of bread, company is coming.

The heel of a loaf is always reserved for a member of the family because it means good luck.

A loaf of bread turned upside down means death or a ship in trouble. (Some think this superstition comes from the legend of an English general who, centuries ago, gave the signal to attack the Scots by turning a loaf of bread upside down.)

BREAK A LEG

See THEATRICAL FEARS

BREAKING GLASS

See WEDDING CUSTOMS

"BREAD AND BUTTER"

Ever walk with a friend and have a tree, a park bench, or another person come between the two of you—then hear your friend mutter, "bread and butter"? It's a common superstition that something else will come between the two of you to end your friendship, unless you quickly say "bread and butter" and cross your fingers.

Saying "bread and butter" is a symbolic way of uniting the two parts of a whole, just as bread and butter go together to form a complete unit.

There are other things to do that are supposed to prevent friendships from breaking up, but this is considered the most effective and certainly the most popular.

BRER RABBIT

Brer Rabbit, along with Brer Wolf, Brer Squirrel, and other animals of the forest, belongs to a group of tales told by Uncle Remus to a small white boy who is enchanted by the humanlike creatures.

These are all characters in stories written by Joel Chandler Harris. Harris was a white man who grew up on Southern plantations and never forgot all the tales he heard from the blacks who worked the land, or the impact the stories made on him and on the black children who listened with him.

The Uncle Remus stories recall the African tradition of humanizing animals. They also serve as an allegory of slavery: The powerless rabbit is victorious within his own "neck of the woods." Brer Rabbit (probably "Brother Rabbit" in Southern dialect) is the "cunningest critter in the forest" and always outsmarts Mr. Man. *Uncle Remus and His Friends, Old Plantation Stories, Songs, and Ballads with Sketches of Negro Character*, by J.C. Harris, are the sources of this imaginative story.

BRIDAL BOUQUET

See WEDDING CUSTOMS

BRIDE

See WEDDING CUSTOMS

BRIDESMAIDS

See WEDDING CUSTOMS

BRIDGES

Since bridges are man-made structures, built to circumvent nature, many superstitions have grown up around them. For example: while driving under a bridge, if a train passes overhead, put your right hand against the roof of the car and make a wish.

It is also believed that if you say good-bye to someone while standing on a bridge, you will never see that person again.

Here's my favorite: If you make a wish at one end of a bridge, then close your eyes, hold your breath, and navigate your way to the other side, your wish will come true. (On the other hand, you could fall off the bridge and drown!)

BROKEN HEARTED

Long ago, people thought that the feeling produced by overstimulation of the emotions (a rapid heartbeat), meant that the heart was breaking. Don't you believe it!

BROKEN MIRROR

See MIRROR

BROWNIES

See SMALL FOLKS

BRUNETTES

See HAIR COLOR

BUCKEYE

See CHESTNUT

PAUL BUNYAN

The legend of Paul Bunyan, America's most famous logging hero, is based on pure fiction. James MacGillivray used the name Paul Bunyan in tales of his own youth as a backwoodsman around 1910. The advertising executive W. B. Laughead took that figure and made him into the larger-than-life hero we think of today, by using him in an advertising campaign for Red River Lumber.

The legend of Paul Bunyan began to grow, and this fictional character took on a quasi-real background. It was said that Bunyan was born in Maine and that his cradle, too big to remain on land, was floated on the river—every time the giant baby turned, he caused a tidal wave. Once, when the baby was sleeping so soundly he couldn't be wakened, the British Navy was called in to help. The sailors spent seven hours shelling the giant in an attempt to wake him. When finally he did awake, he stood up, and in so doing sank seven warships!

Lore takes Bunyan on to Minnesota and Wisconsin with his ox, Babe, who could haul 640 acres' worth of logs at one time. Paul's crew, all giants, ate in a legendary way: It was said that they ate pancakes and biscuits so big that when a biscuit dropped, an earthquake occurred on the other side of the world!

BURIAL BEHAVIOR

See CEMETERY
DEATH CUSTOMS
WAKES

BUTTERCUP

The buttercup might have gotten its name because cows that feed on buttercups give yellower milk. A lunatic may be cured by merely hanging a cloth bag holding the flowers around his or her neck. Herbalists believe that an ointment made from buttercups should be used to draw a blister.

Buttercups are mainly used to tell if a friend is honest, angry, or in love. The test works this way: Hold the flower under your friend's chin, and if the yellow from the bloom shines on the skin then everything your friend says is true. The same reflection can indicate that the friend is angry or in love—or even jealous. It's difficult to decide. You might have to ask your friend what he or she is feeling!

BUTTONS

See also AMULETS

In ancient times buttons were worn strictly as ornaments, until, in the thirteenth century, someone was clever enough to invent the buttonhole. Buttons predate the buttonhole by perhaps fourteen hundred years.

Buttons were thought to be good luck when given as a gift. They were worn as amulets after being exchanged among friends.

Black magic has its own uses for buttons—particularly black ones. If someone thinks that an illness is caused by an evil spell, then the sick person must leave a black button where someone else can find it—thus passing on the illness.

In the United States kids say:

Rich man
Poor man
Beggarman
Thief
Doctor
Lawyer
Merchant
Chief

as they count their buttons, to see which profession will be theirs or which profession their spouses will have.

In England, the rhyme is a little more complicated:

Tinker, tailer, soldier, sailor
Gentleman, apothecary,
Plowboy, thief.
Soldier brave, sailor true
Skilled physician, Oxford blue
Learned lawyer, squire so hale
Dashing airman, curate pale.
Army, Navy,
Medicine, Law,
Church, Nobility,
Nothing at all.

This version of the rhyme takes a lot of buttons.

Finally, there's the common superstition that if you put your button in the wrong buttonhole, bad luck will follow you all day long. To stop such evil goings-on, take the piece of clothing off and start all over again.

CANCER

See also ZODIAC

There is a legend that the Crab, the symbol for Cancer, bit Hercules and was rewarded by one of the warrior's enemies, who made the Crab into a constellation in the sky. Astrologers believe that when the sun is in Cancer, any storms that occur will be catastrophic. People who are born between June 21 and July 22 are Cancers.

Cancers are ruled by the moon. They tend to be homebodies and are often oversensitive in their relationships with other people, causing great emotional problems for themselves. Cancers are highly emotional about most things. They are warm and giving, but sometimes too overwhelming as lovers. They withdraw when criticized and eat when depressed. People go to Cancers with their troubles because they are often sympathetic and kind.

Cancer, the crab

Fridays are good days for Cancers. Eight and three are their lucky numbers; and silver and white are their good colors. Their birthstone is the ruby.

Cancers love to entertain at home. They are torn between their home and possessions and leaving for some new adventure. Those who are friends with Cancers find them tenacious and stubborn—traits they get from the Crab.

Famous Cancers are Louis Armstrong, Nelson Rockefeller, Neil Simon, Ernest Hemingway, James Cagney, and Lena Horne.

CANDLES

To understand the importance of candles in the world of superstitions, one has to understand something about evil spirits. Whenever people are having fun, whenever there is an important occasion of any kind, evil spirits are going to come to help you celebrate whether they are invited or not. An evil spirit isn't likely to come into a brightly lighted room. It prefers dark, intimate places. Throughout most of history, candles were the only means of lighting. They became all-important for keeping evil spirits away from happy events.

In Ireland, people light twelve candles in a circle around a dead body, as double protection: Evil spirits can't cross into a circle (*see* CIRCLES), and they are deterred by the lighted candles.

A couple of ideas that have grown up around the original association of evil spirits and candles:

> A blue light from a candle indicates good spirits nearby (or ghosts).

When a candle goes out during a religious service, evil spirits are lurking in the corners.

Wax candles are used in church because bees come from Paradise. (*See* BEES)

Candles are very big in witchcraft and voodoo. A candle can be used as a voodoo doll. To call a lover, for instance, stick two pins through the candle. Then say this little incantation:

It is not this candle alone I stick
But my love's heart I mean to prick.
If (name) *be asleep or* (name) *be awake*
I'll have (name) *come to me and speak.*

By the time the candle burns down to the pins, your lover should have arrived (it didn't work for me).

CANDLEMAS DAY

See GROUNDHOG DAY

CAPRICORN

See also ZODIAC

Capricorn, the sign of the Goat, is one of the oldest constellations of the zodiac. It is governed by Saturn. People born between December 22 and January 19 are Capricorns.

Capricorns are practical, reliable, conservative, and above all ambitious. They tend to have an insatiable appetite for success. Saturn makes them both frugal and ambitious for power, and their symbol, the Goat, allows them to make great sacrifices in pursuit of their goals.

Capricorns appear cold, even snobbish. Those who know Capricorns well, however, find in them great inner resources of love and affection. Capricorns make highly responsible leaders, with a realistic if not pessimistic outlook. They should eat hot, invigorating foods; they tend to be melancholy but are good in a crisis.

Saturday is a good day for Capricorns. Seven and three are their lucky numbers, and dark green is their color. Their birthstone is the garnet.

Capricornus, the goat

Some famous Capricorns are Muhammad Ali, Marlene Dietrich, Howard Hughes, Benjamin Franklin, Carl Sandburg, Richard Nixon, and Joan of Arc.

CARNATIONS

There is a legend that the carnation is known as a flower of rejoicing because it was first seen on earth at the birth of Christ. This idea is very romantic, but there is evidence that the flower existed prior to the birth of Christ.

During those centuries, in fact, it was believed that the carnation could preserve the human body and keep away nightmares. Some also believed it grew at the graves of lovers, and it became popular in funeral wreaths. Others said it cured melancholy. Pliny the Elder wrote in *Natural History* that carnations were used to spice various drinks.

The carnation is the flower of those born in January. It stands for fascination and a woman's love (*See* CHARTS, Victorian Language of Flowers). Carnations may be worn on Mother's Day (*See* MOTHER'S DAY): A white carnation indicates "in memory"; a red one, "in honor of the living."

In Korea they believe the carnation can tell important things about a person's future. If a cluster of three flowers on a single stem is placed in the hair and the top flowers die first, the last years of life will be hard; if the bottom one dies first, there will be bad luck early in life; and if they all die together then the person is in for trouble throughout life.

CAT

See also BLACK CAT
THEATRICAL FEARS

If a cat crosses the street it is a sign of bad luck.
(Aristophanes, *The Ecclesiazusae* 393 B.C.)

The cat, as an object of worship, had its heyday around 3000 B.C. The Egyptians worshipped the cat-headed goddess Bast, the protector of pregnant women. This goddess enjoyed music, dancing, and the good life. When a cat died in ancient Egypt it was mummified and buried in Bast's temple. The law in those days said that when a household cat died, the members of the family had to shave their eyebrows and mourn. Killing a cat, even accidentally, was a capital offense.

In ancient Burma and Siam it was believed that a holy man's spirit entered a cat after his death; when the cat subsequently died, the holy man's spirit went to Paradise.

Now you have to admit that even with all the love and attention modern cats get, they are not *that* well off. Then again, it was rougher for cats during the Middle Ages, when witchcraft blossomed and cats were thought to be the Devil incarnate. (*See* BLACK CAT)

To straighten out some old wives' tales: Cats do not have eyes that shine in the dark, nor can they see in the dark. They do not suck away the breath of sleeping children, or invalids, although they do like to snuggle close and examine the human face. The idea that they suck away breath comes from beliefs about witchcraft in the Middle Ages. (*See* BABIES) The idea of being "nervous as a cat" is also erroneous—cats aren't nervous; they merely have good reflexes.

Some other superstitions are:

A restless cat means a storm is brewing.
When a cat licks its tail, rain is coming.

If you throw a cat overboard, there will be a storm at sea.
Never kick a cat or you'll get rheumatism.
Never drown a cat or the Devil will get you.
If a cat meows on board ship, it will be a difficult trip.
It's good luck to have the family cat at your wedding.
In New England they say they can tell time (and tides) by the pupils of a cat's eyes.
When a cat puts its tail toward the fire, bad weather is coming.
When a cat licks itself clean, it means fair weather, rain, or company.
If a cat jumps over a corpse, the corpse will become a vampire. You must stop the funeral immediately and wait until the cat has been caught and killed. (Since killing a cat is also bad luck, you can't win when this happens, no matter what you do.)

CEMETERY

See also DEATH CUSTOMS

One of the crying needs of the time is for a suitable burial service for the admittedly damned!

(H.L. Mencken, *Prejudices,* Sixth Series)

These are things that *should not* be done when near a cemetery:
Pointing at a grave (it will make your finger rot).
Counting the cars of a funeral procession.

Here's what you *should* do:
Bury your loved one on the south side (it's holier ground).
Try to pick a rainy day—then the departed spirit will go to Heaven.

Whenever you shiver it means someone is walking over the spot where your grave will be.

CHAMPAGNE

See WINE

CHARMS

See also ABRACADABRA
AMULETS
BITE YOUR TONGUE
EVIL EYE
TALISMAN

Charms are mostly chants recited by priests and other believers. They are thought to bring good luck or protect against bad luck when used with words or gestures.

The term good luck charm, when applied to tangible things, is inaccurate. Those are good luck amulets or good luck talismans. (*See* AMULETS; TALISMAN) Modern usage, however, permits all three terms to be used.

An example of a charm used for burns:

Two angels came from the west
One brought fire,
The other frost.
Out fire, in frost.

CHESTNUT (BUCKEYE)

Got a backache; a headache; pain from rheumatism? Then carry a chestnut. Chestnuts, boiled with honey and glycerine, are supposed to cure asthma and other chest problems. (Oh that old chestnut!)

On Hallowe'en (*See* HALLOWE'EN) chestnuts are always left on the table for any souls of the dead who come back to visit on that night.

CHILDREN

See BABIES

CHRISTMAS

See also HOLLY
MISTLETOE

Although Christmas is a celebration of the birth of Christ, it might have had some connection with the Birthday of the Unconquerable Sun, which was a holiday celebrated by the Romans in ancient times to honor the sun god.

There are several superstitions and traditions which surround the Christmas holiday. For instance:

Christmas morning—The first person through the door of a household must be a man. (In olden times groups of men would go from house to house carrying the image of Christ, to insure that a man was the first to cross the threshold.)

Christmas decorations—All must be down by January 6 (Epiphany) or bad luck will follow all year.

Gifts—Although we think that gift-giving at this time of the year started with the gifts of the Magi, it actually started much earlier, during the Roman holiday that commemorated Saturn and the planting of the seeds. This period was also a day of equality, when Roman rulers abdicated their control and the "elected" peasants ran the land.

Santa Claus—The jolly fellow with the red suit and whiskers who climbs down chimneys was introduced by Thomas Nast in a series of Christmastime cartoons which ran in *Harper's Weekly* from 1863 to 1866. Santa Claus was based on Saint Nicholas, who has been traced to Asia Minor around 350 A.D. The feast day for Saint Nicholas was December 6, and gifts were exchanged on that day. Saint Nicholas was combined with Kris Kringle, a nineteenth-century German creation who brought gifts for the children. Kris Kringle was a corruption of the Christ Child, the original gift bearer of German legend.

Christmas cards—Sending cards at Christmas was a custom introduced in the 1840s by John Calcott Horsley. It didn't become popular in the United States until a printer named Louis Prang took up the idea and marketed the cards, making a bundle of money for himself and creating havoc for the postal service.

Christmas carols—The Christmas carols we know today come mostly from the 1800s. Caroling, however, was an old English custom; it means to sing joyously.

Christmas trees—There are several stories about the possible origin of this custom. There is a legend that Saint Boniface, a missionary who worked among the German druids, was cutting down an oak tree (*See* OAK TREE) that was holy to the pagan druids and found hidden behind the oak a lovely little evergreen tree. More modern legend concerns Martin Luther, who is supposed to have used an evergreen tree to illustrate to his wife and children how wondrous he thought the winter night sky was. He put candles on the tree to indicate the stars.

Poinsettia plants—This is a relatively new custom. Dr. Joel Poinsett, a United States minister in Mexico, first brought back the plant in 1828 after hearing a story about a little Mexican boy who was too poor to put a gift into the poor box on Christmas Eve. The boy prayed outside of the church and on the spot where he knelt a beautiful red plant bloomed. The boy immediately gave the plant to the statue of the Christ Child. In Mexico the plant is known as *Flor de la Noche Buene*—"Flower of the Holy Night."

Feasting—Turkey has been a favorite at Christmastime for a long while. In an 1890 publication, *Statesmen's Dishes,* the following recipe was found by Robert Myers in his "Celebrations":

> *To prepare a turkey for Christmas dinner: the turkey should be cooped up and fed well sometime before Christmas. Three days before it is slaughtered it should have an English walnut forced down its throat three times a day and a glass of sherry once a day. The meat will be deliciously tender and have a fine nutty flavor.*

Eating mince pie dates from the Middle Ages, when the Crusaders acquired a taste for exotic spices. The English wassail bowl, which comes from an expression meaning "be of good health," contains ale, roasted apples, eggs, nutmeg, cloves, and ginger. Finally, there is the idea that the abbreviation Xmas is sacriligious. It isn't: It comes from the Greek X, Chi. In earlier times, X–mas was used as a sacred symbol because it associated the cross with the holiday.

CHRYSANTHEMUMS (MUMS)

The word chrysanthemum comes from the Greek for golden flower. The chrysanthemum is considered a lucky flower for those born in the month of November.

Although it was a popular flower in ancient Greece and Egypt, the chrysanthemum was more important in China and Japan. It became Japan's national flower. During the War of the Dynasty in 1357 (also known as The Chrysanthemum War), each Japanese warrior wore a yellow chrysanthemum as "a golden pledge of courage." For the Japanese, the flower is the symbol of human perfection and the ancient emblem of the Mikado. The coveted Order of the Chrysanthemum is the highest honor the Japanese can bestow.

In China, the flower and its images were considered sacred and were used extensively in temples and in the decorative arts. A story goes that there is a stream in China set in a bank of chrysanthemums, and if you drink from that stream you'll live for one hundred years.

In ancient times the chrysanthemum was considered a symbol of longevity and perfection. Drinking water with chrysanthemums was supposed to prolong life.

The Chinese eat chrysanthemum petals in salad. The Italians use it as an herb. The Koreans boil its roots to make headache cures. It is believed by some to be a good antidote for opium, vertigo, and melancholy. In New England, it was once used in church bouquets to help keep the congregation awake during long sermons.

CIRCLES

Superstitions about circles began in the very early days of the sun worshippers. It was believed that life and all eternity were ruled by the cycle of the sun, which was thought to move in a circle. People did things clockwise (east to west), the way the sun moves, to honor the sun. The expression "going around in circles" probably dates back to very early days.

It was believed that evil spirits could not enter a circle, since the circle represented the sun and the sun was all-powerful. The first use of lipstick also comes from sun worship. The mouth was thought of as the entrance to the body, and a red circle drawn around the mouth kept the soul inside—and the Devil outside.

Later, witchcraft took full advantage of this theory. Witches cannot cross into a circle, nor can outsiders get into a circle a witch has drawn for herself and her coven.

A powerful countercharm to bad luck is turning completely around in a clockwise direction three times. (*See* COUNTERMAGIC)

CLERICAL COLLAR

See KNOTS

CLOCKWISE

See CIRCLES

CLOTHING

Centuries ago people believed they could outwit death by wearing their clothing inside out, as a kind of disguise. This led to our current belief that if you accidentally put a piece of clothing on inside out, and you leave it that way all day, it will bring you good luck.

If you get a new coat you had better put a coin in the right pocket to insure good fortune. Never have your clothing mended on you or you'll soon die (or more likely, your brains will be sewed up by the tailor). If you want to counteract these two dread possibilities, chew on a piece of thread during the whole operation.

COINS

See AMULETS
MONEY
WISHING WELLS

COOK ONE'S GOOSE

This expression, which means ultimately to ruin one's prospects, comes from the Middle Ages. There are two schools of thought about it, both involving King Eric XIV of Sweden.

Eric is supposed to have answered that he wanted to "cook your goose for you" when a town he was trying to capture asked about terms for a truce. The town, not sufficiently impressed by Eric's army, had hung out a goose for Eric's soldiers to shoot at.

The second version of the same legend is that the town had hung out the goose as an insult because the goose is such a stupid and futile creature. This gesture of contempt so angered Eric's army that it burned the town to the ground, effectively "cooking their goose" and ruining the town's prospects of victory.

COPPER BRACELETS

The miraculous curative powers of copper have been rediscovered by both the gullible and the desperate in recent years. In India, copper earrings are believed to ward off sciatica.

The belief that copper bracelets are a protection against the pain of arthritis and rheumatism started in the Middle Ages. Sending your money for a strand of copper to wear is sure to do only one thing—turn your wrists green.

CORAL

See also AMULETS

For many centuries people have worn pure, red coral as an amulet for protection against the evil eye. The Romans, and later the Italians, felt that coral effectively warded off demons and witches as well as the dread evil eye. (*See* EVIL EYE)

Coral is believed to prevent arguments in the home; hung on a bedpost it will prevent nightmares. The Chinese believed that it was protect against madness.

In the Middle Ages, ground coral that was crushed in a mortar made of marble was considered a medicinal aid for almost every ailment.

CORN (MAIZE)

Corn is a product of the Americas, and the legends surrounding it come from this part of the world.

Many Indians believed that the Great Spirit had invented corn during a famine and that anyone who wasted it was doomed to hunger.

There are many Indian legends which tell how corn came to North America. The following is typical of the genre:

> A young warrior dreamt that a lovely yellow-haired squaw appeared to him and instructed him to burn a prairie covered

by grass and then drag her by the hair across the burning fields. Each time he paused, new ears of corn sprouted, and on each ear a small tuft of the squaw's yellow hair grew. These tufts of hair were to remind the warrior that she would never forget her people.

CORNERSTONES

If you think your landlord wants his pound of flesh each month, he's only operating in accordance with the ancient assumption that the gods of the earth must be paid for their land. In the centuries before Christ, people paid for the privilege of building every new building or construction. The payment was in the form of a sacrifice, and more primitive peoples often offered children as payment.

As the custom evolved coins were placed within the walls of the structure as payment. This in turn became the use of a cornerstone to mark the beginning of the building and to bless the construction.

There is a superstition connected with cornerstone-laying ceremonies, which states that an unmarried woman shouldn't attend these ceremonies; if she does, she won't marry for another year. That's silly, but the initial reasoning behind the idea was sound: A virgin was thought to be the best sacrifice to the vengeful protector of the earth.

COUNTERMAGIC

When bad luck is unmistakably coming your way, there are some countercharms which, I am assured, work. Saliva is probably the best weapon. (*See* SALIVA) Spitting over your left shoulder, for instance, is supposed to usually work. Turning around three times (if the threat is very serious, seven times) often precedes spitting. (*See* CIRCLES) Turning around is an attempt to reverse the order of things, in this case to reverse the direction of the bad luck headed your way.

The other, reasonably effective thing to do is to pull your pockets inside out. (*See* CLOTHING)

It's also helpful, whenever in doubt, to cross your fingers. (*See* CROSS YOUR FINGERS)

CRACKED MIRROR

See MIRROR

CRICKETS

See also WEATHER

Almost everybody thinks the cricket is good luck. This is one of the few talismans that from ancient times on have been universally accepted as fortunate to have around. The cricket is a "house spirit" and brings good luck to the dwellers of a home. The cricket is also believed to take his luck with him; when he leaves the house, so does the luck.

The Japanese, the English, and the American Indians all believe in the inherent luck of this little creature. It is thought to be bad luck when he leaves by your chimney; it is terrible to kill one; it is also dangerous to imitate his chirp.

The cricket is a super fortune-teller and knows if rain is coming, if death is near, and even if a lover is coming back.

CROCKETT, DAVY

Davy Crockett always said: "Be always sure you're right, then go ahead." He was called the "coonskin congressman" and he told crowds he was "so ugly that his grin could bring a coon down from a tree."

Crockett, an American hero turned legend, was born in 1786 and served as a frontier scout under Andrew Jackson during the Creek Indian War. He was later elected to the U.S. Congress from Tennessee for three terms. After losing his congressional seat, Crockett fought in the Battle of the Alamo and became one of the few survivors of the battle, only to be shot by Santa Ana in 1836.

The legend of Davy Crockett was in large part perpetuated by Davy himself. He loved to repeat tall tales about himself, like the one where he shot 105 bears in eight months. He reveled in his reputation as a comic backwoodsman with homespun intelligence and unending courage. Because of Crockett we have the coonskin cap, a permanent part of the American wardrobe to be taken out every now and then and paraded in public.

CROSS-EYED

See also EVIL EYE
RABBIT'S FOOT

Don't enter a card game with a cross-eyed partner—you're sure to lose.

If you meet a cross-eyed person of the opposite sex, you'll have good luck; but if you meet one of your own sex, that's trouble. To counteract this bad luck, spit through your fingers and outstare him or her.

CROSSED KNIVES AND FORKS

See KNIFE AND FORK

CROSSES

See also ANKH
CROSS MY HEART AND HOPE TO DIE
CROSS YOUR FINGERS

The sign of the cross as an emblem for a religion, an individual, a country, or as a signature, has been used since the beginning of human time. By the time Christ was crucified on the cross, it had been a symbol for centuries.

Some of the more popular forms of the cross are indicated above.

Latin or Roman

Greek

Saint Andrew's

Russian Orthodox

Maltese

Monogrammed Cross (An example of a personal emblem; this one was used by the Emperor Constantine.)

CROSS MY HEART AND HOPE TO DIE

When you cross your heart (and hope to die) while making the sign of the cross with the index finger of the right hand, you are obviously telling the truth.

Long before Christianity, the cross was a sacred symbol. Since it was assumed that the heart was the seat of all wisdom, it was believed that the heart would know if you were lying.

CROSS YOUR FINGERS

Crossing your fingers is the simplest way of making the sign of the cross. As a charm against the Devil, this sign is particularly effective. A cross has perfect unity: It represents the four directions, and the four elements.

When two lines cross, a wish (or a lie) can be held at the point where the lines meet. A wish will stay at the crossed point until it comes true; a lie will stay at that point, kept away from the evil spirits and from starting trouble.

In early times, two people crossed their index fingers together while the first person made a wish and the second hoped it would come true. This was later simplified into one person crossing his or her own fingers.

So, please keep your fingers crossed for me, will you? I'm hoping for my wish to come true.

CROWS

See also BIRDS
SCARECROWS

There are several little rhymes that tell the story of the crow's place in the world of superstition.

One's bad luck
Two's luck,
Three's health,
Four's wealth,
Five's sickness
Six is death.
(English saying)

One means anger
Two means mirth
Three a wedding
Four a birth
Five is heaven
Six is hell
But seven is the devil's own self.
(English saying)

Do you know the one about how the crow came to be black? The crow was originally white, legend says, but he turned black after eating snake eyes.

CUPID

This wimpled, whining, purblind, wayward boy,
This senior-junior, giant-dwarf, Dan Cupid's
Regent of love-rhymes, lord of folded arms,
The anointed sovereign of sighs and groans,
Leige of all loiterers and malcontents.
(Shakespeare, *Love's Labour's Lost III:1*)

Cupid, or Dan Cupid as he was often called, was the son of Mercury and Venus in Roman mythology and the god of love (known as Eros in Greece). Cupid, as the personification of love, stands for desire, passion, and all the problems associated with the affliction of this emotion.

Cupid is usually depicted as a beautiful, naked boy with wings, carrying a bow and arrow, and sometimes blindfolded. His arrows are supposed to shoot love through his targets.

The name Dan Cupid is another form of "Don Cupid"; don is a title meaning sir or master.

CURSES

See also CHARMS

Everyone curses; it's the way we wish evil onto others. Today when we curse someone, we usually don't mean it—and even if we do mean it, nothing usually happens. But there was a time when a curse was a very serious matter. Families were cursed in perpetuity; their cows couldn't give milk; chickens couldn't lay eggs.

To curse someone is to invoke the wrath of a great power against which the accursed has little protection. A curse is spoken magic; if you don't believe in magic, then a curse becomes blasphemous or meaningless.

DAISIES

"The smile of God" or "day's eye," as the daisy was called in ancient times, opens its white petals with the sun and closes them again at nightfall.

The centuries-old children's game of foretelling one's love life by pulling the daisy's petals and saying, "He loves me, he loves me not," was legitimized by Marguerite in Goethe's *Faust I* (1808).

In the Language of Flowers, the daisy means innocence; the common garden-variety daisy actually says, "share your sentiments," or "I reciprocate your attentions."

Knights in medieval times wore daisies into battle. It was widely known at the time that any knight using the double daisy (a stem with two blooms commonly found among garden daisies) as his emblem was declaring that he loved a lady and she loved him back. If, during those days, a lady couldn't make up her mind, she wore a wreath of white daisies to say "I'm not sure."

Wordsworth said that the daisy was "the poet's darling" (*To the Daisy*) because of all the romantic legends which have grown up around the flower. In sixteenth-century France there was an Order of the Daisy, and some say it was Mary Magdalen's tears that originally created the daisy.

Some common notions are:

If you want to dream of your lover, put daisy roots under your pillow.

Step on the first daisy you see or it will grow on your grave within the year.

> The spirits of babies who died at birth have scattered daisy seeds in abundance to help cheer their unhappy parents.

Medicinally, daisies are supposed to be very useful: If insane people drink a potion containing daisies for fifteen days, they'll be cured (or dead). As a charm, daisies cure wounds, gout, fevers, and ulcers, remove warts, and change gray hair to black. Daisies are also used as an ointment for eye infections. The juice of the English variety was once thought to cure migraines.

DANDELIONS

The dandelion is one of nature's purest diuretics and was known during the Middle Ages as "piss-a-bed." (A current superstition is: If you pick dandelions you'll wet your bed.)

In the sixteenth century the flower was called "lion's tooth" in England, and "dent-de-lion" in France, which later became anglicized to the current dandelion.

Known by children and lovers throughout the world as "love's oracle," the dandelion foretells romance when its puffball of seeds is scattered helter-skelter by someone blowing and reciting the eternal chant, "He loves me, he loves me not."

As a weather forecaster, the dandelion has limited use, but it is believed that if the flower doesn't open in the morning it will rain; and if it blooms in April and July, it will be hot and rainy throughout the summer.

The flower is thought to be good for the liver or against rheumatism, and to purify the blood. Dandelion juice is rumored to work as a tonic on any number of organs.

DEATH CUSTOMS

See also CEMETERY
MOURNING VEILS
WAKES

If you've wondered, as I have, why people are buried so quickly, the reason is that ghosts shouldn't be encouraged to hang around too long.

A classic superstition about death and dead bodies is the notion of putting coins, or coppers, on the closed eyes of the deceased. People still do that today, for numerous reasons. Centuries ago, people thought that the soul left the body through the hollow eye sockets and since the Devil could enter that way, the coins were to close the holes. The Book of Genesis (46:4) says, "Joseph shall put his hand upon thine eyes," along the same lines. In classical Greece and Rome, the coins were payment to the god of the Underworld for his chore of ferrying the soul to the land of the dead.

Schoolbooks say that a doctor laid silver half-dollars on Lincoln's eyelids after he died—but then, he was the President. Most of the time the coins used are of copper.

Before someone dies:

A dog will howl
An owl will screech
A picture may drop
There will be a rapping sound at the window
The sound of church bells will ring in your ears
You'll see a falling star
You'll dream of muddy water

—or any other of a thousand such signs; but there will always be a sign.

About cremation: An early idea was that evil spirits lived inside dead bodies and had to be burned. The Greeks, however, thought that by cremating the body the soul was liberated, and that the souls of buried bodies were outcasts and would probably never get to Paradise.

We know that in ancient Egypt mummification was the preferred burial procedure. The Egyptians believed that the dead spirit would return to the body, and it was preserved for future use.

The Romans didn't believe in an afterlife and didn't bother with mummification.

Today's tombstone is used to mark the site of a grave and to say the person's been in this world. Yesterday's tombstones were designed to keep in the ground the evil spirits that inhabited the dead body.

That was also the reason for the extremely thick, heavy coffins we see in museums.

Remember, in ancient times people firmly believed that death was contagious (*See* BLACK FOR MOURNING), and most of their efforts were to keep the dead from getting at the living and taking them away with them.

DEATHWATCH

This unhappy turn of a phrase refers to a species of beetle known as the deathwatch beetle, whose favorite pastime is tapping on wood. It is believed that if you hear that tap, there will be a death in your family.

Another explanation for the expression comes from the centuries-old tradition of waiting and watching for the ghost of a dead person to come back and cause trouble shortly after the funeral (*See* WAKES)—like Hamlet's father's ghost, who stirred up all those problems because he wouldn't stay dead and buried.

DEMOCRATIC DONKEY

See also REPUBLICAN ELEPHANT

In 1870, the popular *Harpers Weekly* cartoonist Thomas Nast drew "A Live Jackass Kicking a Dead Lion." The sketch, which showed a donkey kicking a carcass, had the U.S. Capitol building and the U.S. eagle in the background. The carcass was generally understood to be the Secretary of War Edwin M. Stanton and the donkey the treacherous Copperheads (the Northern backers of the Southern cause at the time of the Civil War).

The Democrats and other Copperhead opponents loved the cartoon. The donkey became permanently associated with the Democratic Party, even though it was used mostly as a sign of ridicule against them. The Democrats decided to use the donkey as their symbol and turn the ridicule to their own advantage.

The Democratic Party is like a mule.
It has neither pride of ancestry nor hope of posterity.
(Ignatius Donnelly,
Speech in the Minnesota Legislature, 1860)

DEVIL

The Devil and me, we don't agree;
I hate him; and he hates me.
(Salvation Army song)

An apology for the Devil—it
must be remembered that we have
only heard one side of the case.
God has written all the books.
(Samuel Butler, *Notebooks*)

Talk of the devil, and he'll appear.
(Erasmus, *Adagia XVII*, 1500)

The Devil is the Hebrew "Satan," the Old Testament's version of the archenemy of mankind. Translated into Greek Satan was *diabolos*, which means accuser, from *diaballein*, which literally means "to throw across" and in the idiom of the day meant to slander or accuse. The Devil means the accuser of the soul.

In Christian and other mythologies, the Devil is the fallen angel who became God's enemy. All evil is accredited to the Devil.

"To give the devil his due," a common expression, comes from *Don Quixote de la Mancha* by Cervantes.

Should you have an idea that you want to trade your soul with the Devil for some special wish, then the best time and place to meet him is at a crossroads at midnight.

DIAMONDS

See also CHARTS, BIRTHSTONES

A sixteenth-century writer once remarked that the diamond

makes the wearer unhappy because
its brilliance irritates the soul
just as an excess of sun irritates the eye.
(Jerome Cardan)

Of course, that's an irritation most of us don't mind, since the diamond is the most precious of gems.

The Persians believed that diamonds were sinful and an invention of the Devil. Some people believed that diamonds were the result of thunderbolts. In the Middle Ages, it was thought that two diamonds could produce a third.

As one of the most honored of gems, the diamond was regarded by the Romans as insurance for victory in battle when worn as an amulet on the left arm, and as protection against the evil eye for women of beauty if worn close to the face.

The stones have been considered lucky throughout history with the exception of very large diamonds, which have long and bloody histories. The extreme value of the stone added to the problems of keeping and wearing large diamonds.

Until recently, diamonds could only be cut by other diamonds. Today, the laser is used in industry for the cutting of this stone. It is completely false that diamonds cure leprosy and insanity and prevent nightmares. They are symbolic of virtue, purity, and innocence. They have also been proved effective as an amulet against the plague and against sorcery. The diamond is the birthstone for those born in April.

DIMPLES

It's the Irish who first said:

Dimple on the chin,
The Devil within.
Dimple on the cheek
A soul mild and meek.

Or:

A dimple on your cheek
Many hearts you will seek;

A dimple in your chin,
Many hearts you will win.

The ancients believed that people with dimples had special, magical powers, because so few had them. In more modern times, people believed that a man with a dimple in his chin or cheek would never commit murder. (Did you hear that, Kirk Douglas?)

Some people say that a dimple, in a smile, implies a loose character; but Shakespeare had a better idea. He claimed that "Love made those hollows" (*Venus and Adonis* L. 242).

DIVINING ROD

A divining rod is a forked branch, usually made of hazelwood or from a willow tree and is supposed to locate water and other minerals under the earth. It is also known as a dowsing rod.

The stick is said to dip in the presence of water. Many diggers, in fact, won't start a well unless the rod has indicated the right spot. This method of divining water is very old, but it is still widely used.

DOG DAYS

This does refer to the fact that dogs go mad, get fleas, and often howl during certain weeks, but that's not how dog days got their name. The dog days are those between July 3 and August 11 when the Dog Star Sirius rises with the sun, adding its heat to the sun's to produce the hottest time of the year. The Romans were the ones who named the period, calling that part of summer the "days of the dog."

There is a belief that it is an unhealthy time of year. (It might well have been for the Romans, who had no air conditioning.) There is no reason why it is considered unhealthy to swim during the dog days, although many mothers of (unhappy) children still believe that superstition.

According to legend, the flies increase, the rain seldom falls, dogs go mad, and snakes go blind and strike at any sound during the dog days.

DOGS

Love me, love my dog
(Ancient Latin proverb)

We say that some canines are good watchdogs; the Greeks in ancient times called them psychic. They were supposed to howl at approaching evil. Today, we credit a dog's sharp instincts and split-second reflexes for his ability to respond to danger. The Greeks, however, didn't understand those instincts and credited the dog with the power to see evil coming.

One legend suggests that howling dogs were a sign that the god of Wind had summoned death, and that the spirits of the deceased would be carried away because of the howling. Even today, many believe that a howling dog means someone will die, or that the dog has seen death and is howling at it.

The notion of the dog as a faithful companion or as "man's best friend" dates back to Homer's *Odyssey*. The dog, Argos, watched many years for his master, Odysseus (Ulysses), to return, and when Odysseus did, even though he was heavily disguised, Argos recognized him and, wagging his tail with joy, died of happiness (and old age).

DRAFTS

Hot or cold drafts are a sign that evil demons are in the air and are about to pull some nasty prank on good people. Even Kublai Khan believed that.

Many Europeans, Africans, and Americans believe that the warm air of a draft comes from Hell during a summer evening.

DREAMS

We are such stuff as dreams are made on.
(Shakespeare, *The Tempest* IV:1)

We all dream; what those pictures mean, and what causes a certain dream on a given night—that's a whole other kettle of fish.

Interpretations of dreams are different from place to place, and often from generation to generation, although judging by the bookstalls there are plenty of people who think they understand what it all means and can offer a simple list of explanations.

Dreams are as individual as the people who have them. The Romans and Greeks believed that dreams were warnings of future happenings, and the Egyptians thought the gods spoke through dreams. The Hebrews thought that dreams were very important, and often carried out their affairs according to the dictates of a dream. By the 1400s it was believed that "dreams go by contraries"—whatever you dreamed meant the exact opposite.

Edwin Radford, in his *Encyclopaedia of Superstition* (1949), gathers up a list of what many dreams are supposed to mean:

> Dreams of sickness mean marriage among young people.
> Dreams of angels indicate happiness coming quite soon.
> To have dreamed of standing before an altar implies sorrow and misfortune.
> Dream of being angry with someone special and that person will turn out to be your best friend.
> Dreams of fishing mean that for each fish you catch, a friend will die soon.
> Dreams of dancing indicate good fortune is coming.
> Dream that you've gone bald and you'll lose friends or property.
> A dream of a wedding means death.
> Dreams of bees or bee stings are very bad luck.
> Dream of fruit out of season and you'll wake up in tears.
> Dreams of marriage and brides always mean death and illness.

I believe it to be true that dreams
are the true interpreters of our
inclinations; but there is art
required to sort and understand them.
(Montaigne, *Essays* III.13)

DROWNING PEOPLE COME UP THREE TIMES

When you see someone drowning, don't wait for the third time—because there might not be one.

Most drowning people rise because there is air remaining in their lungs (and other parts of the body) creating a buoyancy. Some people never rise back to the surface at all. Sometimes, if the person has some control during the crisis, he or she may rise several times before all the air is expelled. In this case three is not a lucky number.

DWARFS

See also: SMALL FOLKS

According to countless legends, dwarfs are good at blacksmithing and baking, at tailoring and at prophesy-making. They can disappear whenever they want, and they love to attend parties. They often give good advice, but they are known to be thieves and can't be trusted. Dwarfs are so very special, that any gift gotten from a dwarf may turn into a pot of gold.

Dwarfs are tiny, often misshapen people who, during early times, had their own kingdom, ruler, and laws. They dwelt in caves and were considered guardians of the minerals and precious metals found within the earth. The dwarf is a lucky symbol and an important character in mythology.

The main difference between dwarfs and other small folks is that dwarfs are real people—there are no leprechauns, are there?

DYBBUK

A dybbuk is the Jewish version of an evil spirit, the soul of a dead person, or a demon that takes possession of a healthy person and lives through that person. Leo Rosten, in *The Joys of Yiddish*, says that the idea of the dybbuk is as old as demonology, as old as man himself.

The dybbuk is a colorful, corrupt, and often vicious spirit that must be exorcised by a holy man who commands the spirit to leave the person and go to its eternal rest. It is believed that a tiny drop of blood remains behind on the small toe of the right foot to indicate that the spirit has exited the body. You'll also find a small crack in the window by which the spirit has left the room.

EAGLES

There is an ancient Egyptian belief that every ten years the eagle soars through the fires of Hell and then plunges into water and sheds its feathers, thereby acquiring a new life. (This is a good explanation of molting.)

There's another ancient legend that Adam and Eve didn't die but were actually turned into eagles and went to live on an island off the coast of Ireland. (Guess who thought that one up!)

Through the centuries the eagle has come to stand for strength and swiftness, and it was the emblem of many royal houses. The Romans began using the doubleheaded eagle when their empire merged with the Germans', joining their eagle to that of the Huns.

In folklore the eagle has always been a helpful bird, bringing warnings of trouble to the heroes.

EARRINGS

Throughout history, earrings have been effective amulets against the evil eye (*See* EVIL EYE) and were worn by both men and women. Kings, poets, and sailors all wore earrings for superstitious reasons. Sometimes only one was worn; sometimes two were needed.

Sailors believe that wearing earrings protects them from drowning, and that it helps their eyesight.

In ancient times the ear was thought to be the center of intelligence. The custom of pulling a child's ear to make him remember his school lessons comes from that ancient belief.

Earrings for men have come back in style in recent years. For some it's simply an ornament; for others it's a symbol of sexual preference.

EASTER

See also: EASTER EGGS
GOOD FRIDAY
RABBIT'S FOOT

Easter is the first Sunday after the full moon after March 21. According to Christian religion, it was the day that Christ was resurrected.

As with other Christian holidays, there was also a holiday in ancient times that was celebrated at about the same time. In this case, it was the celebration of the vernal equinox—the tribute to the goddess of spring, Eastre. Eastre was an Anglo-Saxon goddess who is reputed to have opened the gates of Valhalla for the slain sun god Baldur, thereby bringing light to man. Easter also refers to the rising of the sun in the east.

Some traditional Easter beliefs are:

The sun will dance as it rises on Easter Sunday.

It's good luck to wear new clothes on Easter Sunday in honor of nature's new green clothes.

The Easter lily is a symbol of purity, which became a popular holiday flower in the United States in the late 1800s.

The eating of ham on Easter Sunday is said to have survived from an old English bit of nastiness—showing contempt for the Jews by eating pork (a non-kosher food).

If it rains on Easter Sunday, it will rain for the next seven Sundays following.

If there was a white Christmas, then Easter will be green (and vice versa).

In the American Ozarks, people believe that an Easter morning bath taken before daybreak will help cure rheumatism.

EASTER EGGS

See also: CHARTS, Holidays
EASTER
RABBIT'S FOOT

Easter eggs come from a very early Germanic custom that honored the goddess Eastre, who was supposed to consider the hare or the rabbit sacred. It was believed that on Easter Eve the hare would lay eggs for all good children.

The egg is a universal symbol of life. It symbolizes the start of spring, the return of life to earth. Colored Easter eggs represent the return of the flowers to the land. A vestige of ancient fertility rites, Easter egg traditions also involve egg-rolling contests, a very old custom that we continue today. The last child to break his or her egg will have good luck during the year.

EENY, MEENY, MINY, MO

This popular children's rhyme (currently out of favor because of the racial slurs implicit in the modern version), actually comes from the ancient druid version (about the first century B.C.):

Eena, meena, mona, mite
Basca, lora, hora, bite,
Hugga, bucca, bau;
Eggs, butter, cheese, bread,
Stick, stock, stone dead—O-U-T!

It is thought by the theologian Charles Francis Potter (*Standard Dictionary of Folklore, Mythology and Legend*, 1949) that this chant, in one form or another, was an ancient magical rhyme used to choose human sacrifices for druidic ceremonies.

Today when we say eeny, meeny, miny, mo, it is a way of choosing up sides, or deciding who will be first to do a task.

ELEVENTH HOUR

The phrase at the eleventh hour means, of course, at the last minute. It comes from the Bible. The Book of Matthew says that all those hired at the eleventh hour were to be paid just as much as those laborers who had worked in the vineyards all day long.

ELVES

See SMALL FOLKS

EMERALDS

See also CHARTS, Birthstones

Who first beholds the light of day
In Spring's sweet flowery month of May
And wears an emerald all her life
Shall be a loved and happy wife.
(Anonymous)

Since ancient times there have been legends about emeralds. The Romans believed the stones lost their brilliance in the presence of evil. The Peruvians worshipped an emerald as large as an ostrich egg. In India, they believed it was possible to learn the "knowledge of the soul of the Eternal" by giving emeralds to the gods as offerings. Almost everyone believed in the wonders performed by the emerald as an amulet. When an emerald is worn near the head, it is believed to provide protection from the evil eye (*See* EVIL EYE); when it is in a ring, it can detect poison. An emerald can also be worn as a symbol of victory by warriors after a battle.

The emerald is invaluable to people who do close work, because it is supposed to relieve eyestrain. It can blind snakes. It is the symbol of love, prosperity, kindness, and goodness. It is the stone of those born under the sign of Taurus.

The emerld has been known to keep people honest, to make them clever and funny, to improve their memories, and to make them excellent public speakers. The emerald also was a cure-all at ancient pharmacies.

ENGAGEMENT (ENGAGEMENT RINGS)

See WEDDING CUSTOMS

ETERNITY

The Chinese symbol for eternity combines rivers, and flowing water. The small splash mark at the top implies a higher meaning. It signifies that eternity is a higher order than life on earth has been.

Eternity

EVIL EYE

See also CROSS-EYED

The concept of the evil eye is probably the most powerful and the most pervasive superstition in the history of the world. "For every one that dies of natural causes, ninety-nine will die of the evil eye," says the ancient Jewish Talmud.

Imagine the terror of a primitive man when he looked into another's eyes and saw a reflection of his own image in those eyes. In ancient times, people believed that the eye was the window of the soul and that certain people had devillike powers to cause illness, death, or bad luck. Man reasoned, therefore, that if his reflection showed in another man's eyes it could only be for evil purposes. Since the ideas of evil and envy are very similar, it was believed that a person with the evil eye coveted your belongings and your good luck.

You've probably heard Jewish friends say, *kayn aynhorah*. It means (loosely translated from Yiddish), "May no evil eye harm you." It is always said after someone has bragged or called attention to a fortunate member of the family. It's a charm against the evil eye, and since it's been in constant use for over four thousand years, it must be considered pretty potent.

This kind of protection against the evil eye is especially necessary for children. The forces of the Devil are especially interested in beautiful, bright, talented children. Whenever someone praises a child, someone else must quickly say a *kayn aynhorah* for protection. The Chinese don't say *kayn aynhorah*. Their method is to say things in reverse: For example, if the child is especially pretty, they will insist that it is really an ugly child; or they will comment on the stupidity of a really intelligent child.

The Greeks count on their blue worry beads for protection, and the Arabs are very careful not to say *el ain* out loud. Today in the Mediterranean countries you can see a plastic blue eyeball suspended from car mirrors or on key chains as an amulet.

Besides charms and amulets, saliva is good protection; spitting over your left shoulder, for example, should be helpful. (*See* SALIVA) The ability to outstare the evil eye often works. In Egypt mascara (or kohl) drawn as a circle around the eye was believed to protect against the power of the evil eye.

There is some evidence that as long ago as prehistoric times, people believed in monsters with eyes of emeralds that could charm their enemies. In ancient Rome there were professional witches who were hired to bewitch enemies. This practice persisted and spread throughout Europe. By medieval times, the fear of the evil eye was so great that people were burned at the stake for possessing one.

How can you recognize the evil eye? Recognition factors vary greatly from region to region. In some places cross-eyes are a sign;

cowlicks or being left-handed is another way to tell; the ability to tame animals scares a lot of people; and if you have different-colored eyes than the average person in the area, you're suspect.

Hypnosis is the closest thing to the evil eye. Svengali was a perfect example of a person who possessed the evil eye.

Honi soit qui mal y pense.
(Make sure the evil is not in the eye of the beholder.)

(motto of the
Order of the Garter 1349,
established by Edward III.)

EYE COLOR

Blue eyes are said to indicate intelligence and even divine powers (because blue is the color of Heaven). Gray eyes mean a person is calculating. Green eyes indicate creativity.

Brown-eyed people like to flirt before marriage but are loyal afterwards.

EYELASHES

See also BEARDS
FINGERNAILS

The eyelash is a valuable talisman to wish upon; if one falls out, put it on the back of your left hand, make a wish, place your right hand (palm down) over the lash, and press. If the lash sticks to the right palm your wish will come true.

There are silly people who blow away the lash after making a wish. This is not a good idea, since eyelashes belong to that group of bodily growths that are believed to be of great use to witches who want to cast an evil spell on you. Remember, always save them.

EYE SHADOW

See also CIRCLES
EVIL EYE

Eye shadow was used in ancient Egypt to draw a circle around the eyes (as lipstick was used to draw a circle around the mouth) so that the evil eye couldn't enter through them. The Devil has no powers against the strength of a circle.

FAIRIES

See also SMALL FOLKS

Fairies are small, often invisible creatures. They can either be helpful or a nuisance; evil or whimsical; and they always live very close to human beings. They're everywhere, in almost every culture. They are almost always associated with the color green (*See* CHARTS, Colors) and sometimes with white.

There are two main types of fairies. One belongs to a nation of fairies who live in a group in fairyland, with a king and queen. The best part of fairyland is that there is no concept of time. Nobody ever grows old or dies.

The second sort of fairies are independent little creatures, who often attach themselves to a household. They check around carefully choose a house and family they want to live with, and then simply move in.

There is of course, a third type of fairies who aren't quite as invisible as some would like them to be. Some say they live in groups and cause trouble. But the latter assumption has never been proven.

Playwright James M. Barrie believed that every time a little child says, "I don't believe in fairies," an unhappy fairy dies just a little bit.

FATAL LOOK

At first glance (pardon the pun) the terms fatal look and evil eye may appear to be the same; but in the history of superstitions they're quite different. Fatal look, as in "if looks could kill," means a look that can literally kill someone instantly. The evil eye, on the other hand, is merely malevolent, causing bad luck or, at the worst, death over a prolonged period.

In Egyptian folklore, the story goes that Isis killed a spy with a single look when he was spotted watching her mourn the dead body of Osiris, her lover.

Throughout the centuries there have been people and creatures who were believed to have the power to kill with a glance. Some snakes, for instance, were believed to poison the air with their eyes.

FATHER TIME

The old man with long beard, scythe, and hourglass who turns up every New Year's Eve is the descendant of Saturn, who represented time. Father Time, as we know him, still carries the scythe to show he can destroy anything at will, and the hourglass is a sign of the unstoppable flow of years. The figure with the scythe has also frequently been pressed into service to represent death; hence the name of the Grim Reaper.

FEAR OF THE DARK

See also NIGHT AIR IS BAD FOR YOU

Having a fear of dark places isn't considered instinctive; it's a response to conditioning. Grandma used to say that night air was unhealthy because it would fill the house with poison. Grandma was only repeating an idea that *her* grandma had been taught: that air after sundown is harmful.

Long ago people thought that at night evil air rose from the ground and floated around, poisoning the atmosphere. Many sun worshippers hid until the sun came out because when it went away they felt they no longer had its protection.

Primitive man had every right to be afraid of the dark, since he lived in forests and caves where darkness often concealed wild animals. To help allay those fears, he created a god of darkness.

FERN SEED

Want to be invisible? Carry a fern seed in your pocket and the world won't be able to see you! Here's how it works: The fern seed is so tiny it was once thought not to exist at all. In early times it was believed to be invisible. Later, it was thought to exist only on Midsummer's Eve. If, on that night, you caught the seed in a white cloth before it touched the ground, you could gain its power of invisibility. (If you're interested in trying this, Midsummer's Eve is the night before summer begins.)

FIG SIGN

Try this: Make a fist, putting your thumb up between your index finger and your middle finger. You've just made a rude gesture that would make many Italians fighting mad!

In witchcraft, this gesture, known universally as the fig sign, is used against the evil eye. It can also be used when you've done something good and don't want the Devil to notice.

It's one of the more ancient finger gestures, and it is closely related to the more popular middle finger in the air that drivers the world over are so fond of using.

There's a legend that Barbarossa first used the sign to indicate contempt for the people of Milan after a battle.

FINGERNAILS

See also HAIR

Don't cut your fingernails on Friday or Sunday. Monday is a very good day to cut them, as this old English rhyme points out:

Cut nails on Monday you'll get good news,
Cut nails on Tuesday will bring new shoes,
Cut nails on Wednesday and you'll travel,
Cut nails on Thursday and you'll get more shoes,
Cut nails on Friday and there's money (or a toothache, or sorrow),
Cut them on Saturday and you'll see your lover on Sunday.
But cut them on Sunday and the Devil will get you.

(Anonymous)

In any case, it's a very chancy thing to cut your fingernails at all, because you never know who'll get hold of the parings and use them in a charm against you. Like other parts of the body, fingernails are believed to be parts of your soul that can be used by witches and other evil beings in potions and charms against you. (*See* BEARDS)

If you must cut your nails, and you've chosen the day of the week that's best for you, don't cut them in order—that invites death. (The nails on a corpse are usually cut in order.)

Dragon Lady nails, the very long kind the actress Gale Sondergaard made famous, were popular from ancient times right through to the nineteenth century in China, where they were considered a sign of great beauty and a symbol of distinction. They indicated that a person was an aristocrat and never worked with their hands. To protect the nails, these people wore sheaths of silver or gold on their fingertips.

There's another superstition, involving the white specks (calcium spots, probably) that many people have on their nails. The rhyme goes:

Specks on fingers,
Money lingers.
Specks on the thumb,
Money comes.

White specks on your fingernails indicate good fortune; on thumbnails means you'll get a gift; on the forefinger, the number of friends; on the middle finger, the number of enemies; on the ring finger, a letter coming; and on the little finger, a journey.

Remember, the total number of white specks on your nails shows the world how many lies you've told!

FISH

See also TALISMAN

Better finish your fish—it's brain food. In the nineteenth century it was discovered that both fish and the human brain contain quite a lot of phosphorus, and since the concept of like makes like is ancient and still well received in many quarters, fish was considered brain food that would make you smart.

If you do eat fish, eat if from head to tail for good luck.

Some beliefs of fishermen are:

If you count the number of fish you've caught, you won't catch any more that day.

If the fish aren't biting, throw a fisherman into the water, then haul him out; the fish will come around then, probably out of curiosity.

Throw the first fish caught back; in ancient times fishermen did this as payment to Poseidon (Neptune), the king of the sea.

In certain parts of the world the fish is sacred. In other parts, it is forbidden as food. In some parts of India it is said to be the food of ghosts.

To some people the fish is a symbol of fertility; in ancient Egypt it was believed to have eaten the phallus of the god Osiris. The Chinese consider it a sign of happiness and in most of India it is one of the eight symbols of Buddha and therefore sacred. Early Christians associated it with the Trinity, because Christ fed the multitudes with five loaves and two fishes.

FIVE

See also NUMBERS

To the Greeks, five was the symbol of the world. They believed that the five-pointed star represented the secrets of life.

There is a rumor that Pythagoras invented the five-pointed star as his symbol of geometric perfection. In the Middle Ages it became the wizard's star and was worn on cloaks as an emblem of the mysteries of the universe.

According to numerology the number five represents fire and love and marriage.

FLAGS MUST NEVER TOUCH THE GROUND

During a battle, a flag bearer always stayed near his king. The troops knew if he lowered the flag to the ground it meant the king was either dead or dangerously wounded. The superstition that it's bad luck for the flag to touch the ground comes from those early wars.

FLIP A COIN

Flipping a coin is an ancient way of deciding a point. People believed that when they flipped a coin, it was the gods who would be deciding the winner as the coin turned over and over in the air on its way to the ground. Later, Fate or Lady Luck made the decision.

The idea "heads you win, tails you lose" comes from Julius Caesar's day, when the emperor's head was on coins. If his head came up when you called it, the decision was made once and for all—Caesar always won.

FLOWERS CAUSE DEATH

Flowers in a hospital room were once thought to be unhealthy or fatal. This was a notion of the Victorians who believed that picked flowers decomposed to such a degree that they made the air unfit to breathe, causing illness and even death.

To illustrate this point the Victorians concocted a morality story, called "The Revenge of the Flowers," about two young girls who took a nap after picking flowers all afternoon. While they slept the spirits of the flowers asked why the girls had killed them and then breathed decomposing, noxious breaths into the sleeping girls' faces until the two were dead. The last line reads: "The flowers are avenged."

In the Language of Flowers, (*See* CHARTS) gathered flowers say, "we die together."

FLOWER LANGUAGE

See also CHARTS, Language of Flowers

> *Flowers are the sweetest thing that God ever made and forgot to put a soul into.*
> (Beecher, *Life Thoughts*, 1858)

As clever as "Say It With Flowers"—the slogan for the Society of American Florists—may be, it represents anything but a new or an original idea. Flowers have had secret meanings (and often mystical ones) attributed to them ever since Greek mythology. A language of flowers, one in which every flower expressed a special word or sentiment, was first popular in the Orient. It reached Europe during the early 1700s via Lady Mary Wortley Montagu, who went to Turkey with her husband and began sending back to England these first interpretations:

Clove—I have long loved you and you have not known it.
Jonquil—Have pity on my passion.
Pear Blossom—Give me some hope.
Rose—May you be pleased, and your sorrow mine.
Straw—Suffer me to be your slave.
Cinnamon—My fortune is yours.
Pepper—Send me an answer.

The Language of Flowers gradually became very popular in England, and to some degree throughout the Continent, but it wasn't until Victorian times that it became all the rage. The intricacies of the game were overwhelming.

Explained by Claire Powell in *The Meaning of Flowers*, a flower, when presented upright, meant something good; when the flower was inverted, something bad. For example, a rosebud with thorns and leaves meant, "I fear you're never going to love me, but I still have hope." Returned upside down, it meant, "You must neither fear nor hope." If the rosebud was stripped of the thorns it meant, "There is everything to hope for"; stripped of its leaves, 'Your fears are correct." Inclined to the right, the flower implied a personal comment about the recipient; for instance, an inclined white lily said, "You are all purity and sweetness." Inclined to the left, the flower represented a personal message about the one who gave it.

This language gets more involved. If the flower was placed in the hair, it might mean caution; in the cleavage, rememberance or friendship; or over the heart, love. All of this depended upon the flower chosen. For example, if a marigold was placed in the hair, it expressed sorrow of mind; in the cleavage, boredom; and near the heart, pangs of love. It might well have taken an Oxford education to have understood what someone was trying to say with a simple daisy!

FORGET-ME-NOTS

sweet forget-me-nots,
That grow for happy lovers
(Tennyson, *The Brook*)

Forget-me-nots have been a symbol of undying love since antiquity. They have also been used as tokens of friendship and remembrance.

Claire Powell, in her *Meaning of Flowers*, tells of an Austrian legend in which two lovers, on the eve of their wedding, walk along the banks of the Danube. A blue flower rides in on the waves, and the young bride sighs because she knows the flower will be swept away. Her lover jumps into the river to retrieve it, but a strong undertow drags him down. With his last effort he flings the flower onto the shore and drowning calls, "Love me. Forget me not!"

There's another, less romantic superstition that steel dipped into the boiling juice of the forget-me-not becomes hard enough to cut through stone. The juice is also said to cure both rabies and sore eyes.

The flower is traditionally given on February 29 as a symbol of remembering a day which only appears once every four years.

FORGETTING THINGS

There's a very popular superstition that if you've forgotten something it's bad luck to go back for it. The reason is that if you interrupt your journey by returning for the item, you'll have broken a circle (*See* CIRCLES), and that's very bad luck, indeed. There is help for this situation. Go back to where you left the forgotten object, sit down, make a wish or say a magic charm like this effective one:

If I sit, bad luck will flit

—and then count to ten.

Some people believe that women forget things more often than men. One African tribe, in its funeral procession for a dead woman, returns to the deceased's home for an hour so that the ghost can have a chance to collect the things she may have forgotten. (This saves the ghost a trip later.)

FOUR-LEAF CLOVER

One leaf is for hope, and one is for faith,
And one is for love, you know,
And God put another in for luck.
(Higginson, *Four-Leaf Clover*)

The druids were the first to believe in the power of the four-leaf clover. They were certain that possessing one allowed them to see evil spirits, like witches, which they could then avoid.

It was believed by some Christians that Eve took a four-leaf clover with her when she left the Garden of Eden.

The Irish say, "he's in clover," when someone is doing very well. Finding enough real, uncultivated four-leaf clovers to wear would be difficult, though—they are freaks of nature, a mutation of the common three-leaf variety.

FRIDAY

Alas! you know the cause too well;
The salt is spilt, to me it fell.
Then to contribute to my loss,
My knife and fork were laid across
On Friday, too! the day I dread;
Would I were safe at home, in bed!
(John Gay, *Fables*, Pt. I, 37, 1727)

Now Friday came. Your old wives say,
Of all the week's the unluckiest day.
(Anonymous)

It has often been pointed out that everything bad in history has happened on a Friday. Adam and Eve were supposed to have fallen from grace on Friday; the Flood started on a Friday; the Temple of Solomon fell on a Friday; and Christ was crucified on a Friday.

So, it's bad luck:

to be born on a Friday
to start a new job on a Friday
to cut your nails on a Friday
to visit the sick on a Friday
to start a voyage on a Friday
to change the bed linen on a Friday
to open a new play on a Friday

Criminals say if you're sentenced on a Friday, you will receive a stiffer sentence. Friday was known as Hangman's Day because executions often took place on it.

Ships never set sail on Friday. That may be since H.M.S. Friday, a ship whose construction began on a Friday, set sail on a Friday, and was never heard from again!

Several major financial panics have contributed to the coining of the term Black Friday. The first happened on a Friday in England in December 1745, and was followed by a number of other similarly disastrous Fridays on the English stock market.

He who sings on Friday will weep on Sunday; then again, if it rains on Friday, it will be clear on Sunday.

FRIDAY THE THIRTEENTH

See also FRIDAY
NUMBERS
THIRTEEN
THIRTEEN AT TABLE

Friday the Thirteenth is an especially unlucky day.

Twelve witches are necessary for a meeting—plus the Devil equals thirteen—and they always meet on Friday. The combination of Friday, a generally unlucky day, with thirteen, an especially unlucky number, can be terrifying.

FRINGE

See KNOTS

FROGS

See also WARTS

A favorite and powerful amulet of the Egyptians, Greeks, Turks, Italians, and others, the frog held an important place in early times. It was a symbol of inspiration and of fertility (due to the large number of eggs it lays at one time). The frog was so important in ancient Egypt that when one died, it was embalmed. It was a popular Roman mascot (*See* MASCOTS) and is still believed to be very good luck in the home. The first-century botanist Pliny the Elder believed a frog had powers that attracted friends and inspired lasting love in those who possessed one.

Others, however, believe that the frog holds the souls of dead children.

Even today the frog is still considered a fairly reliable weatherman because of its susceptibility to atmospheric changes, which cause it to croak whenever the barometers go down.

The ancients found the frog an inspiring creature and often carried amulets made in its image. Today we close our coats with frogs (ornamental cloth closings consisting of a button and a loop through which it passes), perhaps a holdover from an early French custom where frogs were embroidered onto clothes for good luck.

GAMBLING

Since almost all gambling is a matter of luck and good fortune, there are few if any explanations for the following superstitions:

If you're having bad luck at cards, rise and turn three times with your chair in your hand—your luck will change.

It's bad luck for any of the players, when a player places a matchstick across one that you've previously placed in the ashtray.

It's bad luck for a woman to touch your shoulder while you're playing; it's also bad luck to meet a woman while on your way to the gambling room. (These are old superstitions, from times when most gamblers were men.)

No hand with the four of clubs is ever lucky. (The four of clubs is called the Devil's four-poster bed.)

Borrowed money can't lose.

Beginner's luck is indisputable.

It's bad luck to drop a card during the game.

If you sing while you play, your partner will lose.

If you are angry while playing, you'll lose.

He who borrows money to play will win; but he who lends money while playing will lose.

Rub dice on a red-headed person for luck.

Carry dice in a pocket all the time, and you'll have good luck.

If you find a die a certain way it means:

(one spot up) an important letter coming
(two spots) a long and good trip ahead
(three spots) a big surprise; sleep in a strange bed
(four spots) very unlucky—big trouble ahead
(five spots) unfaithfulness from your lover
(six spots) very lucky—you'll get money

Touch a hunchback for luck (*See* LUCK) (the gambler's word hunch comes from hunchback).

Always "back your luck"; in other words, stay with a lucky streak.

GARLIC

Garlick makes a man
Winke, drinke, and stinke.
(Nashe, *Unfortunate Traveller*, 1594)

According to legend, a string of garlic bulbs worn to bed at night could protect you from vampires. As vampires became more and more infrequent night visitors, it was believed that the wreath of garlic bulbs in a house with illness could draw the disease away from the stricken person and absorb the illness.

Today many people tend to think that the garland of bulbs hung on the mantel will bring good luck.

Garlic cloves, we are told, can do wonderous things. They can stop bed-wetting, help toothaches, and when rubbed into the gums of a horse restore its failing appetite.

In classical Greece garlic was left as dinner for Hecate, the goddess of the dead. Since Hecate taught witchcraft and sorcery, garlic became closely associated with the world of the occult.

Certainly the strong aroma of garlic prolonged its life as a mystical substance. It is thought, for instance, that garlic is a good antiseptic —it isn't. In some places, the smell of garlic is believed to keep evil spirits away. Elsewhere it is believed to indicate the presence of evil.

Roman soldiers ate garlic to make them courageous. There are still South American bullfighters who take garlic cloves into the bullring with them, believing that the smell will stop the bull from charging. (It probably would stop a bull or an enemy from charging, if enough cloves were used!)

GARNETS

See Also CHARTS, Birthstones

By her who in this month was born [January]
No gems save garnets should be worn;
They will insure thee constancy,
True friendship and fidelity.

(Anonymous)

Centuries ago Poles began wearing garnets as protection against illness. Later the stone became an amulet for world travelers. In Asia, garnets were used as an ancient form of bullet, and the Persians considered them the only stones good enough for royalty.

The Old Testament says that a large ruby-colored garnet was Noah's only source of light in the Ark during the Flood.

During the Middle Ages a garnet was especially important if it had a lion's head engraved on it. It was then thought to be blessed and to impart good health and honor to its wearers. This was a very rare amulet indeed, since garnets are hard and brittle and usually shatter during engraving.

GEMINI

See also TWINS
ZODIAC

Gemini, a constellation first noted as early as 6000 B.C., is made up of the twin stars Castor and Pollux. It is the sign of those born between May 21 and June 20. Gemini, the Twins, has been associated throughout history with twins like the Roman rulers Romulus and Remus and the Chinese twin principles yin and yang. (*See* TWINS)

The sign of the Twins denotes a multifaceted personality, flexibility, and adaptability. Geminis are likable, impetuous, and good company to have around. They are clever, charming, and attractive. They are ruled by Mercury, which makes them fast and willing learners, but they tend to skim subjects rather than learn them. They are people of the air and changeable.

Gemini, the twins

Geminis also tend to be intelligent, but they procrastinate. They have strong nerves but get depressed by boredom, keep cool in a crisis, and should be involved in self-expressive pursuits.

Wednesdays are good for Geminis, the numbers three and four work well for them, and silver or gray are their good colors. Their birthstone is the agate.

Some famous Geminis are Judy Garland, John F. Kennedy, Marilyn Monroe, and Henry Kissinger.

GHOSTS

Ghosts do fear no laws
Nor do they care for popular applause.
(Unknown, *Thomas Nash His Ghost*, 1600)

Either ghosts are human souls after death, or they don't exist at all. They might be products of an overactive imagination. They might also be people who have died a violent death or who have some unfinished business to do, who return to earth as ghosts to settle their affairs.

Whatever ghosts are, here's what you can do to protect yourself from them:

Make a cross out of two pieces of fruitwood, tie it together with a piece of red string, and wear it between the lining and the fabric of your coat.

When you meet a ghost, spit on the ground between you and the ghost and say, "In the name of the Lord, what do you wish?"

A ghost probably won't do you any harm, though, if you haven't harmed it during its lifetime.

From ghoulies and ghosties,
Long-leggity beasties
And things that go Bump in the night
Good Lord deliver us.
(Scottish prayer, 1800)

GIFT HORSE IN THE MOUTH

We all know the famous Welsh proverb, "always look a gift horse in the mouth." Or is it "never look a gift horse in the mouth?" The second adage probably goes back to Homer, who said, "It is not good to refuse a gift." He may have been thinking of the tale about the Trojan horse—but to have looked that particular gift horse in the mouth would have been a very good idea.

Looking a horse in the mouth means counting its teeth to determine its age and, therefore, its value. Rabelais wrote: (*Works*, I, xi) "He always looked a given horse in the mouth," meaning you never get anything good for nothing, or something bought is often cheaper than a gift.

Perhaps the reverse is true. It's bad manners to look too carefully at a gift; it's an obvious attempt to determine how much it's worth.

Either look or don't look; but never turn your back on a gift horse (or a donkey).

GIVE A PIN, GIVE A PENNY

See PINS

GNOMES

See SMALL FOLKS

GOBLINS

See SMALL FOLKS

GOLDFISH

The first thing you need to know is that goldfish, those tiny, glittering things you win at the penny arcade, are members of the carp family and as such, have a long and happy tradition as a good luck talismans. The Egyptians kept them as mascots (*See* MASCOTS) and thought they were particularly helpful to lovers and insurance of peace and harmony within a household. The Greeks and Romans thought pretty much the same thing, extending the good luck to cover courting and marriage.

In the Orient, a yellow or gold carp was a mascot that brought good fortune as well as good luck to lovers. The goldfish is also an emblem of Buddha.

Claudia De Lys, in *A Treasury of American Superstitions*, tells about the Egyptians who believed that the carp had the perserverance to leap the great waterfall and to reach the chariot made of clouds, which would carry them to Heaven. Other fish didn't have the patience to persist, and simply drowned. That's probably why the carp became good-luck mascots.

GOOD FRIDAY

See also EASTER
EASTER EGGS
HOT CROSS BUNS

Good Friday is celebrated on the Friday before Easter. Like most of the other days of the Christian Holy Week, it became a holiday—for the day Christ died—in the fourth century. Prior to that, all the days of that week were treated as parts of the same celebration.

Some of the many superstitions connected with Good Friday are:

A loaf of bread or buns baked on Good Friday will not get moldy and will bring good luck for a year.

Bread baked on Good Friday and kept is a good luck talisman against shipwrecks.

Babies should be weaned on Good Friday.

The sun doesn't shine as brightly on Good Friday as it does on other days.

Breaking a dish on Good Friday is a good luck sign; it says that no damage will come to that house all year long.

Planting anything on Good Friday will produce a plentiful crop.

A ring blessed on Good Friday will bring protection for the year.

In Scandinavian countries, to insure against evil spirits, witches, and snakes, branches of mountain ash are placed around the door of homes. This will only work if done on Good Friday.

GOOD HEALTH

The Egyptians had a magical sign to bring them good health and to ward off illness and disease. This symbol was also believed to counteract the effects of the evil eye. The mere representation of an eye was used for protection. It might be painted on the prow of a ship to guard it against shipwreck or it might be painted on a cup or a plate to prevent breakage.

GOOD LUCK—GIVE OLD SHOES

See WEDDING CUSTOMS

GO TO THE DICKENS

This little euphemism has nothing to do with Charles Dickens and probably originated long before he was born. Shakespeare used it in the *Merry Wives of Windsor*: "I cannot tell what the dickens his name is."

According to the language expert Lewis B. Funk, dickens probably came from devilkins, meaning little devils. Today "go to the dickens" usually means "go to Hell."

GREMLINS

See SMALL FOLKS

GROOMS (BRIDEGROOMS)

See WEDDING CUSTOMS

GROUNDHOG DAY

If Candlemas Day be fair and bright,
Winter will have another flight.
But if Candlemas day brings clouds and rain,
Winter is gone and won't come back again.
(Old English proverb)

On February 2, it is said, the groundhog wakes up from its long winter snooze and pokes its head out from its underground hiding place. If it sees its shadow, there will be six more weeks of winter (if the sun is out, he will see his shadow and be frightened back to his nest). If there's no shadow, summer is on its way, and the groundhog will stay out and play. This modern superstition has more to do with harvesting and planting than with weather forecasting. Six more weeks of winter can't be very good for the farmer who would prefer a mild February.

This modern superstition has early origins. February 2 used to be Candlemas Day (and still is in some countries). Candlemas was the time when the candles were blessed for the year in observance of the Purification of the Virgin and the Presentation of Christ at the Temple.

An even earlier observance of this day was in Rome where it was the day to honor Venus.

GYPSIES

There are many theories about who gypsies really are. One legend says they are people condemned to wander the earth without rest because they refused hospitality to Joseph, Mary, and Jesus on their flight to Egypt. The word gypsy is a corruption of the word Egyptian. At least that was the medieval notion of the origin of the name gypsy.

The gypsies refer to themselves as Romany for many contradictory reasons that outsiders can't seem to fathom. They maintain an aura of mystery as a separate nation, it is thought, by following strict taboos against intermingling and revealing any of their secrets.

Some believe they migrated from Northern India in about 100 A.D. and proceeded to Central Europe by the 1400s. Others say they are descended from the outcasts of the temple of Thoth in Egypt.

Even today, the gypsy is characterized as a moon worshipper who is born, lives, and dies in the outdoors. They are fortune-tellers and horse handlers. They can read the tarot cards (*See* TAROT) and although they often believe in their own ability to see the future, they hardly ever give a real reading to an outsider. Gypsies believe that until a dead person has been burned, the soul roams the earth unable to get free of this world, remaining a constant accuser of those who have not burned the body.

HAIR

See also BEARDS
HAIR COLOR

Is your hair curly? Then you must be lucky! Since curly hair resembles the wavy rays of the sun, the sun god likes you and will protect you. You say your hair got curly by eating bread crusts; but have you heard about pouring rum or grapes on your head in order to make your hair grow curly?

Down in the southern states they say that combing your hair at night makes you forgetful.

Since ancient Egyptian times people have believed that hair clippings can be used by a witch in casting evil spells. The Egyptians believed that a potion made of hair, nail clippings, and human blood could produce absolute power over another person.

A hairy chest has always been a sign of strength; and cut hair has been, since Samson and Delilah, a sign of weakness. There's a myth that hair changes color overnight; and that hair and fingernails continue to grow after death, they don't; and be careful—whistling causes your beard to grow.

HAIR COLOR

There was never a saint with red hair.
(Old Russian proverb)

If you pull out a gray hair, seven will come to its funeral.
(Old Pennsylvania proverb)

Gray hair is a sign of age, not wisdom.
(Old Greek proverb)

A chaste woman ought not to dye her hair yellow!
(Menander, 310 B.C.)

Trust no man
Even your own brother
Whose hair is one color
And beard is another.
(Anonymous)

Blondes are fickle and make false friends. Brunettes are sincere and have good health. Redheads are emotionally unstable and have terrible tempers.

Redheads really come in for criticism. Since red is the color of fire, Romans, Egyptians, and Greeks all thought that redheads were particularly unlucky people to have around. In the Middle Ages, it was believed that redheads were witches and therefore deserved to be burned.

Some superstitions about hair color are:

It's good luck to run your hand through the hair of a redhead.

It's lucky if a dark-haired person crosses the threshold first on New Year's Day.

Bees always sting redheads.

Redheads are good conversationalists.
Curly black hair means neatness; straight black hair, extravagance; and thick black hair, good health.
Brunettes have the best chance of survival except when disease strikes—then blondes fare better.
Men with dark hair are deceitful.

HALLOWE'EN

See also CHESTNUT

October 31 was, according to the Celtic calendar, the last day of the year and a time to honor the dead. The druids celebrated that day by inviting the souls of all the evil people who, upon death, were sentenced to live in the bodies of animals to a feast to see if their deeds since death earned them redemption. On that night, ghosts and witches walked the earth; some said they journeyed for two days. You can imagine what sort of havoc was created and why it was unsafe for decent folk to be out on that terrifying night. It was called All Evil Day, or something similar.

In 837, Pope Gregory IV changed the name to All Saints Day and discouraged the common belief that witches and evil spirits went partying. The word hallow was an old English one meaning saint or holy man, in common usage until the fifteenth century. All Hallow Eve became All Souls Eve or All Saints Eve. Hallowe'en is simply a contraction of the original. (In Italy and other places it is still All Saints Eve.)

Symbols and superstitions of the day are:

Ghosts and hobgoblins roam the earth, but since they are invisible, children dressed as adults perform their mischievous pranks for them.
Black cats (*See* BLACK CAT) are Satan in disguise and can be seen cavorting with witches.
Pumpkins, a symbol of harvesttime, are cut to resemble faces, and a candle lighted inside the pumpkin pays tribute to the sun at harvesttime.
The disagreeable custom of chalking a person's back comes from

merry old England, where they drew white circles on backs to indicate that summer was over and the rule of the sun was coming to an end for another year. (*See* CIRCLES)

Fortune-telling is very popular on this night and started when witches got together to feast during druid times and told each others' futures.

HANDS

See also PALMISTRY

The old German proverb that says "Warm hands, cold heart" or "cold hands, loving spirit" (whichever seems appropriate) is little more than a compliment for people with cold hands.

If your right palm itches you'll get money, but if your left palm itches you'll have to pay money out.

Don't wash a baby's hand, or you'll wash away its luck.

HANDSHAKE

Two things happen when you shake hands: You make a cross with your clasped palms; and you effectively put your weapon hand (unless you're left-handed) out of commission, so it can do no harm.

Handshaking is not a worldwide custom. In the Orient, for instance, people, in meeting, have joined their own hands in front of them for centuries; they are effectively proving their good intentions by placing their hands where they can be seen, away from their weapons.

Do you think muggers could be taught this simple little gesture?

HAT ON BED, HAT ON TABLE

A hat on a bed is bad luck and will cause a fight in a household.

In the Orient it was believed that placing a hat (or turban) where another person might lay his or her own head was a dangerous business. This was a way one might contract bad luck from the evil eye. (*See* EVIL EYE)

Long ago people believed that evil spirits lived in the hair. They based this idea upon the crackling sound they sometimes heard, which is caused by static electricity.

The superstition about putting a hat on a table is like the superstition about putting your hat on a bed. Putting your hat on a table, however, will let evil spirits get into your food.

HATS

For good luck, turn your hat front to back; it's symbolic of changing the natural order of things and is supposed to be quite helpful in warding off bad luck. Jockeys do it, as do ball players sometimes.

Remember, wearing a hat indoors causes a headache.

HELL

> *The descent to hell is easy; the gates*
> *stand open night and day; but to reclimb*
> *the slope, and escape to the upper air,*
> *this is labor.*
>
> (Virgil, *Aeneid*, VI)

Hell appears to be what bores or frightens each person the most. It is also known as Hades or any one of an endless string of euphemisms—even New York is called a helluva town.

Hell is the opposite of Paradise. It is where mortal souls go because of bad behavior during life. Hell is also God's way of condemning a mortal for lack of faith.

In the world of superstition, Hell is the place where the Devil and his playmates come from, it is where the evil eye resides, and it is the most frightening place a superstitious person can imagine. At the mention of Hell such a person will spit three times, turn around seven times, knock on wood, touch iron, or perform any other countercharm that might help avoid a meeting with the Devil, and going to Hell.

HEXES AND HEX SYMBOLS

The Pennsylvania Dutch are fond of putting gaily painted symbols on their barns to frighten away the Devil and to protect their animals from the evil eye. The symbols usually contain a good deal of red, a color that frightens witches. The word hex does not come from hexagon, although hex symbols are frequently six-sided.

Hexes are spells, the stuff of which witchcraft is made. A witch, or more often a warlock, will place a hex on a person or thing to cause trouble. Even today there are professional hex doctors, who make a hex to order. There are spells for all the evils that can befall people. (It can be a very lucrative business.) The spells come mostly from medieval times and are contained in two very important handbooks, *Seventh Book of Moses* and *The Long Hidden Friend.*

HICCUPING

Sneezing will stop a hiccup.
(Hippocrates, 400 B.C.)

Aristophanes suggested holding your breath or gargling with water as a cure for it. Later, more sophisticated people suggested remedies like spitting on the forefinger of the right hand, crossing the front of the left shoe three times, and repeating the Lord's Prayer backward.

Hiccups were said to have been caused by the evil eye. (*See* EVIL EYE) When they happened in Church, it was an indication that a person was possessed by the Devil.

If you can say the Lord's Prayer backward on one breath and then repeat three times:

Lick up, hiccup
Stick up, hiccup
Trick up, hiccup
Begone, hiccup

—you will be cured.

Scaring someone out of their hiccups works, generally, because it breaks the tension that may have caused them in the first place. (You might also believe that scaring the person often scares the evil spirits away.)

To stop hiccuping, you can:

hold your nose, tilt your head back, and take a sip of water for each year of your age
drink a glass of water from the far edge of the glass or through a napkin
bring your little fingers as close together as you can without having them touch
hold your breath and say hiccup nine times

If none of the above work, start over again from the top.

HOBSON'S CHOICE

Hobson's choice isn't a superstition. At the end of the sixteenth century Thomas Hobson owned a livery stable in Cambridge, England. Each customer who came to rent a horse had to take the animal nearest the door; there was no picking or choosing. Hobson's choice, of course, was no choice at all. Hobson is reputed to have explained that his policy was to have every customer alike well served, and every horse ridden with the same justice.

A common expression goes, "where to elect there is but one, 'tis Hobson's choice—that or none."

HOCUS-POCUS

Hocus-pocus may have started with a Norse sorcerer named Ochus, Bochus. It may also be a shorter version of hokuspokus-filiokus, which was a popular sacrilegious mockery of part of the Catholic mass, *hoc est corpus filii* ("this is the body of the Son"). It certainly has a important place in the history of witchcraft.

Thomas Ady's *A Candle in the Dark; or A Treatise Concerning the Nature of Witches and Witchcraft* (written in 1656) contains an early description of hocus-pocus:

> *I will speake of one man . . . that went about in King James' time . . . who called himself 'The Kings Majesties most excellent Hocus Pocus', and so he was called, because that at the playing of every Tricke, he used to say, 'Hocus pocus, tonus talontus, vade celeriter jubeo,' a dark composure of words, to blinde the eyes of the beholders, to make his trick pass the more currantly without discovery.*

Early in the seventeenth century, magicians, conjurers, and jugglers went by the name Hocus Pocus or variations of it, as they traveled around the countryside practicing their trade.

Today, hocus-pocus simply means flimflam, nonsense, or charlatanism.

The word hoax is a shortened version of hocus-pocus.

HOLLY

Before Christ, holly was loved in Rome. It was an emblem of friendliness and goodwill and was sent to friends as gifts during midwinter celebrations.

In Northern Europe people hung holly at their doors so that wood spirits could shelter themselves against the chilly winds in it, and to insure good luck.

Holly became a Christmas symbol almost immediately after the death of Christ. Some people believed that the Cross was of holly wood, which was punished by being turned into a scrub. Others say that Christ's crown of thorns was made of holly and that the berries had been yellow, until the crucifixion, when they turned red from his blood.

Two superstitions are that bringing holly into the house before Christmas Eve will provoke family quarrels and that holly must be burned after the twelve days of the holiday.

Witches hate holly. That's why some people grow it around their windows.

Holly is supposed to repel lightning, and the syrup from its bark, to cure coughs.

HONEYMOONS

See also WEDDING CUSTOMS

> *When a couple are newly married, the first month is honeymoon, or smick-smack.*
> (John Ray, *English Proverbs*, 1670)

It was once customary for a newlywed couple to drink a potion containing honey (usually mead mixed with honey) on each of the first thirty days of their marriage. During such a period the moon goes through all of its phases and then disappears. The affection of newlyweds was therefore regarded as waning like the moon. Hence the term honeymoon, a combination of the honey potion and the phases of the moon.

It has been reported that Attila the Hun drank so much of this honey and mead concoction that he suffocated to death during his honeymoon.

"The first month after marriage . . . there is nothing but tenderness and pleasure," said Samuel Johnson (in *Boswell: A Life*, 1755). That might well have been what the early bridegroom had in mind when he stole his bride and hid out for a month to escape the wrath of his new in-laws. In merry Old England, the most common way for a man to get married was for him to simply steal his bride, something he learned from his predecessor, the cave man. Some say that's how the honeymoon period got started, with the name and the pleasure coming later.

> *The honeymoon is the period during which the bride trusts the bridegroom's word of honor.*
> (Anonymous)

HORSESHOES

See also IRON

There are dozens of reasons why the horseshoe is lucky. First of all, there's the shape. Since the earliest of recorded times, man has believed that the crescent or u-shape was the most powerful protective sign. We see this in the arched windows of old churches, temples, and mosques (as well as in the arched doorways of modern churches and public buildings). These were built in that shape as a protection against evil.

Second, the horseshoes are nailed with seven iron nails. (*See* IRON, SEVEN) Seven is probably the most important number in the world of superstition. The nails used by blacksmiths have magic as powerful as the horseshoe itself. Rings made from these nails are said to give the same protection against evil as the horseshoe.

When you put a found—it's important that you find it—horseshoe over the doorway, or on the front of a barn as an amulet, you are following in the footsteps of the Greeks and the Romans who believed firmly in the protective powers of the horseshoe. Some say that if you hang the horseshoe with the prongs up, should the Devil get too close he will be sucked in and destroyed. Others say that if you hang it with the prongs down, the magic pours out and prevents the Devil from coming in at all. Whichever way you hang it, the important thing is that it's hung securely or its magic won't bring good luck; instead it will send you straight to the hospital.

There's a legend about Saint Dunstan, a blacksmith in England who later became the Archbishop of Canterbury. He was asked by a figure in a cloak to shoe him instead of his horse. Remembering that Satan had cloven hooves, which needed shoes, Dunstan nailed Satan to the wall and poked him with a red-hot poker until Satan agreed never to enter a house with an inverted horseshoe protecting its door.

HOT CROSS BUNS

Crosses made on food, either with a knife or icing, were thought to keep evil spirits away from the house. These foods became especially popular during holidays, when evil spirits liked to cause problems for the celebrants.

There is a superstition that hot cross buns baked on Good Friday will stay fresh for a year; and as long as they remain fresh, the men of the house who are sailors will not drown. (They may, however, get very bad cramps from these old buns.) (*See* EASTER)

One a penny, two a penny,
Hot cross buns,
If you have no daughters,
Give them to your sons;
If you have none of these
merry little elves
Then you may keep them all
for yourselves.
(The street vendor's cry)

HUNCHBACKS

See TOUCH A HUNCHBACK FOR LUCK

IF LOOKS COULD KILL

See EVIL EYE
FATAL LOOK

ILL WIND

Over four hundred years ago John Heywood wrote (in *Proverbs*, II:VII) "It's an ill wind that bloweth no man to good." Shakespeare wrote, "Ill blows the wind that profits nobody." (*Henry IV*, II:V)

The Old Norse language meaning of the word ill was evil, and that's the meaning implied in the phrases about an ill wind.

The idea that someone profits from every disaster has followed us through the centuries. Some modern lyricists have often used the expression.

INSIDE OUT

See CLOTHING

IRIS

For wheresoev'er thou art in this world's globe,
I'll have an Iris that shall find thee out.
(Shakespeare, *Henry VI, III:2*)

Iris was the goddess of the rainbow, and her name has been applied to the flower because of the many colors in which it blooms. Legend has it that Iris was the bearer of news from Zeus, who hardly ever left Mount Olympus.

In the Victorian Language of Flowers, an iris meant, "My compliments, I have a message for you."

IRON

See also HORSESHOES

Imagine how surprised primitive man must have been when he discovered iron. It could break wood and stone and withstand fire. It must certainly have been a gift from the gods, if a frightening one. To make iron, ancient man cut pieces of meteorites which were believed to have come from Heaven.

Long ago people believed that if iron touched a plant that was used for healing purposes, the plant would lose its powers. Iron was used for healing, though. A piece of iron was nailed to a tree after having been applied to the part of the body where the illness was. Their belief was that the illness would be trapped in the tree.

If you nail an iron amulet to the outside of your house, it will protect you from witches. Witches are terrified of iron. They have no powers against it at all, since iron is a special gift of the gods.

ITCHING

See also HANDS

What causes an itch? People in ancient times couldn't explain it except to consider it a craving. Thus the following sayings were born:

About an itching ear—Left for might,
Right for spite.

Left or right,
Good at night.

About an itching nose—If your nose itches.
Your mouth is in danger—
You'll kiss a fool,
And meet a stranger.

If your nose itches,
If your nose itches,
A stranger is coming
With a hole in his britches.

About an itching eye—If your left eye itches it's
itching for bad;
If your right eye itches good luck
is coming.

About an itching palm—You'll receive money from an
unknown source.

You'll take a bride.

In either case, as Shakespeare pointed out in *Julius Caesar*, it's not a good trait:

Let me tell you, Cassius, you yourself
Are much condemned to have an itching palm.

IVY

Ivy was popular in Victorian times. It represented friendship and fidelity. The Victorians often wore brooches that showed ivy growing around a fallen tree, with the inscription "Nothing can detach me from it."

The plant was sacred to the Egyptians, and the Greeks always used it to make victory crowns for their conquering armies.

Like most plants, ivy has been considered useful as a cure for all sorts of afflictions, especially sunburns, wounds, and swelling.

JADE

Much of the jade deposits of the world are found in China, and so it's not surprising that the Chinese have always worn jade as an amulet. It is thought to protect against stomachaches, to bring rain, and to help make men fertile.

In China, a piece of jade placed in the mouth or on the eyelids of a deceased person was believed to help bring back the spirit for another life on earth.

JINX

See also CHARMS

Many of us believe that people or things can be jinxed. Most of us use the word and believe in the idea that a continual run of bad luck is a jinx. It doesn't effect us much, though, unless we're gamblers, actors, accident-prone people, or sailors. Certain ships and theaters are said to be jinxed. (*See* FRIDAY, THEATRICAL FEARS)

There's little a person can do about a jinx, except to ride it out and be careful.

JUNE WEDDINGS

See WEDDING CUSTOMS

JUNIPER BERRIES

Today many people think that juniper berries help ward off bad vibrations when mixed in a potion called gin. The more we drink of this potent magic, the better the protection.

The Greeks appeased the gods of the underworld by burning the berries and the branches as incense. They also burned the berries at funerals to ward off evil spirits.

Today juniper berries often provide the perfume used in air purifiers. Some people also think that the perfume from the berries is a good luck charm against evil.

In the Victorian Lanugage of Flowers the juniper berry, given as a gift, meant protection and asylum from enemies.

KAYN AYNHORAH

See EVIL EYE

KEEP YOUR FINGERS CROSSED

See CROSS YOUR FINGERS

KISSING

> *Lord! I wonder what fool it was that*
> *first invented kissing.*
> (Jonathan Swift, *Polite Conversation*, 1738)

There's a story that the fool who first invented kissing did it as a way to see if his wife had been drinking wine behind his back. It is said that the Romans also used to press their lips on their wives' to test for sobriety.

Later the kiss was used to sign the deal after a marriage contract had been arrived at; hence the betrothal kiss. (*See* WEDDING CUSTOMS)

Kissing is thought to have started in Asia Minor as an expression of loyalty, and of sentiment—a mother kissing her child.

KNIFE AND FORK

You're aware that crossed knives and forks at the table after dinner will cause an argument, aren't you? There's a long-standing fear of crossing dangerous objects, which dates back to the use of crucifixion as a popular form of punishment.

Some other superstitions about knives and forks are:

Knife falls, gentleman calls;
Fork falls, lady calls;
Spoon falls, baby calls.

Dropped knife means good luck.

When the knife falls and the blade sticks in the ground, company is coming; if it doesn't stick, you'll probably have a fight.

Beware when giving a knife (or a pin) as a gift. (*See* PINS) This can be very unlucky. Some say that giving a knife means love will end; others say if you give a penny in return for the gift, then say:

"If you love me, as I love you
No knife can cut our love in two."

—the relationship will be safe.

KNOCK ON WOOD

They'd knock on a tree and would timidly say
To the Spirit who might be within there that day;
'Fairy fair, fairy fair, wish thou me well;
'Gainst evil witcheries weave me a spell!'
(Nora Archibald Smith, 1900)

I rarely like to be any considerable distance
from a piece of wood.
(Winston Churchill)

I always knock on wood before I make my entrance.
(Will Rogers)

The concept of knocking on wood or touching wood is almost as ancient an idea as man himself. It is supposed to ward off punishment for bragging ("He who talks much of happiness," goes an old proverb, "summons grief"). Long ago people believed that if you pointed out your good fortune to the evil spirits, they would be jealous and take it away; and so they knocked on wood.

In the days of the druids, it was believed that good, helpful gods lived in trees. People would touch the bark and ask a favor. When the request was granted, they would come back and knock again on the bark to say thank you.

Knocking three times has the additional magic or frightening away the evil spirits by the noise, preventing them from hearing of the good fortune. (*See* THREE)

KNOTS

To catch an evil spirit, and stop it from causing trouble, make a knot. Evil spirits are known to get caught at the spot where a knot is tied. The best example of this widespread belief is the clerical collar. It was feared that evil spirits would be caught in the priest's tie if it were knotted and hide there and cause trouble during religious ceremonies.

Some people tie knots on their kitchen aprons for protection, and some tie pieces of red string around a wart as a cure.

In medieval times if a knot was tied during a wedding it meant that the couple would never have children. This was considered such strong magic that the only way to break the spell was to untie the knot. Today, in many parts of the world, all knots in a house are untied during childbirth to make the delivery easier and safer.

The knot remains of paramount importance in wedding ceremonies. It symbolizes love and duty and an indestructible relationship. It is often represented by knots in the bridal bouquet. (*See* WEDDING CUSTOMS) In India, all knots in the clothing of the bridal party are loosened before a marriage, and retied at the wedding ceremony as a symbol of eternal unity.

Knotted fringe has a special place in religious ceremonies. The fringe keeps evil away by entangling the spirits, and when the fringe is knotted, it adds an extra layer of protection by confusing the spirits and then trapping them in the knots. Orthodox Jews add an extra bit of protection from any spirits that might get trapped by knotting the fringe in such a way as to spell out one of the sacred, unspoken names of God.

LADDERS

See WALKING UNDER LADDERS

LADYBUG, LADYBUG

Ladybug, ladybug, to your home you must turn,
Your house is on fire and your children may burn.
(Early English nursery rhyme)

Never, never kill a ladybug; it will inevitably bring bad luck.

If a ladybug lands on you when you are ill, she will take the illness away with her.

The ladybug, a member of the beetle family, came to earth via lightning, according to an ancient Norse belief. It was closely associated with the goddess of love and beauty.

There appears to be an uncanny empathy between children and ladybugs. Children tend to talk to them and expect an answer.

LAST SUPPER, THE

See THIRTEEN AT TABLE

LEFT-HANDED

See also RIGHT SIDE OF THE BED

Left-handed people have always been discriminated against, since a left-handed person was thought to be a messenger of the Devil. It's common knowledge that the Devil lives on the left side of the body. In most modern languages terms like left-handed mean indirection, insincerity, and even treachery.

Although about seven percent of the population is born left-handed, as children many are taught to be right-handed. It is believed that if left-handed people are made to use their right hands, they will stutter.

Left-handed people are supposed to be clumsy. The Book of Judges, however, says: "There were seven hundred chosen men left-handed; everyone could sling stones at a hair breadth, and not miss." Leonardo da Vinci painted the *Mona Lisa* with his left hand.

The Moslems have always been very concerned about left-handed people. They say that each Moslem has two guardian angels on earth; the one who lives on the right side takes note of all his or her good deeds; while the one on the left notes all the bad things. They also believe that at the Creation, God threw one fistful of dust to the right, creating people who would be happy, and one fistful of dust to the left, creating people who would be unhappy. The Moslems reserve the use of the left hand for all unclean acts: for example, petting dogs, which are unclean to Moslems.

In the Middle Ages, being left-handed was a sure sign of witchcraft and a reason to be burned at the stake.

LEO

See also ZODIAC

Leo, the Lion, is the astrological sign for those born between July 23 and August 22. Ruled by the sun, Leo is also a fire sign, which gives people born under it enthusiasm. There is an Egyptian legend that tells of Leo as the symbol for the heat of summer, the time when the Nile was said to overflow and lions to appear.

Leos tend to be generous, gracious, gregarious, and courageous. They like to do things their own way. Sometimes they are egotists, and they may marry for money instead of love.

Leos have active imaginations and can easily please an audience. They give orders easily and are impulsive leaders who try to rule. They are most effective in a crisis.

Sunday is a good day for Leos; five and nine are good numbers for them; and golden tones are their best colors. The birthstone for Leo is the sardonyx.

Some well-known Leos are Mae West, Alex Haley, Jacqueline Kennedy Onassis, George Meany, Benito Mussolini, Cecil B. De Mille, and George Bernard Shaw.

♌ *Leo the Lion*

LEPRECHAUNS

See SMALL FOLKS

LET GEORGE DO IT

This peculiar expression was first credited to King Louis XII of France who, in a sarcastic moment, said it about his minister Georges d'Amboise (later Archbishop of Rouen) who was a true Renaissance man and excelled in many areas. In fact, Louis was hard put to find something that George couldn't do. An expanded version, "Let George do it; he's the man of the Age," was quite popular among the people of France in the late fifteenth century.

The expression popped up in America when the cartoonist George McManus called his comic strip "Let George Do It."

There's also a saying that no man named George has ever been hanged (the accuracy of which I could not guarantee).

LIBRA

See also ZODIAC

Libra is the sign of the balance, or justice. It implies a strong love of justice and an eternal search for balance. This is the sign for those born between September 23 and October 22.

Libras are ruled by Venus, which makes them unusually artistic and creative, as well as sensitive. Their overriding concern in life is beauty and art. They love beauty in all forms. They are independent, intolerant, and idealistic; they often complain. They also search for justice and follow intellectual pursuits. Libras are under the influence of the air sign and are changeable, though clearheaded and quick to act once they make a decision. They are easily disturbed emotionally, and they need incentives to succeed.

Libra, the scales

Fridays are good days for Libras, and the numbers six and nine are usually lucky for them. Blue is their color. Libras are beautiful people, often charming and amusing, and always survivors. Their birthstone is the opal. They often have back problems.

Some famous Libras include David Ben-Gurion, Truman Capote, Carole Lombard, Dwight Eisenhower, Sarah Bernhardt, and Barbara Walters.

LIGHTING THREE CIGARETTES ON A MATCH

See THREE ON A MATCH

LIGHTNING

What did you hear? that God was bowling and the thunder was the sound of the pins falling and the lightning was a strike? did you hear the one about God being angry? If you heard the second explanation, that's the one that ancient civilizations believed. They were terrified by lightning and would fall on their faces in the dirt, believing that evil spirits were in the air all around them. They thought that lightning must be either a god (Thor, Jupiter, Zeus) or that it came from a god. When fires were started by lightning, they thought it a punishment from Heaven and wouldn't allow it to be extinguished.

There is a Hindu saying that goes, "Lightning strikes the loftiest tree first;" that makes sense. But the saying, "lightning never strikes the same place twice," isn't even logical. It's true that the law of averages is against it striking twice, but since lightning strikes an exposed area or structure, it can and has hit the same spot several times.

Some superstitions about lightning are that: A dog's tail draws lightning to it—don't believe it; acorns are supposed to be good protection against lightning when placed near windows (*See* ACORN); holly used at Christmastime is good protection (*See* HOLLY); and lightning turns milk sour unless a rusty nail has been added.

Weather watchers say:

Forked lightning at night
The next day clear and bright.

An Old English rhyme goes:

Beware of an oak:
It draws the strokes;
Avoid an ash:
It counts the flash;
Creep under the thorn:
It can save you from harm.

LILY OF THE VALLEY

Were lilies of the valley created by Eve's tears as she was expelled from the Garden of Eden; or are they, as the Irish say, fairy ladders that the small folks (*See* SMALL FOLKS) run up and down on, ringing the bells?

One legend says that Saint Leonard, a young Christian, went out to fight a dragon (which was really evil in disguise). Saint Leonard won, but he was wounded. As his blood fell to the ground the fragile little lilies grew to commemorate the battle Saint Leonard won for Christ (and the forces of good).

The lily of the valley stands for purity and humility. In early Christian times, it was dedicated to the Virgin Mary. In fact, in Great Britain and in France it is also known as "Our Lady's Tears."

The flower is believed to be able to cheer up a sad person; to help with a weak memory; and to stop the pain of gout.

Today the flower is still popular in bridal bouquets and actually is used in some medical treatments for heart disease.

It is the flower for those born in the month of May, and means the return of happiness.

LIPSTICK

See CIRCLES

LITTLE BIRD TOLD ME, A

See BIRDS

LOVE APPLE

See TOMATO

LOVE AT FIRST SIGHT

She lovede right fro the firste sighte.
(Chaucer, *Troilus and Criseyde* II, 1375)

Loving comes by looking.
(Latin proverb)

Whoever loved that loved not at first sight?
(Marlowe & Chapman, *Hero and Learning* I, 1598)

To the Greeks, it was common knowledge that each of us was part of someone else; we had been divided at birth, and the search for a mate was a search for the other half. When the two natural halves met they would know each other. This would be love at first sight.

The little purple flower known as love-in-idleness is said to have been pierced by Cupid's arrow and then reclaimed by the fairies. It is the flower most closely associated with love at first sight; as Shakespeare observed in *A Midsummer Night's Dream* (II:1):

Yet mark'd I where the bolt of Cupid fell:
It fell upon a little Western flower,
Before milk-white, now purple with love's wound,
And maidens call it love-in-idleness.
Fetch me that flower; The herb I shew'd thee once:
The juice of it on sleeping eyelids laid
Will make or man or woman wildly dote
Upon the next live creature that it sees.

LOVE-IN-IDLENESS

See LOVE AT FIRST SIGHT

LUCK

See also THEATRICAL FEARS

All the world of superstition is based upon luck—good luck and its sinister twin, bad luck.

"Luck is a mighty queer thing," wrote Bret Harte in *The Luck of Roaring Camp*. "All you know about it for certain is that it's bound to change." Luck is something that's always around. It can't be willed, bought, or bribed, although the ancients spent most of their waking hours trying to do just that. When they spoke of evil spirits, they meant bad luck. It was their way of explaining why things went wrong.

A senior Rothschild once advised:

> *Never have anything to do with an unlucky place or an unlucky man. I have seen many clever men, very clever men, who had not shoes on their feet. I never act with them. Their advice sounds very well, but they cannot get on themselves; and if they cannot do good to themselves, how can they do good to me?*

A momentary spate of bad luck is often cured by:

- holding your right hand to your face and spitting three times through your forefinger and middle finger
- making the fig sign (*See* FIG SIGN)
- pulling out your pockets and turning around in a circle (clockwise) three times (*See* CIRCLES)

Beginner's luck is usually good luck because there is some good magic in anything new.

A lucky man is rarer than a white cow.
(Juvenal, *Satires* VII)

It is a very bad thing to become accustomed to good luck.
(Publilius Syrus, first century B.C.)

LUCKY AT CARDS, UNLUCKY IN LOVE

This proverb simply means you can't have everything. It probably dates to the Italians, who often shrugged about bad luck. They would say: He who is lucky in love should never play cards.

LUCKY BREAK

See also WISHBONES

Lucky break is a poolroom term referring to one or more balls going into pockets during the player's first turn. Circus people use the reverse; they say, "it broke bad," if the weather is bad, causing business to be off and bringing bad luck.

All this goes back to primitive times when a tribal member would break a stick in the middle to make a noise that would frighten away evil spirits. If things went well it was a lucky break.

A vestige of that is snapping your fingers at the mention of trouble to scare away bad thoughts or evil spirits.

MAKE A WISH

See BANANA
MOON
ROBIN REDBREAST
STARS
WISH ON A STAR

MANDRAKE ROOT

The reputed powers of the mandrake root came from the magical shape of the root when it is pulled out of the earth. This is one of the few natural objects that has the shape of the human body. It looks, in fact, as though it has been carved to order. The root so strongly resembles a man (or woman) with his legs spread that it has had sexual connotations since early Greek times. The Emperor Julian is said to have drunk a solution of mandrake root soaked in wine each night as an aphrodisiac before going to bed. Today, in some sections of Greece, young men carry a piece of the root in their pockets as a love charm. John Donne, referring to the root's attributes as a fertility drug, (in *Go and Catch a Falling Star*) wrote, "Get with child a mandrake root." (*See* APHRODISIACS)

Given the shape of the root, it isn't surprising that the ancients, who believed strongly in the concept of like makes like, believed that the root had the power to make a woman pregnant. It is also not surprising that, because of its shape, it would take on frightening and mystical powers. The Greeks believed in its powers for relieving pain. It was supposed to scare away the Devil. During the Middle Ages it was used as an anesthetic. In Arabia it was called the devil's candle because it was supposed to shine in the dark.

These roots were believed to grow under the gallows of murderers, nourished by the body's drippings. When these roots were torn up from the ground, they were said to utter piercing shrieks that brought death to those who heard them.

The mandrake root is mentioned in the Bible; Thomas Newton (18th century) in his *Herball to the Bible*, wrote, "It is supposed to be a creature having life, engendered under the earth, of the seed of some dead person put to death for murder." With this belief rampant, it is no wonder the root, which is rare and difficult to unearth, would have such magical qualities attributed to it.

MAN IN THE MOON

Shakespeare (in *A Midsummer Night's Dream*) called it "This man with lantern, dog and bush of thorns, presenteth moonshine." Some people saw the moon's face as that of a peasant (usually with a bundle of twigs on his back, and sometimes with a dog). It was also supposed to be Judas, sent to the moon as a punishment for betraying Christ.

In Panama, the man in the moon was sent there as punishment for incest.

MARIGOLDS

Pick a marigold you didn't plant
And chances are you'll take to drink.
(Old English saying)

It is said that the marigold took its name from the fact that the Virgin Mary wore it on her breast. It is believed also that the flower sprang up from the blood of those who died during Cortes's conquest of Mexico.

In Western Europe and the British Isles the marigold has some unfortunate connotations. The Germans, for instance, say it is unfavorable to love. It usually represents grief, pain, or anger in other countries.

The marigold is intimately linked with the sun. It blooms all year, following the pattern of the sun, as it opens its petals early in the morning and closes them again by midafternoon.

The romantic Victorians modified the unhappy connotations of the flower somewhat by joining marigolds with roses and giving this meaning: "sweet sorrows of love." Mixed flowers with marigolds mean "changing tides of life from good to ill."

In Eastern countries the marigold, in conjunction with poppies, says, "I will soothe your grief."

MARRIAGE

See WEDDING CUSTOMS

MASCOTS

See also TALISMAN

The word mascot probably comes from a variety of roots associated with sorcery and witchcraft. The most popular modern mascots are those we see on ball fields: real or imitation animals that represent a ball team. It's an ancient and popular belief that a chosen mascot can bring good luck and keep the Devil away.

A mascot can be a person (a batboy), an object, or an animal. The animal mascots of today can be traced to ancient times, when there were animal totem poles endowed with godlike powers. These were believed to have a supernatural sense of good and evil and to be effective protection for man.

Zeus had an eagle as his mascot, which became the emblem of the Roman legion and later the symbol of Napoleon's army. Assyrian kings had trained lions and leopards, which accompanied them into battle, and cats rode with Egyptian warriors as their mascots. There is the Russian bear, the Chinese dragon, and of course, Uncle Sam. (*See* UNCLE SAM)

MATCHES

See THREE ON A MATCH

MAY DAY

You must wake and call me early, call me
Early, mother dear;
Tomorrow 'ill be the happiness time of
all the glad New Year—
Of all the glad New Year, mother the
maddest, merriest day;
For I'm to be Queen o' the May, mother
I'm to be Queen o' the May.
(Tennyson, *The May Queen*)

May Day celebrates new crops and flowers. It marks the time when cows are able to start feeding on fresh grass again. It's traditionally the day when dairymen are able to start making cheese and everything is in abundance. The Romans celebrated the day with parades followed by prayers of thanks to their gods.

The druids, who celebrated May 1 as the beginning of the new year, considered it their second most important holiday. Their most important tradition was to start a vast bonfire, as a symbol of the spring sun. Then they had the cattle walk through the fire to be purified. Lovers walked in the smoke of the fire for luck.

The maypole comes from the Romans, who cut a pine tree, stripped it of its branches, and then wrapped it in violets. During the Middle Ages, people incorporated the druid's custom of dancing around the bonfire into dancing around the maypole in honor of the spring crops. When the Puritans took over England in 1649 they ended the maypole traditions, calling them devilish instruments and forbidding the poles to be raised at all. After the Puritan reign, the maypole and the dancing returned.

Some popular superstitions about May Day are:

- in England, the belief that to wash your face with dew from the grass on May Day will make you beautiful
- choosing the Queen of the May
- "bringing in the May," a tradition of going into the forest and bringing back flowers and branches to decorate the house

During the twentieth century, May Day has taken on yet another connotation. It is strongly associated with the communist and socialist philosophies. That concept started in 1889, when the French Socialists declared the day to be devoted to the workers. By the 1920s Russia had appropriated the day to celebrate communism.

MAYPOLES

See MAY DAY

MEZUZAH

See TALISMAN

MILK

See SPILLED MILK

MINT

This plant is associated with the legend of a young nymph named Minthe, who allowed Pluto, the god of the Underworld, to make love to her. Proserpine, Pluto's wife, became so jealous that she changed the nymph into a plant. That is why, it is said, mint loves damp ground and grows in profusion near streams.

Mint is believed to be good for stomach disorders and is used in baths to calm nerves.

MIRROR

Mirror, mirror tell me
Am I pretty or plain?
Or am I downright ugly
And ugly to remain?
Shall I marry a gentleman?
Shall I marry a clown?
Or shall I marry old knives and scissors
Shouting through this town?
(Old English saying)

All mirrors are magical mirrors;
Never can we see our faces in them.
(Logan Pearsall Smith, *Afterthoughts*)

The Devil's behind the glass.
(J.C. Wall, *Devils*)

Woe unto thee who breaks a mirror: Seven years' bad luck without fail will follow. This belief is an old and important one in the history of superstition.

Long before there were mirrors, shiny surfaces were used to reflect images. These were considered tools of the gods. When ancient man saw his reflection in a lake or pond he thought he was seeing his soul or his other self. When that image was disturbed or broken he thought it was some kind of injury to himself. The Romans, at the beginning of the first century, added the time of seven years, the amount of time they believed it took for life to renew itself. (*See* SEVEN)

To break the cycle of seven years' bad luck, bury the pieces of the broken mirror.

Another popular superstition about broken mirrors is that someone will die in the family within a year. This idea may have come about from the basic belief that the mirror was a gift of the gods. A broken mirror prevented you from seeing the image of death. This was a way the gods had of trying to protect you from knowing that something bad was going to happen.

This belief has spilled over into several others, including:

Don't let a baby look into a mirror for a year or it will die.
Cover all mirrors in a house where someone has died so that his soul won't get caught in the mirror and be detained. It has also been said that the ghost of the deceased would gather the souls of all those who were reflected in a mirror and take them with it.
When a mirror falls off the wall of a house, it means death.
Witches and vampires are not reflected in mirrors because they have no souls.
A mirror which is framed on only three sides has been used by a witch to see over long distances. (I'd get rid of that one, if I were you.)

MISTLETOE

Kissing under the mistletoe at holiday time is one of the most pleasant customs we have retained from ancient times. The most pleasing explanation of this custom comes from Norse mythology. It tells of the goddess Frigga, whose tears turned into pearls on the mistletoe when her son, Baldur, came back to life. In gratitude for this miracle, Frigga put mistletoe under her protection, preventing it from ever being used for evil purposes. Since Frigga was the goddess of love and marriage, a kiss under the mistletoe symbolized her protection over the love of the two people kissing.

Mistletoe was sacred to the ancient druids, who worshipped it because it grew near the oak tree. (*See* OAK TREE) They believed that it reached the oak through a stroke of lightning from heaven. They used mistletoe in ceremonies, but only if it had been harvested with a golden sickle. They thought that the mistletoe would lose its holy powers if iron touched it. Because the druids worshipped mistletoe, it became bad luck for Christians to use it in churches.

There is another old tale about mistletoe that tells of priests harvesting it, never letting it touch the ground. It thereafter was hung over doors and arches as a sign of welcome to priests and as protection against witches.

To the Victorians mistletoe meant "surmounting all obstacles." That meaning came from another Norse legend in which Baldur received invulnerability from Frigga, except she neglected to protect him against mistletoe. An enemy gave the blind god, Hodur, a piece of mistletoe and told him to throw it at Baldur. He did, and Baldur died. Thus a blind man was able to surmount all obstacles and kill an invulnerable man with mistletoe.

MOLES

See BIRTHMARKS
CHARTS, Language of the Mole

MONEY

Money makes everything legitimate, even bastards.
(Hebrew proverb)

Coins date at least to the tenth century B.C. Carrying a lucky coin dates back to the time of Croesus ("as rich as Croesus").

We have these beliefs about money:

Finding a penny means that more will follow.
A jar of pennies in the kitchen brings good luck.
If a bridegroom gives his bride a coin and she wears it in her shoe at the wedding, it will bring a happy marriage.
Turn a piece of silver in your pocket on seeing a new moon and your wish will come true.
Put the first money you receive each day into an empty pocket; it will attract more coins (in English marketplaces this custom is still very popular; that coin is called handsel). This is an example of the popular theory of like makes like. (This is the same superstition as framing the first paper money received by a new store or restaurant.)
Another example of like makes like is the old saying that seeing someone with polka-dot clothing will bring you money.

Place a coin in a new coat, pocketbook, or wallet for good luck.

Coins with holes in them are thought to be especially lucky. This comes from the ancient belief that shells or stones (once used as barter) with holes were worn by the gods of the sea and were especially helpful in keeping evil spirits away and especially in preventing drowning.

Finally, there's the superstition about beggars and money. In the old days beggars used to stand outside the churches after services and ask for money from the worshippers. If someone didn't give, he was cursed by the beggar. It is believed, even today, that beggars have the power to curse people.

MOON

See also MAN IN THE MOON

If you think I'm going to tell you the moon isn't made of green cheese, have no fear. That theory is too old to tackle; it goes back to the beginnings of language. The expression comes from the meaning of green as new or inexperienced. A new or green cheese resembles the moon in shape and color. In *Proverbs* II John Heywood wrote: "Thinke that the moone is made of green cheese . . . is a dolt and a fool." (1500) Erasmus, in *Adagia*, said the same thing: "He made his friends believe the moon is made of green cheese."

There are hundreds of moon superstitions, probably because the moon, like the sun, comes and goes, and marks the passage of time. It was of great importance to the ancients, and it provided easy and convenient ways for them to explain certain phenomena. These are some which have been passed down to us:

It is unlucky to see a new moon through closed windows or through the branches of a tree.

It is unlucky to see the new moon over your left shoulder but lucky to see it over the right shoulder; luckiest of all—to see it straight ahead.

A robbery committed on the third day of the new moon will fail.

If you become ill on the eighth day of the new moon, you'll die.

If the rays of the full moon fall on your face while you're sleeping, you'll go crazy.

The best marriages begin on the full moon, or a few days before.

A child born at the full moon will be strong.

A ring around the moon means rain or snow.

A crescent moon when increasing means good luck for travelers and lovers (an effective amulet for travelers is a crescent-shaped pendant).

It is a generally accepted fact that a full moon affects people in peculiar ways. There tends to be more fighting, more crime, more murders, and more fires during a full moon. In New York City in 1981 the police chief blamed a rash of bomb threats on a full moon.

The phrase dark side of the moon refers to the time from the full moon to the new moon. It's also known as the waning moon and has been particularly harmful to people throughout history.

The light side of the moon, traditionally a better time for people, is the period from the new moon to the full moon. During this time the moon is waxing, or increasing.

There is moon madness; you can be moonstruck or be mooning. The word lunatic derives from moonstruck and comes from the Middle Ages, when it was believed that the moon blanched the brains and particularly affected mentally ill people.

If you want to wish on the moon you can try these phrases:

I see the moon, the moon sees me
The moon sees somebody I want to see
(Presumably you'll name someone
and then you'll get to see that person soon.)

New moon, true moon,
Star in the stream;
Pray tell my fortune
In my dream.

MOTHER-IN-LAW

Mothers-in-law did not originate with stand-up comedians. They were a serious problem even in early times. The Zulu say, "man should not look upon the breast that has nursed his wife." In old civilizations, mothers-in-law were completely shunned. Some American Indians have been particularly cruel to their mothers-in-law, since they believed that a man would go blind if he looked into the eyes of his wife's mother. One reason for this attitude was the primitive, deep-rooted fear of incest.

Happy is she who married the son of a dead mother.
(James Kelly, *Complete Collection of Scottish Proverbs*, 1721)

I know a mother-in-law who sleeps in her spectacles, the better to see her son-in-law suffer in her dreams. (Ernest Coquelin, 1900)

MOTHER'S DAY

Near the turn of the century, Miss Anna Jarvis, who had dedicated her life to taking care of her mother, decided to launch a campaign to celebrate Mother's Day throughout the United States. The first Mother's Day was held in West Virginia (Miss Jarvis's home state) in 1908, and it was celebrated with a religious service. Miss Jarvis brought carnations to the service to make it more festive because they were her mother's favorite flower. Since it's political suicide even to hint that Mother is not the most important thing in the world, Mother's Day became an official, signed and sealed holiday throughout the nation in 1913 (the second Sunday in May was the designated day).

As early as the 1600s, however, England had been observing Mothering Sunday on the fourth Sunday of Lent. It was the day that all indentured servants and apprentices were given the day off to go home and visit their mothers. It was the custom in those days to bring along a gift.

Around the time that the United States began to celebrate Mother's Day, England took up the American version of the holiday, and changed the day to the second Sunday in May.

MOTHS

White moths are particularly important to the superstitious. For them, white moths (which only fly at night) are the souls of the dead. If the white moth flies around your head, it is the soul of a dead friend saying hello.

The black moth, as befits its color, foretells the death of someone in the house during the year. Perhaps this belief led to the invention of the camphor ball.

MOURNING VEILS

See also DEATH CUSTOMS

A widow is supposed to wear a veil during mourning to hide from death and to prevent others from catching death from her. Since the widow was the person closest to the deceased, she probably has some death vibrations around her, so she'd better conceal them. Death, you know, is contagious.

There is an old idea that close relatives must not wear jewelry or appear in public places for weeks after a death in case the evil spirits that had taken the deceased were still around. Today we observe some part of this tradition out of respect for the deceased.

MOUSE

Roasted mice help cure: measles; colds; sore throats; fever; and mixed with honey, make an excellent mouthwash, according to Pliny the Elder (77 A.D.).

The mouse is a bad luck omen in most places. If it eats your clothing, that is a sure sign of death.

It is believed that the Devil created the mouse in the Ark; or, that they fell to earth from the clouds during a storm. In Germany, witches made mice. In most places, mice make trouble.

Consider the little mouse, how sagacious an animal it is which never entrusts its life to one hole only.
(Plautus, *Truculentus*, IV:1, 190 B.C.)

When a building is about to fall, all the mice desert it.
(Pliny the Elder, *Natural History VIII*)

It was discovered in modern times that powdered mice made a respectable cure for bed-wetting. (I'll bet!)

MOVING INTO A NEW HOME

Three removes are as bad as a fire.
(Benjamin Franklin, *Poor Richard's Almanack*, 1736)

A good luck charm is essential when moving into a new home. The custom is from the early wandering tribes who had a prevailing horror of the unknown and relied heavily upon charms and talismans to help them. Giving a housewarming gift is a remnant of this fear: a new broom, a loaf of bread, a used box of salt, and water must be placed in a new home before the residents can move in.

Other superstitions are:

Never move downstairs in the same building.
Moving on a Saturday means a short stay.
It's bad luck to move on a Friday. (*See* FRIDAY)
Moving on a rainy day means great unhappiness.
It's unlucky to enter a new house through the back door.
Move while the moon is on the increase, for good luck.
Always use something old in building your new house (used bricks or lumber from the old house).

MURDER

Other sins only speak; murder shrieks out!
(Webster, *Duchess of Malfi*, IV:11, 1623)

It is believed that the murderer's image remains on the victim's eyes. This superstition, which has haunted many a murderer through the decades and has been the basis of several novels, comes from the ancient belief that there is a permanent image of the last thing a person sees before dying on the eyes of the corpse (and a dead man always talks.)

MYRTLE

For the Jews in ancient times, myrtle was a symbol of the eyes, and so it became a symbol of atonement for lust (which shone from the eyes). To the Greeks, the myrtle tree was a symbol of love and was dedicated to Aphrodite. It was said that in Rome myrtle groves surrounded the temple of Venus and that after the rape of the Sabine women the Roman soldiers crowned themselves with myrtle in honor of the conquering of Venus.

In England myrtle is considered lucky. In Wales myrtle is planted on each side of a home to insure love and to keep the atmosphere peaceful.

In early Germany brides wore it to prevent pregnancies. (As a birth-control measure it didn't work nearly as well as the Crusades, which kept the men away from home for years.)

NAMES

See also CHARTS, Names

"What's in a name?" you ask. It depends upon whom you ask. At one time in history a person's name was as much a part of that person as eyes or soul. It was considered very bad taste to mention the name of a dead person because you might disturb the ghost. If you did mention the name, you always added, "may he [or she] rest in peace." Some people still say it.

The first Book of Samuel (written about 500 B.C.) says: "As his name is, so is he." Because of that philosophy, people have changed their names to bring themselves better luck and to confuse the evil spirits.

Having seven letters in your name is very lucky (either given name or family name) and people with thirteen-letter names should probably add one letter to bring them better luck.

"Change the name but not the letter, is change for worse instead of better!" (Chambers, *Book of Days*)

It was once believed that naming a child after a living person brought death to that person. Now it's believed that it will bring long life to the child. In some families it is believed that naming another child after a deceased one is dangerous, because the dead child will call the living one to heaven.

Then there's the Bible's instruction (in the Book of Exodus, 20:7): "Thou shalt not take the name of the Lord in vain;" so you see, this name thing can be a very serious business. People believed so strongly in certain ancient gods with several names that when protection was needed they said all these various names. The primary name of the god was saved for very dangerous situations. Otherwise, you would be using the name of the Lord in vain.

NARCISSUS

Narcissus, the flower of people born in December, was named after the boy who refused to love anyone, although many women easily fell in love with him. One version of the legend says that one day Narcissus saw his own reflection in a pond and was so enraptured by his own perfection that he fell into the water in an attempt to touch it. When his body was washed upon the shore, it had been changed into the flower we know as narcissus. The Furies wore these flowers as a crown to indicate their own egotism, the most fatal of all vices.

The word comes from the Greek word for numbness. The Greeks believed that the narcissus gave off a dangerous scent that caused headaches, madness, and sometimes death.

The Victorians failed to dim the unpleasant association of the flower. They acknowledged its meaning of egotism and self-esteem.

As a medicine, the narcissus root has been used in antiseptics and healing wounds. Mixed with honey it is used as a general painkiller.

NEW YEAR'S DAY

There are many customs associated with New Year's Day and its companion holiday, New Year's Eve. The New Year's resolution is the most ubiquitous of New Year's traditions. Wiping the slate clean, paying all old debts, returning all borrowed items, letting the past be forgotten is a very ancient idea.

New Year's superstitions are numerous. Starting with the evening before, here are the things you must do to insure a good new year:

Open the window a few minutes before midnight to let the bad luck out and the good luck in.

All debts and arguments must be settled before midnight.
It's bad luck to let a fire (in the fireplace) go out.
At midnight it's essential to use noisemakers to chase the evil spirits who have gathered in great numbers to celebrate New Year's.
Ringing out the old is an old tradition of ringing the church bells to let everyone know the new year has arrived.
"Auld Lang Syne," the song we all sing, is Scottish for "old long since," or "long ago."

New Year's Day has come superstitions all its own:

If you have no money in your pockets on New Year's Day, you'll be poor all year long.
Nothing should be removed from your house on New Year's Day; it's unlucky.
If you give a gift on New Year's Day, you'll give your good luck away.
Whatever activity you do on New Year's Day you'll do often during the coming year.
The person who drinks the last liquid from a bottle will have good luck.
Babies born on New Year's Day will have lucky lives.
Get somebody to kiss you. If you do, you'll be kissed often during the year.

NIGHT AIR IS BAD FOR YOU

Never greet a stranger in the night, for he may be a demon.
(Talmud, *Sanhedrin*)

Now the night comes—and it is wise to obey the night.
(Homer, *Iliad*, VII)

The night is no man's friend.
(*Old German proverb*)

The night has been associated with evil and death throughout history. Night air is believed to be poisonous and to carry infection. The night is believed to produce evil air.

The Romans believed that the air from the fields surrounding Rome was the cause of malaria. Actually it was the mosquitoes that festered in the ground outside of Rome that caused the disease. Even after the real cause of the disease was discovered and acknowledged, people still refused to go out at night, believing that mosquitoes bit only at night. Many still believed that the night air itself had caused the pestilence.

In Medieval times night was always a dangerous time for Jews. Evil spirits were everywhere. Tuesday and Friday were somehow more dangerous for them than other nights, and so they said special prayers for protection as they returned home from prayers on those evenings.

NIGHTMARES

See also DREAMS

Literally a nightmare was a spirit. The Anglo-Saxon word mare meant an evil spirit or monster that came during the night and sat on the chests of sleeping people. It stopped their breathing, creating bad dreams and an oppressive feeling. In the Middle Ages they were called night-hags. Then Freud got hold of them, and nightmares were never the same again.

NINE

See also NUMBERS

The number nine is frequently associated with magical things because, $3 \times 3 = 9$ and if you multiply it by any number, the answer will always add up to be nine or one of its multiples. For example, $9 \times 3 = 27$—$2 + 7 = 9$.

Nine often appears in ancient amulets. People wore nine stones in a necklace and on breastplates, and they tied nine knots (*see* KNOTS) in their prayer shawls.

Doing something nine times (like knocking on wood, or spitting, or turning in circles) is considered to have very powerful magic for bringing good luck.

Odd numbers in general are very lucky. Leases are often written for an odd number of years, like ninety-nine, for example.

NINE-DAY WONDER

Chaucer referred to a nine-day wonder as early as the fourteenth century, and a popular sixteenth-century proverb goes, "The wonder (as wonders last) lasted nine days." Another saying was: "Wonder lasts nine days and then the puppy's eyes are open."

Someone who causes a great sensation for several days, then goes back into oblivion is a nine-day wonder.

You'd think it strange if I should marry her.
The would-be ten days wonder, at the least
That's a day longer than a wonder lasts.
(Shakespeare, *Henry VI*, III:2)

NOT WORTH A TINKER'S DAMN (DAM)

This old adage seems to have two roots, both meaning that the object or idea referred to isn't worth anything at all.

The first comes from the well-known fact that tinkers (men who fixed things) had notoriously profane language and so when they said damn, it didn't mean much.

The second makes more sense. It refers to a tinker's dam, which was a piece of bread used to help catch solder and to prevent it from running through the holes in the pans being mended. Afterward the bread would be thrown away—since it wasn't worth a damn.

NUMBERS

See also: FIVE
NINE
NUMEROLOGY
SEVEN
THIRTEEN
THREE

There is divinity in
odd numbers, either in nativity,
chance or death.
Shakespeare, *The Merry*
Wives of Windsor, V)

The gods delight in odd numbers
(Virgil, *Eclogues*, VIII)

Throughout history numbers have been assigned special powers. The ancients attributed each number with a life of its own:

One meant reason; God; unity; and the sun.

Two stood for divisibility; opinion; sociability; and the moon.

Three (*See* THREE) had many meanings; perhaps most important, all religious symbols are in threes.

Four was imagined as square; the foundations of all things, as the four seasons and the four points of the compass.

Five (*See* FIVE) stood for fire; love; and marriage.

Six was a perfect number, since it equals the sum of 1+2+3; it represents creation.

Seven (*See* SEVEN) was a magical number, since the world was created in seven days.

NUMEROLOGY

This is a quasi-science, based on the rhythms of a person's birthday and the study of a person's name. According to Pythagoras, there were nine basic numbers and everything else was repetition. All compound numbers could easily be reduced to a single digit by adding up the values. For example, 502 becomes 5+0+2=7. All even numbers were female, all odd numbers were male, and the world was ruled by numbers.

The common conversion chart from letters to numbers looks like this:

1	2	3	4	5	6	7	8	9
A	B	C	D	E	F	G	H	I
J	K	L	M	N	O	P	Q	R
S	T	U	V	W	X	Y	Z	

To find the number for your view of life (the soul urge), add the vowels of your name.

To find the number that is the key to your daydreams, (the quiescent self), add the consonants in your name.

To find what you must do with your life, or fate (the life path), add the date of birth.

Here's an illustration:

C A R O L E = 3 1 9 6 3 5 = 27; 2+7=9
My soul urge=1+6+5=12; 1+2=3
My quiescent self=3+9+3=15; 1+5=6
My life path=1+1+4=6

NUTMEG

If you carry a nutmeg in your pocket
you'll be married to an old man.
(Jonathan Swift, *Dialogues*, 1738)

There's a notion from Michigan that carrying a nutmeg in your back pocket will help rheumatism.

You can also use nutmeg to remove freckles; to improve your eyesight; and around the neck, to prevent boils, sties, and cold sores. But frankly, there's no hard and fast evidence for any of these ever working.

OAK TREE

See also ACORN
LIGHTNING

The oak, struck by lightning, sprouts anew.
(Ovid, *Tristia*, IV:10, 10 A.D.)

It is an ancient belief that lightning strikes the oak tree more often than any other object. People have planted oaks near their homes for centuries, to act as lightning deflectors. Some people simply keep acorns on their windowsills for protection. (*See* ACORN) A smart person, with a slightly superstitious nature, will not make the roof of his house out of oak planks.

The oak tree was worshipped by the druids, who saw it as a symbol of endurance and strength. They believed the gods lived in oak trees. (*See* KNOCK ON WOOD) The Celts also worshipped the oak as a symbol of their most highly prized virtue, hospitality. The Victorians also regarded the great oak as a sign of hospitality.

The Greeks dedicated the oak to Zeus, since it was an oak that shaded his cradle when he was a baby in Arcadia.

The Romans made crowns of oak leaves, which symbolized bravery and humanity. The oak-leaf crown was their highest award, given for killing the enemy, winning a battle, or saving the life of another Roman. Our military honors continue to reflect this tradition. "With oak clusters" signify that an award is for greater recognition than one without clusters.

OLD SHOES

See also WEDDING CUSTOMS

> *nowe for good luck, caste an olde shoe after mee.*
> (John Heywood, *Proverbs*, I:9, 1546)

The shoe is a symbol of fertility in many cultures. For the Scots and the Irish, the throwing of old shoes at all new ventures insures a fruitful conclusion to the endeavor.

In China a childless mother borrows a shoe from the altar of the mother goddess to make her fertile.

In ancient Palestine, Hebrews sealed the purchase of land when the seller gave the buyer a sandal as a sign of luck and fertility.

Today, at weddings, old shoes are tied to the back of cars to indicate that the rights of the parent over the child have ended, and that the husband is now responsible for the bride's debts.

ONCE IN A BLUE MOON

See BLUE MOON

ONIONS

See also GARLIC

Onions, like garlic, come from bulbs and have encouraged many of the same superstitions. In sickrooms, for instance, they are believed to draw the illness away from the patient; and cut in half and placed under the bed, they remove fever.

One sixteenth-century writer claimed, "The juice of onions anointed upon a pild or bald head in the sun, bringeth the haire again very speedily." Another early writer said onions "inclined one toward dalliance." Alexander the Great fed onions to his troops to increase their passion for war. Indian gurus think the onion induces tranquility.

The Egyptians frequently took oaths with their right hand on an onion, which was a symbol of eternity because of its spherical shape. (*See* CIRCLES)

Some people believe that dreaming of onions means good luck; others, that a sliced onion stops the itching of an insect bite.

The onion is famous as a weather forecaster:

Onion skin very thin
Mild winter coming in;
Onion skin thick and tough,
Coming winter cold and rough.

Sydney Smith offered this eighteenth-century salad dressing:

Let onion atoms lurk within the bowl
And, half suspected, animate the whole.
(Recipe for Salad Dressing)

OPALS

See also CHARTS, Birthstones

October's child is born of woe
And life's vicissitudes must know;
But lay an opal on her breast,
And hope will lull woes to rest.
(Anonymous, *Note and Queries,* 1889)

An emblem of hope and the gemstone of those born in the month of October, the amorphous, color-changing opal was highly valued in Roman times. There's a tale about Mark Antony exiling a Roman senator who refused to give up his hazelnut-sized opal. Opals were set in the crown of the Roman emperors to guard their royal honor, and so Antony was being cautious of the senator's ambitions.

The opal, like many symbols popular in ancient times, was a good luck amulet or an attraction for evil spirits. The Orientals even believed it was alive because of its changing colors. Long before Christ, the opal was among the most prized of gems. Around the fourteenth century, during the time of the Black Plague, it became an evil omen. It was said that an opal turned a brilliant color and then lost its luster when its owner died of the illness.

King Alphonso XII, of Spain, is said to have given his bride a beautiful opal. Shortly after she started wearing the gem, she died. The deaths of the king's sister and sister-in-law followed. The king started wearing the opal himself, and he, too, died shortly afterward.

In the nineteenth century, Sir Walter Scott's novel, *Anne of Geierstein*, which was about a young woman who disappeared after holy water fell on her opal, again gave the gem a bad reputation. Queen Victoria tried to bring the stone back into fashion by frequently wearing opals. With Victoria, however, it was as much an economic gesture as an expression of admiration. Australia was just then opening vast opal mines, but the stones weren't selling because of the bad luck stories attached to them.

The medicinal powers of the opal involve the eyes. They are thought to strengthen sight, cure eye diseases, and even make the wearer invisible. This last belief has turned the opal into the stone of the underworld, for obvious reasons. It is also believed that the opal is like the evil eye, and can invade the wearer's privacy.

The opal is said to turn pale in the presence of poison, to protect its wearer from contagion, and to dispel melancholy and sadness.

A blonde will stay blonde longer by wearing an opal necklace.

OPEN SESAME

This is a magical formula used to open rocks, doors, trees, and mountains. It's found in the story of Ali Baba and many other folktales throughout the world. Open sesame usually refers to the opening of something that offers wonderful treasures within, as in this early poem written for children:

Dear little child, this little book
Is less a primer than a key
To sunder gates where wonders wait,
Your 'Open Sesame!'

(Rupert Hughes, *With A First Reader*)

ORANGE BLOSSOMS

See WEDDING CUSTOMS

OWLS

The owl was the constant companion of Athena, the Greek goddess of wisdom. Because Athens was her city, the owl was sacred in the city, and Athens was almost overrun by them.

The Romans, on the other hand, hated the owl, considering it a bad omen and believing that its hooting meant death.

Today it is believed that the owl can foretell the future and if one hoots or hollers near your house disaster is not too far off. It may even be foretelling a death in the family. So wear your clothes backward, pull out your pockets, throw salt over your left shoulder, and put a knot in your handkerchief.

There was an old owl liv'd in an oak,
The more he heard the less he spoke;
The less he spoke the more he heard;
O, if men were all like that wise bird!
(*Punch*, Vol. LXVIII, 1875)

OYSTERS

See also APHRODISIACS

The oyster is believed to be one of nature's true aphrodisiacs. Along the lines of the ancient belief that like makes like, it was thought that the oyster, which resembles the female reproductive organs, would stimulate sexual interest. Casanova believed this and prescribed oysters as a sexual aid. Lord Byron in *Don Juan*, II, said, "Oysters are amatory food."

"It is unseasonable and unwholesome in all months that have not the letter 'R' in their name to eat an oyster," said Henry Buttes in 1599 (in *Dyet's Dry Dinner*). This idea came from King Edward III, who in 1375 restricted the farming of oysters from May to September for conservation purposes. (During months without an R, oysters lay their eggs.) The idea that oysters are unwholesome in the summer comes from the simple fact that lack of refrigeration doubled the spoilage rate of the oysters and invariably brought food poisoning to the indulger.

Oysters are fascinating creatures. They are female during the spawning season, and then they become male.

PALMISTRY

The general idea of palmistry is to read hands in order to gain knowledge about a person's personality, past history, and likely future. If you're any good at this art, you can probably divine quite a lot of interesting information. Palmistry may be as old as the Stone Age, although the modern practice is based upon concepts from ancient India that reached Europe via wandering gypsies, who have always claimed an uncanny ability to foresee the future. (*See* GYPSIES)

Some notable people who have believed in palmistry were Alexander the Great, Eleanor Roosevelt, Aristotle, Balzac, Mark Twain, and Pope Leo XIII.

Palmistry is based on the shape of the hand, its size, the configuration of the fingers, and the lines and the protuberances of the palm.

PARSLEY

Parsley grows better for a wicked man than a good one.
(Old English superstition)

Because Romans decorated their graves with parsley, we believe that a gift of parsley will bring bad luck, illness, and even death to both the giver and the receiver.

Transplanting parsley is a bad idea. It brings bad luck, particularly when transplanted from an old home to a new one. Leave it for the new owners.

The Greeks, although they also sprinkled parsley on graves, believed it brought cheerfulness and good appetite and are said to have worn wreaths of parsley at banquets. This idea was revived in the Victorian Language of Flowers, which lists parsley as meaning to feast.

PASSING A PRIEST

The French, in some ways a curious people, believe it is unlucky to pass a priest on a country road. If they do, they will touch iron (*See* IRON) to ward off evil spirits.

PEACE SYMBOL

Perhaps our generation's most famous symbol has been the one for peace and disarmament. It is comprised of the signal flags for the initials N (nuclear) and D (disarmament). When you combine the two signals they form an ancient sign of man upside down, which means the death of man. Put that in a circle, which represents the unborn child, and you have a symbol that says dead children.

"No Nukes"

PEACOCK FEATHERS

See also THEATRICAL FEARS

Peacock feathers are notoriously bad luck. The guy upstairs from me swore he had seven years' bad luck after he brought some peacock feathers into his living room.

The reason they are such bad luck is that at the end of each feather is an eye (some believe the evil eye) that watches you even in your own home, causing terrible things to happen.

The source of this common belief is rooted in the Greek legend of Argus, the hundred-eyed monster who was turned into a peacock, with all his eyes in his tail, never to be able to shut them again. The story goes that Hera turned Argus into the peacock because he fell asleep while on a spying assignment for her.

In the sixteenth century, peacock feathers were bestowed upon liars and cheats to signify they were traitors.

In India, the peacock was lucky because it warned of approaching evil, and it was thought to be magical. The Indians say the peacock "has an angel's feathers, a devil's voice and a thief's walk."

Since the earliest recorded times, the peacock has been highly esteemed in China and Japan. It was an indication of rank and a reward for achievement from the rulers.

The Peacock Throne became the symbol of the first Shah of Iran, Reza Shah Pahlavi (1877–1944). Nobody had told him that it was a bad luck charm, which his son, the second Shah, Mohammed Reza Pahlavi, discovered when his government was overthrown in 1978.

PEARLS

See also CHARTS, Birthstones

Liquid drops of tears that you have shed,
Shall come again transformed to orient pearl.
(Shakespeare, *Richard III*, IV:4)

The unique gem that grows in the sea as a result of an irritation in the shell of an oyster, the pearl is believed to shine at night, to watch over the affairs of people, and to forecast danger, sickness, and even death by its loss of luster and its increasing brittleness.

The pearl is drenched in superstition and has at least two popular origins. The first is in the Scandinavian tale of Baldur, the god of light, who was slain with an arrow of mistletoe. It was believed that the tears of his mother, the goddess Frigga, brought him back to life. Her Tears congealed, and became the pearls on the mistletoe. (*See* MISTLETOE)

The second legend comes from the Orient. The pearl was associated with the fullness of the moon, which overflowed with heavenly dew, drawing the oysters to the surface of the sea. The oysters opened their shells and received the dewdrops, which hardened into the all-perfect pearls. (At least this acknowledges the participation of oysters in the production.)

Although it is doubtful that Cleopatra really dissolved her pearls in wine, pearls can dissolve in an acid wine or vinegar; however, this is a very slow process.

The Hindus believed that pearls grew inside elephants and were therefore holy.

Throughout the East, the pearl is a main ingredient in all love potions.

PHOTOGRAPHS

See also MIRROR

In many parts of the world it is considered very unlucky, even fatal, to be photographed. It is understood that a person's soul is in the image, and to take a photograph, creating a duplicate of that image, will allow the Devil to take possession of the soul.

When three people are photographed together, it is thought that the middle person will die.

It is unlucky for an engaged couple to be photographed together; why tempt the evil spirits into breaking up a good thing?

It's especially bad luck to be photographed with a cat, (*See* CAT) since that cat could contain the spirit of a witch.

To bring a curse upon someone, turn the person's picture to the wall, or upside down. If you turn it both upside down and backward, it's much stronger magic!

PEPPER

In ancient Rome, pepper was a status symbol. In the fifth century B.C., it was prescribed for female complaints and taken as an antidote for hemlock poisoning.

Spilled pepper isn't as dangerous as spilled salt (*See* SALT OVER YOUR LEFT SHOULDER), but it could indicate that an argument is brewing between two good friends.

If you want an unwelcome guest to leave, surreptitiously place a pinch of pepper under his or her chair.

If you have a fever, don't eat pepper—it will increase.

Pepper can't be all bad; Yale University was partly subsidized by monies made on the trading of pepper. According to *Reader's Digest Stories Beyond Everyday Things*, Elihu Yale worked for the East India Company, made millions of dollars, and then started Yale University with the money.

PINEAPPLE

To the Victorians, this pungent fruit stood for perfection. Given as a gift, it is thought to mean "you are perfect." It was enormously popular during the Victorian era, and a piece of prose from those days calls it:

> *So beautiful that it might seem*
> *to be made solely to delight the eye;*
> *So fragrant that we might be induced to*
> *cultivate it for its perfume only.*
> (Anonymous)

Although the pineapple may have declined in value as a precious gift in everyday life, the custom was revived temporarily for the 1966

Broadway musical *Cabaret*, which featured the pineapple as a token of love in the song "It Couldn't Please Me More."

PINS

See a pin and pick it up
All the day you'll have good luck

Pick up a pin, pick up a sorrow.

Pass up a pin, pass up a friend.

See a pin, let it lie, all the day
You'll have to cry.

Lend a pin, spoil a friendship.
(Halliwell, *Nursery Rhymes*)

Black-headed pins must not be used when fitting a dress.
Finding a safety pin is good luck.
A pin shouldn't be used to remove a splinter. Use a needle instead.
It's bad luck to give someone a brooch unless the receiver *doesn't* say thank you.
Pins stuck in a wax image of a person will cause that person pain.

Pins, because they have always been made from shiny materials, have always been considered magical. In early times pins were made from thorns and fowl and animal leg bones that had been sanded and shined. They were used to make clothing and tents.

Commonly associated with pins is spilled blood, because of a pin's ability to draw blood with the slightest touch. Primitive man was the first to put a pricked finger to his mouth to catch the blood. He probably wanted to get the blood before the evil spirits could.

Most superstitions about pins mention picking them up. That idea dates to the days of witchcraft, when it was believed that witches used odd bits of metal to cast magic spells. If you didn't pick up a fallen pin, a witch might.

PISCES

See also ZODIAC

Pisces is the astrological sign for people born between February 19 and March 20. Their symbol is two fish swimming in opposite directions, representing the extremes in a Pisces's character. The constellation Pisces contains two widely separated stars connected by streams of small stars. There is a legend that Pisces was created when Venus and Cupid became two fish in order to escape the fury of terrible Typhon.

Pisces are often dreamers and idealists, and because they are influenced by Neptune or Jupiter they are optimistic. They are also very adaptable, since Pisces is a water sign. The Pisces person is very imaginative and intuitive, and often a great creative talent. Pisces tend to be emotional and sensitive, and to worry about things.

The aquamarine is the birthstone of Pisces. Friday is their lucky day; five and eight are their lucky numbers; and lavender is a good color for them.

Some famous Pisces people are George Washington, Albert Einstein, Rudolph Nureyev, Frederic Chopin, Joanne Woodward, and Ted Kennedy.

Pisces, the fishes

POINTING

We've all been taught that it's bad manners to point. This comes from a time when it was believed that it was bad luck to point. People with the dreaded evil eye were thought to point at their victims.

Witches begin their chants by pointing at the accursed person. In fact, whole chants are recited while witches point at hapless victims.

POPPIES

Poppies are those innocuous little flowers from which opium is made. Can you imagine what ancient man made of this phenomenon? The extraction of opium from poppies was discovered in Persia and Asia Minor and was kept a secret from other areas for centuries. The narcotic effects of the poppy were understood early and are reflected in the ancient legend of Somnus, god of sleep, who placed poppies around a resting goddess who grieved for her lost daughter. When the goddess awoke, she picked some poppies, ate them, and then slept again, this time losing her sadness and sense of bereavement.

Opium was used for headaches and as a sleeping pill. Sometimes poppy juice was given to children as a sleeping aid.

In early times, the poppy was often offered to the dead. It was a symbol of death and its presence was considered a bad omen.

In Flanders Fields the poppies blow
Between the crosses, row on row . . .
(Col. McCrae, *In Flanders Field*)

This 1915 poem refers to a belief that poppies grew on the battlefields from the blood of dead soldiers, a sign that Heaven was angry at the evil deeds of mortals.

The poppy remains a symbol of dead soldiers and is worn each year on Veteran's Day (originally known as Poppy Day).

POT OF GOLD

See RAINBOWS

POTTER'S FIELD

And they took counsel, and bought . . .
the potter's field, to bury strangers in.
(The Book of Matthew)

We know that beggars are still buried in potter's fields. The original potter's field (mentioned above) was a plot of earth outside of Jerusalem bought by the chief rabbi with the thirty pieces of silver Judas received from the Romans for betraying Jesus. The land was set aside for the burial of the poor and the strangers passing through town. It is supposed that potters got their clay from this field before it became a graveyard.

PULLING THE WOOL OVER HIS EYES

Centuries ago the word wool was an accepted slang version of hair. In those days, men wore powdered wigs often made of real wool, and the expression grew out of the habit of jokingly pulling the wig over a man's eyes so that he was unable to see what was happening.

PURPLE

Remember the expression "To the Purple Born"? Since the earliest times purple was reserved for royalty and gods. Priests and other religious leaders also wore purple. It was a difficult color to mix, although it was used in very ancient times, made from the murex shell, which yields a brilliant purple.

The color became an imperial one and indicated divine right because of its association with the gods. Common folk were forbidden to wear purple until recent times.

In 1895 Gelett Burgess wrote this little ditty (called "The Purple Cow"):

I never saw a purple cow,
I never hope to see one;
But I can tell you, anyhow,
I'd rather see than be one.

Burgess and his rhyme became an instant hit. The rhyme was so popular that a few years later Burgess wrote:

Ah. Yes, I wrote the "Purple Cow"—
I'm sorry, now, I wrote it!
But I can tell you, anyhow,
I'll kill you if you quote it.
(Cinq Ans Après)

PYRAMID

The Egyptian pyramids, a wonder of engineering, have always seemed to say to the world, "we hold a secret." Many people have pursued that idea, believing that there was some hidden knowledge or even fortune in the design and construction of the pyramids. Alas, nothing has ever been found to substantiate these claims, but believers still look for clues.

In ancient times the triangle shape was considered the strongest for physical structures and was thought to be a sacred form. Many people took three iron nails, placed them in the shape of a triangle or pyramid, and had them driven into the front door of their homes to protect them from the evil eye, and later, from witches. (*See* IRON, EVIL EYE, WITCHES, AND WARLOCKS)

In modern times, around the outbreak of "Egypt-fever" in the 1920s, when the tombs were being discovered and opened, "pyramid power" and a con game called the pyramid game, took hold. It was believed that the shape of the pyramid prevented decay. People have reported that razor blades placed inside a pyramid lasted longer, that seeds germinated faster and were healthier; some people even claimed to feel better just being near a model pyramid. "Pyramid power" has never been completely disproved, nor has it been accepted as a scientific fact. The con game mentioned above was based on a chain-letter theory. People gave money, and as the base of the pyramid grew, the people at the top would receive huge sums of money. When museums began exhibiting the tomb of Tutankhamen in the late 1970s, both surfaced once again, proving only that people will believe anything if it's packaged right.

RABBIT'S FOOT

See also AMULETS
EASTER EGGS

In the world of superstition the hare and the rabbit have become interchangeable, particularly in America, where the hare is hardly ever separated from its relative.

The special magic of these animals lies in the fact that they are born with their eyes open, which invests them with special powers over the evil eye.

The rabbit's foot—which, to be truly lucky, must be the left hind paw from an animal which has been killed at the full of the moon by a cross-eyed person—holds a lasting place in the hearts of the superstitious as a good luck charm. Incidentally, it should always be carried in the left pocket.

The rabbit's foot is considered a powerful charm against evil forces because the rabbit's strong hind legs touch the ground ahead of its front ones (a most unusual way for an animal to walk). To the ancients, this technique was so remarkable that they ascribed magical powers to those feet.

The hare was worshipped in the British Isles long before Christianity reached there. In witchcraft, the rabbit, like the black cat, was thought to be a witch in disguise. It was believed that a witch who disguised herself as a rabbit could be killed only with a silver bullet.

Even today, farmers watch rabbits as a weather indicator: a thick coat means a hard winter; a thin coat indicates a mild one. This has, however, proved to be as inaccurate a way of predicting the weather as checking onion skins or watching for the groundhog.

RADISH

John Seymour in his *Gardener's Delight*, tells us that the Egyptian pyramid builders were fed vast quantities of radishes, along with garlic and onions. "Maybe," writes Seymour, "they moved those huge stones into position with their *breath*."

The Greeks adored the radish. They even placed a golden radish inside their temple at Delphi.

The Romans, however, found it vulgar, treated it as a food, with disdain, and are thought only to have extracted the oil of the radish for medicinal purposes.

RAINBOWS

See also SMALL FOLKS

Rainbow at night; Sailor's delight;
Rainbow in the morning; Sailors take warning.
Rainbow to windward: Foul all the day;
Rainbow to leeward: Damp runs away.
(R. Inwards, *Weather Lore*)

Coleridge called the rainbow: "That gracious thing made up of tears and light." Some think that a pot of gold can be found where the end of the rainbow touches the ground.

Others believe that the rainbow is the bridge over which souls are taken from earth to Heaven (or Hell). Some think that anyone walking under a rainbow will be transformed into the opposite sex.

RAM

See ARIES

REDHEADS

See HAIR COLOR

RED-LETTER DAY

The significance of red-letter days probably originated with the Book of Common Prayer. Saints' days and religious holidays of all sorts were printed in red in all early prayer books and church almanacs, and they still are. Our own calendars often have holidays and weekends in red, indicating that a day is special.

In ancient Rome, important days were noted down with chalk. Regular days were noted with charcoal, making them into black letter days. Red, a color which frightened the Devil and his witches, means good luck. Black-letter days are considered not quite so lucky; in fact, they're just ordinary days.

RED PEPPER

See AMULETS

RED RIBBONS

See also CHARTS, Colors

My mother claims she tied dozens of red ribbons to my carriage when I was a baby because so many people stopped to say how pretty I was. The ribbons were supposed to protect me from the evil eye, which spent a lot of time hanging around pretty babies.

The superstition is that for each person who praises the baby, another ribbon must be added. Each ribbon is a *kayn aynhorah*, meaning "may no evil harm you." (*See* EVIL EYE) Judging from the results, I'd say my mother's ribbons weren't large enough.

Another common superstition, following in the same tradition as the one about baby carriages, is to tie a red ribbon somewhere inside a new car for luck. This tradition probably comes from the ancient one of using a red rag to counteract bad luck and to bring good luck. Red, after all, is a color which frightens the Devil.

REPUBLICAN ELEPHANT

In 1870, the cartoonist Thomas Nast introduced a jackass into a cartoon that quickly became the Democratic donkey (*see* DEMOCRATIC DONKEY), and so it wasn't surprising when Nast drew another cartoon, in 1874, for *Harper's Weekly*, which suggested a Republican symbol to the public. The drawing showed wooden boards, which were platform planks, being thrown everywhere by an out-of-control elephant labeled the Republican vote. The cartoon also included a donkey in a lion's skin, which was immediately recognizable to most of the magazine's readers as the Democratic Party.

Like the Democratic donkey, the elephant was not intended as a complimentary symbol at first, but it caught on and has remained with the Republicans.

Some historians point out that Nast did not introduce the elephant to the Republicans, that it was first used in the 1860 political campaign by Abraham Lincoln. In an August 9 advertisement, an elephant was depicted wearing boots and carrying a sign which read "We Are Coming! Clear the Track!"

RICE

See WEDDING CUSTOMS

RIGHT SIDE OF THE BED

See also LEFT-HANDED

Yes, Virginia, there is a right side and a wrong side of the bed!

The ancients believed that the gods lived on the right side of man and the evil spirits lived on his left side. Further, they believed that the human heart was located on the left side of the body, and if you slept on that side, you'd crush your heart and die. You had to climb into bed on the right side and get out on the right side, to complete a circle begun when you went to sleep. Going to the right symbolized following the course of the sun; going to the left, the direction that the Devil and his witches took.

Climbing out of bed on the left side (the wrong side) is considered so unlucky that hotel rooms are often designed so that you can't get out on the left side.

In the right-versus-left controversy, the Romans took the position that right quite literally meant dexterous or adept; left meant sinister, corrupt, or even evil. (Maybe that's why the people who believe in the political right always feel they're on the side of the angels and that the political left are nothing but a bunch of troublemakers.)

You must always enter a place with the right foot first. This was particularly important to the Romans, who believed that when you entered with your right foot, you entered with the gods at your side. If you placed your left foot first, you were in big trouble; the evil spirits were right there beside you. They believed so strongly in this that they stationed guards at the entrance to all public places to make certain that people entered with the right foot.

That's how the custom of carrying the bride over the threshold got started. The groom was trying to make sure that the bride didn't start their married life off on the wrong foot. (*See* WEDDING CUSTOMS)

RIGHT VERSUS LEFT

See RIGHT SIDE OF THE BED

RING OUT THE OLD

See NEW YEAR'S DAY

ROBIN REDBREAST

See also BIRDS

If a robin you should dare to kill,
Your right hand will lose all its skill.
(Old Proverb)

Make a wish on the first robin redbreast of spring and it will come true.

There are two theories about how the robin got its red chest. The earlier version says that he singed it while carrying water to a good god caught in the fires of Hell.

A later belief is that the robin pulled the thorns from the crown of Christ and the blood turned his breast red.

In either case, legend has made the robin a special bird, with special reverence paid it by many people.

ROSEMARY

If you wear rosemary around your neck, you'll have a better memory. If you eat rosemary, you'll have an even better memory because, it has been said, rosemary stimulates the nervous system, which in turn stimulates the memory process.

In ancient Rome, students wore rosemary in their hair to help them remember their studies.

When planted near doorsteps, rosemary keeps witches away:

Run witch run, flee witch flee,
Or it will go ill with thee.
Run witch flee. Begone!
(Rhyme from Middle Ages)

Say that when planting your rosemary, or it probably won't work as well.

As a symbol of friendship and remembrance, rosemary was included in early funeral wreaths. Often it was thrown into the grave to indicate a special memory of the deceased.

A batch of rosemary under the bed at night keeps nightmares away and insures a sound sleep. (Probably because the witches can't get at you.)

Rosemary is such a miraculous plant that it was considered to be a cure-all for diseases of the mind and to make old people young again!

In the thirteenth century, Queen Elizabeth of Hungary claimed her paralysis was cured by something known as Hungary water, whose principal ingredient was rosemary.

ROSES

See also CHARTS, Language of Flowers

There's one tale that the rose first appeared in Bethlehem, when a young woman was falsely accused of a crime and sentenced to die at the stake by burning. Miraculously, the wood changed into roses and prevented her death.

It is also said that the Greek god of silence stumbled upon Venus, the goddess of love, while she was enjoying a romantic assignation. At the same time Cupid, Venus's son, came along and bribed the god of silence with a rose. This god of silence is depicted as a half-naked young man holding a finger to his lips and a white rose in his other hand.

Throughout history the white rose has represented silence and secrecy. The Latin *sub rosa*, under the rose, has been found in writings dating back to 479 B.C. It meant "you are sworn to secrecy." The white rose carved into doorways and above arches in banquet rooms reminded guests that whatever they heard inside was to be kept secret. The same emblem was once carved over the confessional, and sometimes white roses were actually placed near the confessional booths.

Roses have been the unofficial emblem of England since the War of Roses, when Henry VII joined York's white rose with the Lancaster red as the symbol of his unified kingdom.

Roses that bloom beyond their proper season are said to foretell death. The Romans sprinkled rose petals over the graves of the dead. The ancient meaning of roses was joy; today love has replaced that meaning. Every variety of rose, however, has its own meaning.

There's a marvelous legend that Cupid was stung by a bee while admiring a rose, and he became so angry that he shot an arrow into the rosebush, causing it to bleed. The blooms turned red, and thorns were created. Another legend has it that Cupid prankishly spilled wine over the roses, making them red.

RUBY

The glowing ruby should adorn
Those who in July are born,
They will then be exempt and free
From love's doubt and anxiety.
(Anonymous, *Note and Queries*, 1889)

Some early folks thought the ruby had the power of driving out evil thoughts, reconciling arguments, controlling passions, promoting tranquility, and preserving health.

It is said also to bring cheerfulness to its wearers, to protect against fevers, and to repel bad dreams. It is an antidote for certain poisons, protects against the plague, and changes color when an enemy approaches. It can also cause water to boil; but you need ideal circumstances for that to happen.

The star ruby is the most potent of the rubies. The gleam inside is said to be a good spirit.

The ruby is a particularly important stone in the Orient. It figures prominently in the Koran and is highly respected by the Hindus.

RULE OF THUMB

See also THUMBS UP

The rule of thumb is a measurement. It probably dates back to very early brewmasters who tested the heat of the beer with their thumbs to judge its degree of doneness.

RX

The symbol used to identify pharmaceutical prescriptions was originally the Roman astrological sign for the planet Jupiter. In the Middle Ages, when physicians believed that the planets influenced a person's health, Jupiter was thought to be the most powerful of all the heavenly bodies in the curing of diseases.

SAGITTARIUS

See also ZODIAC

Winston Churchill was a Sagittarius; so are Eugene Ionesco, Frank Sinatra, and Jane Fonda. Sagittarians are born between November 23 and December 21. Their sign is that of the Centaur or, more popularly, the Archer, which probably comes from an ancient Babylonian war god. Sagittarians are often enthusiastic, self-reliant, and outspoken, but they are not always as open to others as they may appear to be. They often don't marry, and they have a love for adventure and travel. They have high ideals, and quick tempers, enjoy life, and hate criticism.

Sagittarius, the archer

Sagittarians seem to have well-balanced characters, but they tend to be impractical. They are loyal and law-abiding, and they possess a commanding nature. Sagittarians have a tendency to gain weight and love an active social life. They often have magnetic appeal to others because they are cheerful to have around.

The birthstone of the Sagittarian is the turquoise. Thursday is their day, and the number nine is their lucky number. Purple tends to be their best color.

SAINT SWITHIN'S DAY

Saint Swithin's Day, if thou dost rain
For forty days it will remain.
St. Swithin's Day, if thou be fair,
For forty days 'twill rain nae mair.
(Old Scottish proverb)

On July 15, 862, Saint Swithin, the Bishop of Winchester, was canonized and due for reburial in his cathedral. It rained so hard on that day that the proceedings had to be canceled. It rained for forty days. When the rain did not stop it was decided to leave Saint Swithin where he lay. Actually, Saint Swithin had asked to be buried outside the church grounds, in the fields nearby.

SAINT VALENTINE'S DAY

See VALENTINE

SALIVA

See also CROSS-EYED

Spit on the ground to stop evil from getting around.
Spit loudly three times if frightened and no harm will reach you.
Spit into your hands for luck before a fistfight or any other contest.

Spit into the boat before sailing if you want good winds.
Mothers must spit on their babies whenever anything is ailing them. All complaints can be cured with saliva.
Gamblers spit into their hands for luck (just as the ancient Greeks did).
Spit over your left shoulder—spit into the face of the Devil.

It is believed that Jesus cured blindness using clay made from dirt and saliva. When the Masai tribe in Africa spit to say hello, good-bye, or anything, it means goodwill.

Saliva has always been the most effective countercharm against all forms of evil because it is a body fluid and cannot be faked. Anything found in the body is important in the world of superstition, with saliva holding a special place in lore.

Pliny the Elder said (in *Natural History,* Bk VII, 77 A.D.):

> *All men possess in their bodies a poison which acts upon serpents; and the human saliva, it is said, makes them take flight, as though they had been touched with boiling water. The same substance, it is said, destroys them the moment it enters their throat.*

Any substance that can do all those things to a snake is pretty powerful stuff.

SALT OVER YOUR LEFT SHOULDER

> *Salt is pure and white—there is something holy in salt.*
> (Hawthorne, *American Notebooks,* 1840)

Salt was, for primitive man, a magical substance that could be used for both good and evil. Salt became protective against the Devil when it was realized that salt could preserve food; more than likely, it could preserve people. When salt was spilled, a special guardian spirit was supposed to be warning against some impending danger.

Because the common belief was that good spirits lived on the right side and evil ones on the left (*See* RIGHT SIDE OF THE BED), people threw salt over the left shoulder into the eyes of the Devil.

Salt has many implications. Foremost it is a symbol of sadness. The Norwegians believe that as many tears as necessary will be shed to dissolve spilled salt. In New England people will throw salt on a stove to help tears dry more quickly.

The historical circumstances surrounding salt are important for an understanding of the superstitions about it. Salt was, first of all, very scarce, very expensive, and very difficult to mine. Therefore it was extremely precious and was the favored token to bestow upon strangers as a sign of welcome; hospitality among the ancients was a far more important virtue than it is today. In ancient Greece, strangers were greeted with a pinch of salt placed in their right hands. In Eastern countries salt was given as a pledge against ill will. In Hungary, salt was sprinkled over the threshold of a house so that no witch or other evil could enter. (Remember, Hungary was the home of Dracula and many other vampires, and the people had to be very careful.)

Both the Greeks and the Romans worshipped a goddess of salt, which might seem excessive until you realize that they believed that salt purified the sea. Since they made their living from the sea, they lived by the grace of these gods. At birth, a pinch of salt was placed on an infant's tongue to insure long life and good health.

Because salt was so rare in ancient Rome, it was sometimes used as wages for soldiers, from which we get the word salary and the expression "not worth his salt."

Another common saying, "I'll take that with a grain of salt," comes to us directly from Pliny the Elder (circa 77 A.D.), who prescribed a grain of salt to be taken to prevent poisoning (in *Natural History*). It was his idea that to build up an immunity to poison, one should take a little poison at a time, with a grain of salt to help it go down more easily. Today, when we accept an exaggeration with a grain of salt, it makes the boast more digestible.

"He's the salt of the earth" comes from the Book of Matthew in the Bible, and it means, he's the best there is. When the Arabs say, "there's salt between us," they mean that something sacred has taken place, as a deal signed with salt in good faith.

Salt is still a sign of purification and hospitality in the East. In Japan, for example, salt is sprinkled on the floor of a wrestling ring before a match between sumo wrestlers.

In the West we continue to fear spilled salt, maybe because Leonardo da Vinci, in "The Last Supper," painted Judas spilling salt as a symbol of his betrayal to Christ.

SAPPHIRE

Wytches love well this stone,
For they wene that they may werke certain wondres
By vertue of this stone.
(Ancient witchcraft manual)

The deep blue stone known as sapphire has been worn throughout the centuries by kings, clergy, virgins, Hindus, and by those born during the month of September, under the sign of Virgo.

According to legend, the sapphire protected kings and other royal personages from envy, and it is thought to have attracted divine favor. The Hebrews revered the stone, because it was said to have been set in King Solomon's ring. It is the favored stone for bishop's rings. It is thought to aid in the fulfilling of prayers.

The Greeks dedicated the stone to Apollo and believed it was a good luck charm. It was thought to have the power to kill spiders, to serve as protection to virgins, and to be effective against the forces of evil. The Hindus think it is a very good luck charm and that it brings health, wealth, energy, and the blessings of the gods.

The sapphire stands for wisdom. The star sapphire is particularly helpful against sorcery because of the good spirits that twinkle from within.

A maiden born when Autumn leaves
And rustling in September's breezes
A sapphire on her brow should bind,
'Twill cure diseases of the mind.
(Anonymous, *Notes and Queries*, 1889)

SCARECROWS

See also BIRDS
CROWS

Long ago two pieces of wood were nailed together to form a cross to be used in the fields as protection for the crops. Later, clothing was added over the cross as a disguise to fool the Devil. Modern farmers have continued the practice of clothing a straw figure to scare crows away (in other words, to protect the crops from evil). It is thought that the scent of people, which lingers on the clothing, keeps the crows out of the fields.

There is a saying that a single crow is a sign of evil. Should you see a crow approaching, remove your hat and bow bareheaded toward the crow. It ought to work.

SCORPIO

See also ZODIAC

Scorpio, the scorpion

The symbol of Scorpio is either the Scorpion or the Eagle. Scorpios are those born between October 23 and November 23. They are quiet people with an inner strength, sharp vision, and a deep understanding. They can be secretive loners who resist change. They tend to be courageous, strong-willed and idealistic, with a strong, healthy desire to succeed. They sometimes have perception bordering on the psychic, and they almost always have a deep survival instinct.

Scorpios are influenced by Mars, who loves arguments. They are partial to Tuesdays, the numbers three and five, and the color red. The Scorpio birthstone is the topaz. Scorpios love to love, but since they find fault easily with those they love, they are difficult to love in return. Scorpios are either very, very good or very, very bad. Mystery intrigues them, and they love solving puzzles.

Some well-known Scorpios are Katharine Hepburn, Indira Gandhi, Pablo Picasso, Grace Kelly, and Kurt Vonnegut.

SEVEN

See also NUMBERS

Seven has always been considered a special number. To the Egyptians, for example, the earth was represented by a four-sided house in which three gods dwelled, which added up to seven. This became their lucky number.

Seven really came into its own with the Christian interpretation of Creation:

The world was made in seven days.
There are seven days to the week.
There are seven graces.
There are seven stanzas in the Lord's Prayer.
There are seven ages of man.
Christ uttered seven last words.

Most of the above beliefs are ruled by the different phases of the moon, which change every seven days.

The Romans believed that the mind and the body changed completely and were renewed after seven years. (They also started the seven years' bad luck concept—*See* MIRROR)

Seventh Heaven is an Islamic concept, and it represents the best of all possible places. It is the heaven of heavens, the residence of God and his angels. There is also a very early Islamic belief that there are seven heavens, one lying right above the other, graduating in degrees; depending upon how good a person was on earth, he or she could say, "I'm in Seventh Heaven."

Seven is especially lucky to gamblers. (*See* GAMBLING)

SEVENTH HEAVEN

See SEVEN

SEVENTH SON OF THE SEVENTH SON

See also SEVEN

The seventh son born to the seventh son is thought to be doubly blessed. He is believed to be a clairvoyant, with natural healing powers. Throughout the Middle Ages, the seventh son usually practiced magic and administered the laying on of hands to the sick.

George Stimpson, in *A Book About a Thousand Things*, hints that there might be some basis for this lasting belief. There is a school of thought that believes the children of older fathers are smarter: They inherit the level of intelligence of the father at the time of conception. Coincidentally, many geniuses have been the younger child in a family.

Gypsies seem to be the only people who give equal credit to the seventh daughter of the seventh daughter. They say she always tells an accurate future.

SHAKE HANDS ON IT

See also HANDSHAKE

When two people shake hands on it, they are blessing their deal. The two hands form the sign of the cross, which is thought to bring good luck to any venture. It is also a pledge of good intentions.

SHAMROCKS

See also FOUR-LEAF CLOVER

The shamrock has been the emblem of Ireland since about 433 A.D., when Saint Patrick is said to have used the three leaves to explain the meaning of the Trinity to the pagans and druids.

A three-leafed plant very like the shamrock was greatly honored by the Greeks, who believed that Zeus's horse ate them for protection. It was also found on Roman coins and drawn on walls in Egyptian tombs. Many ancient civilizations honored the plant and believed it brought immortality, riches, and protection from evil.

The word shamrock is Gaelic, or at least comes from the Gaelic *seamrog*; or from the Latin *trefoil*, which means three-leafed.

It is solemnly believed by all good and loyal Irishmen that a real, true shamrock will never grow on English soil.

SINGING BEFORE BREAKFAST

If you sing before seven, you'll cry before eleven.

If you sing before you eat, you'll cry before you sleep.

Sing while eating
Or sing in bed,
Evil will get you
And you'll be dead.

In an interview after her divorce from Eddie Fisher, Elizabeth Taylor complained that he used to sing to her at the breakfast table. Well, no wonder she divorced him! Maybe the superstition that singing at the breakfast table will bring disappointment in love has some validity.

It's also a bad idea to sing while playing cards. You'll probably lose. But if you sing unconsciously while in the bathtub you'll probably have good luck.

Most of these ideas come from the Greeks, who very sensibly thought that being happy early in the day was not wholesome since you hadn't had time to earn your happiness. They also felt that expressing happiness (as through singing) early in the day would bring unhappiness later, because things usually work in reverse.

SLEEP WITH YOUR HEAD POINTING NORTH

This superstition is still pretty commonly followed. It says that the head of the bed should always point north so that the person will be lulled to sleep by the earth's magnetic waves, which are thought to flow from north to south. This idea got started when it was discovered that the earth was round.

If you should be so foolish as to place your bed with the head facing south, or if, like me, you don't know which direction your bed is facing, then place glass coasters under the feet to break the electrical current's flow.

SLEEP ON RIGHT SIDE

See RIGHT SIDE OF THE BED

SMALL FOLKS

See also DWARFS
FAIRIES

For our purposes, small creatures such as gremlins, elves, leprechauns, and trolls will be known collectively as small folks.

The small folks have been around for a very long time. Usually they cause no end of trouble, but something always makes them a welcome sight. In fact, we often go looking for them.

Some common legendary small folks are:

Brownies: from Scotland; a very friendly species of small folks, they live with families and work around the house while the family sleeps.

Elves: from Germany; mischievous and often spiteful creatures who are believed to have no souls.

Goblins and Gnomes: basically French, they are household spirits who have bad tempers; they like to live in dark places but when in the mood are helpful around the house. Gnomes are particularly fond of mines and work there under the earth.

Gremlins: probably English in origin; these are associated almost exclusively with airplanes. They are blamed for all mishaps, no matter how big or small, which befall planes and for all the crews' problems.

Leprechauns: Irish by nationality; these two feet tall shoemakers are believed to be very rich and stingy; have great crocks of gold, which they won't give up; and vanish at the slightest provocation. (*See* RAINBOWS)

Trolls: Scandinavian; they can be either dwarfs or giants and live in caves by the sea or in the mountains. Fishermen are said to be terrified of them, and they are believed to burst when sunlight hits them head-on.

As Batchelor and de Lys in *Superstitious? Here's Why!* explain, "The best part of these small creatures is that they can be blamed for all human ills, bad luck, and accidents which is one of the main reasons why they were invented, of course!"

SMOKEY THE BEAR

The most successful public-service campaign in America has been the Smokey the Bear campaign against forest fires. He was quickly taken into the hearts of Americans and made a part of the culture—even if he didn't prevent too many forest fires.

Smokey was discovered during a massive forest fire in 1950. He was a bear cub separated from his mother, who was rescued and began living quite happily in captivity, which was most unusual for a grizzly bear. They named him Smokey because of the fire in which he was found. Eventually he went to live in the National Zoo in Washington, D.C. and was a great favorite of visitors.

When the U.S. Forest Service decided to use Smokey as a fire-prevention symbol, Congress passed a law making it illegal to give his name to any other animal. That's how Smokey the Bear became a national symbol.

The current Smokey was adopted by Smokey Sr. and his mate, Goldie, when it was discovered they couldn't have baby bears.

SNAKES

Some people eat snakes to keep young. But then, some people will believe anything.

"Snakes change milk into poison," says an old Sanskrit proverb, echoing the age-old belief in the snake as the symbol of evil. After all, it was a snake that tricked Eve out of the Garden of Eden. That in itself is enough to have created a bad reputation for the creature.

The Egyptians couldn't decide about the snake. They both hated and worshipped it. It was the symbol of old age and wisdom; when seen with its tail in its mouth, it formed the eternal circle. (*See* CIRCLES) Because a snake sheds its skin during molting, they believed it was immortal. Combined with the circle shape it forms, that made the snake a pretty important symbol.

In India there has always been a snake cult. They believe that the souls of the dead come back as snakes, and so snakes must never be killed.

Historically, the snake has often been a phallic symbol.

A snake ring has long been a popular amulet. It is believed to bring long life and good health to its wearer.

Finally, the symbol of the medical profession is two snakes wrapped around a rod of harmony, which symbolizes healing.

SNAP YOUR FINGERS

See LUCKY BREAK

SNEEZING

Gesundheit!

How many times have you heard that? In fact, how many times have you said it; or God bless you; or any of the benedictions that save the soul? In case you hadn't realized it, sneezing is a breeze from Satan's wings; the Arabs believed that Allah created the world with a sneeze.

On Monday, sneeze for danger
On Tuesday, kiss a stranger
On Wednesday, receive a letter
On Thursday, something better
On Friday, sneeze for sorrow
On Saturday, see your lover tomorrow

On Sunday your safety seek
Or the Devil will have you the rest of the week.
(Harland, *Lancs. Folk-Lore*)

A newborn child is under the spell
of elves until it sneezes.

The Hebrews say, "God bless you," when you sneeze because they believe that at the moment of sneezing you are as close to death as you can be. They say that Adam sneezed after Eve took a bite of the apple, and he saw his own death.

The Romans used to say, "May Jupiter preserve you!" because they, like many of the early civilizations, believed that the essence of life—man's spirit or his soul—was in the form of air or breath, which lived inside his head. A sneeze could expel that spirit unless the gods interceded. Many civilizations, both old and modern, bow toward the sneezing person, saying, "May your soul not escape," with the gesture.

During the Middle Ages, sneezing was a very serious matter. It was believed that evil spirits were let loose with a sneeze, which could be dangerous to anyone nearby. With the Black Plague lurking around every corner, this fear was "not to be sneezed at." In this respect, sneezing does spread disease, or evil spirits.

The Greeks wrote that Prometheus invented the sneeze.

The Romans were the ones who started the belief that if you sneeze while making a decision, your choice will be right.

In the seventeenth century, sneezing was thought to be a sign of good health, and English hospitals discharged patients who sneezed three times—that, they felt was a sure sign of recovery.

The Book of Genesis says that God breathed into his nostrils the breath of life in its description of the birth of Adam. Well, reasoned early man, if life went in through the nose, wouldn't it be logical to think that it could leave in the same fashion? Couldn't a good, solid sneeze dislodge life just as quickly as it had begun? Therefore shouldn't a blessing be said to prevent the soul from being damned? Gesundheit!

SOMETHING OLD, SOMETHING NEW . . .

See WEDDING CUSTOMS

SPEAKING AT THE SAME TIME AS SOMEONE ELSE

See WISH ON A STAR

SPIDERS

A spider in the morning is a sign of sorrow
A spider at noon brings worry for tomorrow
A spider in the afternoon is a sign of a gift;
But a spider in the evening will all hopes uplift.

(Old proverb)

Spiders have a long and varied history in the world of superstition. It is believed that a spider spun a web to hide the infant Jesus in the manger when the messengers of Herod came to look for him, therefore it is unlucky to kill a spider or to disturb its web.

A spider mixed with syrup will cure a fever (the spider will eat the fever).
A spider in a walnut shell worn around the neck will ward off the plague.
If you run into a spider web, you'll meet a friend.
If you see a spider run down its web in the afternoon, you'll take a trip.
The daddy longlegs spider is especially lucky, and if a plow kills one, the cows will go dry.

It is said that Robert Bruce, the king of Scotland, watched a spider build its web for inspiration while he was held prisoner. Following the spider's lead, he freed Scotland from English rule.

The name of the dance known as the tarantella was derived from an old folk cure for the bite of a tarantula spider.

It was also believed that the Roman gods were jealous of the delicacy of the work that spiders did when spinning their webs.

SPILLED MILK

There's an old Russian saying: "Don't cry over spilled milk. If you would live forever, wash milk from your liver. What is taken in with milk, goes out with the soul."

Some say it's unlucky to spill milk. To counteract this bad luck, say your enemy's name three times.

If bubbles form at the top of a glass of milk (or a cup of coffee) immediately after you've poured it and you can drink the bubbles before they break, you'll get a lot of money.

Any spilled drink, in Ireland, is lucky. The liquid on the ground is a treat for the small folks (*See* SMALL FOLKS), and they'll repay your kindness.

SPILLED SALT

See SALT OVER YOUR LEFT SHOULDER

SPIT

See SALIVA

STAR OF DAVID

See also TALISMAN

The Magen David or Solomon's Seal, as the Star of David is known, is used as an amulet against demons and bad spirits.

In today's world, it is also the symbol of Israel and the symbol of the Jewish religion. It is made by superimposing one triangle, representing the male, over another triangle, representing the female. For very religious Jews, the true name of God must never be spoken, and so this star, which is thought to comprise the four elements of the universe, is used to represent God. Synagogues, religious books, and paraphernalia of the religion are stamped with this Magen David.

STARS

Whenever a mortal falls in sin,
Tears fall from angels eyes.
And this is why at times there
fall Bright stars from out the skies.

All wishes made on a shooting star will come true.

Shooting stars mean someone will die because they clear the path for a soul to get to Heaven.

It's unlucky to point at the stars; stars are really angels, and if you point you might poke their eyes.

Falling stars are the souls of children coming to earth, on their way to being born.

Stars are lucky; because of that, more countries have stars on their flags than any other symbol.

The Star of Destiny is a popular concept in astrology. It is believed that each person receives a Star of Destiny that serves as his or her own guiding star throughout life. It disappears at death. (*See* FIVE)

Stars are also special people with star quality. Noel Coward, when asked if he had something special to say to the star of his new play, answered, "Yes. Twinkle."

STEP ON A CRACK

The most popular form of this grisly saying is:

Step on a crack, break your mother's back.

It comes from the belief that a crack represents the opening of a grave. To step on that crack meant you might be walking on the grave of someone in your family. (This notion predates pavement, with its carefully measured squares that make it almost impossible *not* to walk on a crack.) The whole poem goes:

Step in a hole
You'll break your mother's sugar bowl.
Step on a crack,
You'll break your mother's back.
Step in a ditch,
Your mother's nose will itch.
Step on the dirt
You'll tear your father's shirt.
Step on a nail
You'll put your father in jail.
(Old children's rhyme)

STEP UNDER A LADDER

See WALKING UNDER LADDERS

STONES

See AMULETS

STORKS BRING BABIES

This is such a persistent legend in households that have small children that most of us half believe it even though we really know better—don't we?

The whole superstition is that storks pick up infants from marshes, streams, caves, and other such places where the souls of unborn babies reside and bring them to their new homes.

The stork has been observed to be a good "family man" for many centuries. Aristotle made it a crime to kill a stork in Greece in 330 B.C. The Romans passed the stork's law (*lex ciconaria*), which compelled children to care for their needy parents in old age.

In Roman mythology the stork was sacred to Venus. When a stork couple built their nest on the rooftop of a home it was considered a blessing from Venus and a promise of enduring love.

Veteran stork watchers know that each year the stork returns to build its nest on the same spot, and superstitious stork watchers know that means good luck.

It is a popular custom to say that when a stork flies over a home, a birth is about to take place.

STRING AROUND YOUR FINGER

Unfortunately, many people tie a piece of string around a finger, then forget what it was they were trying to remember. If they had tied the string onto a finger of the left hand it probably would have worked better, because the seat of all knowledge and memory is on the left side—where the heart used to be.

In the old days, the spirit of life often presented itself in the form of a body pain. The pain was thought to stay in one place if a piece of string or other fabric was tied around the spot. This also served as a cure for the pain. The string remained on the spot as a reminder of where the pain had been. That's how come we tie string around our fingers to remind us of something.

STY

An absolutely guaranteed way to cure a sty in your eye is to rub it nine times with a gold wedding ring.

If that doesn't work, press a copper penny to your sty then throw the penny away. The person who picks up the penny (a greedy soul) will pick up your sty. (If that doesn't work, see a doctor.)

SUCKER BORN EVERY MINUTE, THERE'S A

This saying has been ascribed to the circus man P. T. Barnum, and he should have known.

SUN

See SUNRISE

SUNRISE

Sad soul, take comfort, nor forget
That sunrise never failed us yet.
(C. L. Thaxter,
The Sunrise Never Failed Us Yet)

That's the amazing thing about the sun and sunrise—it always comes again the next day. It was a great comfort to primitive man, and so he worshipped the sun and feared its loss. An eclipse frightened him terribly. He believed, as many modern, more sophisticated people do, that the sun hides before every great sorrow.

Sunrise is a time often associated with death. More deaths, even through natural causes, are recorded at sunrise than at any other time. The unwholesome association of death with the sun comes from early man's worship of the sun. Sacrifices to the sun god were always made at sunrise.

Condemned people are traditionally executed at sunrise.

SWAN SONG

He made a swan-like end,
Fading in music.
(Shakespeare,
Merchant of Venice III:2)

Back in 400 B.C., the Greek god Apollo was said to have given his soul to the swan. When swans died they were said to sing in anticipation of the great times Apollo had in store for them after their deaths.

The final piece of work from a creative person is said to be his swan song, the last the world will hear of him.

Incidentally, swans don't sing before they die. They hardly even groan.

SWEEPING THE DUST OUT

This superstition is popular in many countries and probably falls under the old wives' tale heading. It goes: When you sweep the dust out the front door, good fortune goes with it; or if you use a broom to sweep the dust out the front door, you'll sweep your friends away.

One explanation for this is based on the belief that at homes fortunate enough to have good spirits living just outside the front door (for protection against the Devil), you would get dust in their faces when you swept out your home—and that's insulting.

TALISMAN

See also AMULETS
CHARMS
GARLIC
HORSESHOES
MASCOTS

In the world of superstition, people wear amulets and talismans for luck. An amulet is a passive protection against evil and the Devil; the talisman, is a more active form of this sport. You touch it, kiss it, or wave it about to bring good or evil upon yourself or someone else.

The cross and the mezuzah are popular talismans, as are other religious symbols such as the ancient ankh (*See* ANKH), saint medallions, the Polynesian tiki, and the fish emblem.

Let's examine the mezuzah because it, probably more than any other object still in use today, embodies all the principles of a talisman. The mezuzah is a hollow metal tube into which is inserted a small roll of parchment, inscribed with the central vow of Judaism: "Hear, O Israel, the Lord our God is One." This small object is then nailed to the doorpost of a home, consecrating that home and keeping it safe. For more active protection, each person touches or kisses the mezuzah when entering the house. The concept comes from the Book of Deuteronomy: "upon the doorposts of thy house and on thy gates."

Some scholars have indicated that the mezuzah is descended from the Egyptian practice of writing lucky sayings over the entrance of homes and gathering places. The Muslims inscribe the name Allah and print verses from the Koran on their buildings.

All doorways and entrances are invitations to the Devil, and so doorknobs and knockers often represent angels, lions, gargoyles, or other protective images. Throughout history there have been carvings placed over windows and arches for protection, just as garlic (*See* GARLIC) is hung over doorways, or a horseshoe (*See* HORSESHOES) is nailed over the barn door.

Talismans are often simple objects like coins or lucky hats that weekend fishermen wear. We mortals have such little faith in our own abilities and are so susceptible to the influences of the Fates that we will credit any odd piece of cloth or metal with supernatural powers in order to change what will, undoubtedly, be bad luck without divine intervention.

TAROT

The mystical tarot is a deck of seventy-eight cards used primarily for fortune-telling. There are four suits within the pack, divided into the Minor Arcana (of which there are fifty-six cards), and the Major Arcana (twenty-two cards). The Major Arcana are said to have been derived from the world's oldest book. They are pictures with very specific meanings.

The original tarot cards were used to predict the rise and fall of the waters of the Nile, which fertilized the lands around the river and therefore sustained the Egyptians. Some say they came from the ancient Hebrews. Others insist they were a part of Egyptian mythology. A large school of thought believes they came from ancient India and China via the gypsies, who added their own well-known fortune-telling skills to the practice of reading the cards. (*See* GYPSIES)

There is one legend which says that after Alexandria was destroyed, Fez became the most important city in the world. Wise men from all over came to compare knowledge, but they didn't understand each other's language, and so a book of pictures was devised. That book became the Major Arcana of the tarot deck. Eden Gray, in *The Tarot Revealed*, explains that the "symbols are the picture forms of inner thoughts; they have rightly been called the doors leading to the hidden chambers of the mind." According to Gray, "The magic is in the cards."

Actually, the magic is in the reader of the cards. A skilled tarot interpreter can weave wild and wonderful tales of futures filled with money, success, love, and good health or foresee a world filled with death, destruction, and hate. It's all in the eye of the beholder.

The Major Arcana, the important fortune-telling section of the pack, consists of twenty-two different pictures:

1. The Magician or Minstrel: skill
2. High Priestess: mystery
3. The Empress: marriage
4. The Emperor: control or power
5. The High Priest: convention
6. Lovers: choice and responsibility
7. Chariot: success in artistic endeavors
8. Strength or Justice: spiritual over material wealth
9. The Hermit: help and guidance
10. The Wheel of Fortune: good luck
11. Strength or Justice: balanced future
12. The Hanged Man: self-sacrifice

13. Death: change
14. Temperance: adaptability
15. The Devil: black magic
16. The Tower being struck by Lightning: conflict
17. The Stars: hope
18. The Moon: dreams and imaginiation
19. The Sun: material happiness
20. The Final Judgment: renewal
21. The World: assured success
22. The Fool: choice (also numbered as 0)

None of these cards mean anything by themselves. It's up to the tarot reader to put together the meanings of all the cards dealt to the individual. Each card has an auxiliary meaning, depending upon when it turns up during the reading.

The earliest deck of tarot cards still in existence dates to 1390 and can be found in a French museum.

TATTOOS

Sailors with bulging muscles weren't the first to discover the (debatable) art of tattooing. In ancient Egypt, women of high birth wore tattoos on their heads, necks, and breasts. Iranian women still wear them.

Tattoos on women were a fad during World War I, when George Burchett advertised his designs as "Dainty Tints Imprinted on Society Ladies' Cheeks." In 1920, an Egyptian tomb was opened to reveal a vastly tattooed princess, and Burchett's business boomed.

Sailors, of course, do wear tattoos. Some wear them as a sign of virility, and some believe tattoos protect them from drowning. Centuries ago sailors had tattoos done as protection against smallpox. Some sailors have tattoos printed on their skins just because they have nothing else to do some nights.

TAURUS

See also ZODIAC

Taurus, the bull

They say you can always tell a Taurus, whose symbol is the Bull, because Taurus people are strong-willed, with firm, almost unshakable characters. Born between April 21 and May 20, they are conservative people who are loyal and helpful to their friends. They are creative and are often active in the arts.

They are very good at solving problems, although it may take them a little while, since they get bogged down in details.

Taurus is under the guardianship of Venus, which makes the Taurus an attractive person, primarily a homebody devoted to the family. To the Greeks, the Bull was the disguise that Zeus took when he kidnapped Europa. It is the sign that indicates the beginning of spring, and it is an earth sign, making the Taurus melancholy and stubborn.

The best day of the week for Taurus people is Friday, one and nine are their best numbers, and blue is their special color. The emerald is the birthstone for Taurus people.

Some celebrities born under this sign are Harry S Truman, Shirley Temple, Duke Ellington, Fred Astaire, and Salvador Dali.

THANKSGIVING

This holiday was first celebrated about 2000 B.C., when the Mayans celebrated their harvest as our Pilgrims did when they held their party at Plymouth Rock in 1620. The Aztecs also celebrated the holiday, but they were more demanding in their rites than either the Mayans or the Pilgrims: They insisted that a young girl be beheaded as a sacrifice to the gods for the crops.

There is some evidence that the Mayans played games not unlike our football on their Thanksgiving. Even the Pilgrims are thought to have played stoolball on Thanksgiving, and so the football game at the college of your choice on this holiday is really from an ancient tradition.

Most of the ancient civilizations honored the sun and the crops with a fall feast. We know for certain that the Syrians, Egyptians, Mesopotamians, and Hebrews had special days very like our Thanksgiving.

Both Washington and Lincoln proclaimed a day of Thanksgiving in November (Congress made it an official national holiday in 1941), even though people had been celebrating it for years. There is no evidence that turkey, cranberry sauce, or pumpkin pie were ever served at that first celebration (but so what?).

THEATRICAL FEARS

> *A certain woman went to the theater and brought the Devil home with her. And when the unclean spirit was pressed in the exorcism, and asked how he durst attack a Christian, 'I have done nothing,' says he, 'but what I can justify, for I seized her upon my own ground.'*
>
> (Tertullian, 200 A.D.)

It's no contest. Actors are the most superstitious people in the world. (They are followed closely by gamblers, jockeys, and sailors.) Everything they do contains an element of chance. An actor thinks: If it rains the audience will be small; if the playwright doesn't die before finishing the play, I'll get a starring part; if the producer remembers my sister, maybe I can get an audition. If I'd read earlier they would have remembered me better. If I'd been taller (shorter, thinner, fatter) I might have been right for the part. If the taxi had been quicker; if the cat hadn't walked in front of me; if there had only been an apple to eat; if there hadn't been so many people at the reading . . .

An actor always looks for the uncontrollable reason behind why he didn't get the role or why the performance didn't go well. Otherwise he'd have to look to his own talent, and that would be shaking his already questionable faith in himself. Rejection, the bane of the acting game, must be rationalized. There are so many theatrical superstitions that it's amazing the curtain ever goes up.

Let's start with the things which mean good luck:

A bad rehearsal means a good opening night.

It's good luck to start a performance thirteen minutes late.

A cat is good luck backstage. The best luck is for a cat to make a mess backstage before a performance.

A rabbit's foot (*See* RABBIT'S FOOT) ought to be used to put on stage makeup. An actor must never lose the rabbit's foot—that would indicate a loss of talent.

Americans say, "shit," and the French and English use "*merde*" once during a performance, for luck.

It's lucky to fall or trip on your entrance cue on opening night.

The Barrymores believed it lucky to eat an apple before each performance.

A hunchback in the company is good luck. (*See* TOUCH A HUNCHBACK FOR LUCK)

Visitors to dressing rooms must step in with their right feet first. (*See* RIGHT SIDE OF THE BED)

Champagne drunk in large quantities on opening night, or any night, is good luck. If some champagne gets spilled on opening night, you must put your finger into the spilled liquid and dab it behind your ears (they say Tallulah Bankhead originated this one).

Saint Genesius was the patron saint of actors. He was thought to be a martyred actor in ancient Rome.

On Broadway a "gypsy" robe is passed on from musical chorus to musical chorus; each time the robe has some addition from the play it has visited sewn onto it. The custom began in 1949 with *Gentlemen Prefer Blondes* and continues among "gypsies" (chorus dancers) to the present day.

On to some bad luck superstitions:

Whistling in the dressing room means that the person closest to the door will be fired. There are two reasons for this superstition. First, whistling carries and could be heard on stage. Second, during Elizabethean times most stagehands were former sailors who could handle the riggings; they communicated by whistling to each other, and a wrong or confusing whistle might bring a sandbag or a curtain down on an actor's head. (*See* WHISTLING)

The last line of a play must never be spoken out loud until opening night. This line, known as the tag line, would complete the play and tempt the Fates into causing trouble. Because, by saying the last line the play becomes perfect, and perfection is not for mere mortals.

Never open an umbrella onstage. This superstition has been traced to 1868 when an orchestra leader named Bob Williams opened his umbrella onstage just before leaving the theater in the pouring rain. He went directly to the pier, where he boarded a ship to cross the Atlantic, and an hour later the engine of the ship exploded, killing Williams. (*See* UMBRELLAS)

Peacocks and peacock feathers (*See* PEACOCK FEATHERS) are especially unlucky to theater people. Edwin Booth called it "that miserable bird of malignant fate!" when his theater went bankrupt in the 1870s after a friend gave him a stuffed peacock for the lobby as a good luck gift. The old Bijou Theatre had peacocks painted across the auditorium. It was considered a hard luck theater until it was painted over. Then the theater began to prosper.

Yellow is an unlucky color in the theater. (*See* YELLOW) It was worn by the actor who played the Devil in medieval plays. Yellow roses sent to an actor means death to a friend; and a yellow dog onstage means someone in the company will die.

Green is a really unlucky color. It has to do with limelight. In the old days the light used on the leads was green, or lime, and if an actor wore green onstage the colors canceled each other out; in other words, you couldn't see the actor (a terrible tragedy for any actor). Villians were highlighted with lime. Green is also difficult to light, since it turns muddy. It is the color of the fairies, who get jealous if mortals wear it and are apt to cause trouble onstage.

Never wish someone good luck before they go onstage. There are several reasons for this: To wish someone else luck means parting with it yourself; to wish luck is to tempt the Fates. Instead of good luck, one often hears the opposite: break a leg, or fall down backward. Actors are also told to "give 'em hell," "be brilliant," and "enjoy yourself."

Cats are bad luck onstage.

According to the venerable Lunts, it was fatal to a performance to pass someone on a staircase either on or offstage.

The words turkey and bomb are unlucky. Turkey is bad luck because of an early play called *Cage Me a Turkey*, which closed after the first act.

Opening a play at the end of the week is bad luck; opening on the thirteenth is also bad luck. Both will cause a play to end in financial disaster.

Finally we come to the most prevailing of theatrical superstitions. This has to do with *Macbeth*. The play is cursed—anyone who has ever worked in the theater will agree. It is even heresy to utter the name of the play or to quote it while in a theater. If the play must be referred to, it is called "The Play" or "The Scottish Play." It all started on opening night in 1606, when the actor slated to play Lady Macbeth was taken seriously ill and Shakespeare himself had to go on in his place. The play, written on commission from the king, was roundly disliked by its patron and removed from the repertoire for fifty years.

Throughout the history of the theater there has rarely been a production of *Macbeth* that has not been plagued by illness, death, fire, earthquake, or another disaster. In 1849, for example, there was a riot at the Astor Place Opera House where *Macbeth* was playing, and thirty-one people were killed. In the 1937 Old Vic production, the director was in an auto accident, the producer's dog died, the star Laurence Olivier lost his voice; a twenty-five-pound weight fell and almost crushed Olivier; finally, the head of the Old Vic died on opening night. As recently as the 1981 Lincoln Center production, the lead, Philip Anglim, lost his voice right after opening night and couldn't go on for two weeks.

There is an antidote for foolish actors who insist on quoting *Macbeth*: Leave the dressing room, turn around three times, spit over your left shoulder, and then knock three times on the dressing-room door before reentry. Most theatrical scholars agree, though, that there's little hope where *Macbeth* is concerned; it's cursed.

THIRTEEN

See also NUMBERS
THIRTEEN AT TABLE

As in many important superstitions, the number thirteen can be either lucky or unlucky. The Egyptians liked the number and thought it represented the last stage of earthly presence before going to heaven. They pictured a symbolic ladder with twelve steps that had to be climbed during a lifetime, each step representing another on the road to knowledge. The thirteenth step led to eternal life. Death was merely a transformation and not an ending, and death, or the thirteenth step, was to be wished for with some impatience.

There's a theory that thirteen became unlucky because primitive man reasoned that when you added up your fingers you got ten; then add two feet and you had twelve; after that came the unknown, or thirteen. Since the unknown is frightening, thirteen became a scary concept. But primitive man also believed there were thirteen moons (not our more traditional twelve), and if there were thirteen moons it must be lucky. So you see, a quandary formed even in prehistoric times.

The Hindus have been blamed for inventing the idea that thirteen people at a gathering are unlucky. (*See* THIRTEEN AT TABLE)

For Americans, thirteen is officially a lucky number. On the back of a dollar bill there is a pyramid of thirteen steps, there are thirteen leaves and berries on the olive branch, and the eagle holds thirteen arrows (for the thirteen colonies). Unofficially, it's an unlucky number, and that's why so few office buildings and apartment houses have a thirteenth floor.

If you feel uneasy about the number thirteen, then you probably suffer from triskaidekaphobia. It's hardly ever fatal, and it's never contagious. Just take two aspirin and go right on to fourteen.

THIRTEEN AT TABLE

See also THIRTEEN

> *A Riddle: "Why should Pope Leo XIII have been a very unlucky man?"*
> *Answer: "Because he was always the thirteenth at table."*
> (A popular riddle during Leo's reign, 1878–1903)

There were thirteen at dinner during the Last Supper and guess who was the thirteenth? Judas, of course; but the fact is that thirteen at dinner was unlucky long before Christ had a final Passover with his friends. In Norse mythology there is a story of an unlucky dinner with thirteen guests. Twelve gods were invited, but after everyone had arrived, Loki, the god of mischief, crashed the party. During the dinner one of the gods was killed.

If you find yourself at a dinner party with thirteen at the table, the only possible protection against someone dying before the end of the year is for everyone to join hands and stand up together, as one. That should help.

THREE

See also NUMBERS

We all know that things happen in threes. Bad things especially happen in threes. You know that if one famous person dies, two more will. We get three chances, we give three cheers, we have three strikes before batting out, and we usually get three wishes. Why three?

Since the dawn of time three has been a magical number for man. It represents the miracle of birth: Man plus woman equals child. All of life is represented in the number three: birth, life, death; or, beginning, middle, end. These add up to the eternal triangle; the deity; the Trinity. Almost all religions, no matter how ancient, are based on a trinity of some sort. In ancient Greece three important rulers were worshipped: Zeus (the heavens), Poseidon (the seas), and Hades (the underworld).

Trinities can be seen everywhere:

Man is made of body, mind, and soul.
The world is earth, air, and water.
The Christian Graces are Faith, Hope, and Charity.
Nature is made up of animals, vegetables, and minerals.
The primary colors are red, yellow, and blue.
Jonah spent three days in the whale, Daniel met three lions, and Peter made three denials.

Perhaps the two most famous trinities are: the Catholic Church's Father, Son, and Holy Ghost; and Shakespeare's three witches in *Macbeth*.

THREE BRASS BALLS

The everpresent symbol of the ignoble pawnbroker still hanging over the shops of modern dealers was first used during the Middle Ages by the Medici family in Florence, Italy.

Averardo Medici, legend has it, killed a mighty giant during a battle against Charlemagne with a mace containing three gilded balls. Averardo liked the legend and to perpetuate it, adopted the three brass (or golden) balls as his personal and family shield. Not long after, the Medici family, the richest in Italy, became the leading bankers and moneylenders, and the family crest (the three balls) hung outside their business establishments. It's still there.

THREE ON A MATCH

Never light three cigarettes on the same match. That superstition, which appears to come from World War I, is really far older and encompasses more than just cigarettes.

In World War I, soldiers became very shaky about lighting three cigarettes on a match when they realized that while they were lighting up in the trenches, the match glowed long enough to show the enemy where they were. Lighting a quick one or two butts wasn't so bad, but the third was almost certain to cause death since the match led the enemy straight to the last smoker. During World War II, attempts were made to squelch the superstition because there was a sulfur and match shortage.

In very early civilizations, when a chief died, all the fires of the tribe were extinguished except his. The medicine man (or witch doctor or shaman) relighted the fires three at a time with a stick from the chief's fire that was believed to contain his spirit.

Oddly, the Christian church adopted this pagan custom, and during the tenth century the Russian church lighted three candles from one taper at funerals to help the departed soul into eternity. Since these services were conducted only by ordained clergy, it became bad luck or even taboo for others to light any three things on a taper or match.

THREE WISE MONKEYS

See no evil, hear no evil, speak no evil. This idea goes back at least to the eighth-century Buddhists in China, who often depicted this philosophy in a three-monkey drawing. When the Japanese saw it, they loved it and emulated their neighbors.

The Western world took to the idea immediately after meeting the monkeys when the port of Japan opened in the 1800s.

Monkeys were very plentiful in China. Their agility made it seem possible for them to put their paws over their eyes, ears, and mouth.

THUMBS DOWN

See THUMBS UP

THUMBS UP

See also RULE OF THUMB

An itching thumb means visitors.

If your thumb turns back it shows you can't save money.

Pricking the thumb with a pin or a needle is a sure sign some evil is coming your way.

A long thumb implies stubborness; a wide one, wealth

The position of the thumb at birth was a very important sign to the ancients. A baby usually comes into the world with thumb clutched in fist; as the child becomes more alert the thumb slowly gets released and moves upward. This was thought to be a sign of life beginning. Conversely, at death, the thumb relaxes back into the palm, or thumbs down, meaning death.

This brings us to the renowned gladiator games of Roman times. When a gladiator fell, the audience was asked to decide whether he was to live (indicating by thumbs up) or die (thumbs down). There was often peer pressure to vote thumbs down, as Juvenal (*Satires* III) explained: "They win applause by slaying with a turn of the thumb."

To us, the symbol of thumbs up usually means approval, and thumbs down means failure.

THYME

The herb thyme stands for activity and energy. It takes its symbolism from the Greeks, who believed that the perfume restored energy.

Its powers expanded during the Middle Ages to include strength and courage. Ladies-in-waiting and lovers embroidered the image of thyme on the edges of handkerchiefs to be given to knights before battles. The most popular design showed a bee humming around a sprig of thyme and carried the double meaning of activity and strength. (*See* BEES)

If a single lady puts thyme in one shoe and rosemary in the other, and places a shoe on either side of her bed on Saint Agnes' Eve (January 20) or on Saint Valentine's Day (February 14), she'll dream of the man she's going to marry.

TOADS

See FROGS
WARTS

TOASTING WHEN DRINKING

> *Here's to the whole world, lest some damn fool take offense*
> *I drink to the general joy of the whole table.*
> (Shakespeare, *Macbeth:* III.4)

What's all this nonsense about clinking glasses together before drinking? It's to stop the evil spirits from opposing the eloquent toast you've just made.

Originally the custom was to pour a bit of the guest's wine into your own glass and vice versa, to insure that neither had been poisoned. When poisoning began to recede in popularity, the custom was modified to simply clinking glasses to scare the evil spirits, and saying something like "to your good health" as a gesture of goodwill.

The word toasting dates at least to Elizabethan times when a piece of toasted bread was placed at the bottom of a tankard before the ale or wine was poured. The toast was thought to improve the taste, and perhaps it did, because it served to absorb the sediment that collected among the nonvintage dregs.

There is one other assumption about toasting, or raising one's glass in honor of another. The custom comes from a time when the Danish conquered the British Isles. The inhabitants were not allowed to drink alcohol without the permission of their conquerors and had to wait until the Danes lifted their glasses.

TOMATO

The tomato, found as early as 500 B.C. in South America, has a strange history for a common plant. The tomato was first exported to Morocco, where it was later boarded onto ships for Italy and France. It was known as *pomo dei Moro*, apple of the Moors. The French changed it to *pomme d'amour* when they exported it to England, and so by the seventeenth century the English were eating something they called love apples. In Germany the common name for the tomato is still *liebesapfel*, which means love apple.

With this exotic name, how could the tomato avoid being considered an aphrodisiac? The love apple quickly became feared by virgins throughout the world, who often would not eat tomatoes until they were safely married.

Around the 1820s, and for years afterward, the tomato was believed to be poisonous, and the cause of a great many illnesses that couldn't otherwise be explained at the time.

The Italians have evolved an intricate folklore around the tomato. Tomato sauce, they think, brings health and wealth. A large red tomato sitting in the window wards off evil spirits; another on the mantelpiece will bring prosperity. These beliefs of wealth and health have brought us the pincushion in the shape of a tomato as a talisman. Most of us have lived with this kind of pincushion, never dreaming it was supposed to be lucky.

TOPAZ

Who first comes to this world below
With drear November's fog and snow
Should prize the topaz' amber hue—
Emblem of friends and lovers true.
(Anonymous, *Notes and Queries*, 1889)

Cleopatra is said to have liked topaz because it reminded her of honey. Ancient princes wore this stone to bring riches. In the Middle Ages, the topaz, set in gold, was worn as an amulet on the left arm, to ward off enchantments and other annoyances.

The topaz is supposed to help cure gout, get rid of worries, and calm the mind. It is a charm against lunacy and sudden death. It will increase intelligence, brighten a dull wit, and insure faithfulness in your love life. The Hindus even believe it will prevent thirst if worn near the heart.

Scorpios wear the topaz as their birthstone.

TO RAISE CAIN

According to the Book of Genesis, Cain, the brother of Abel, was the world's first murderer and criminal. In early days, the name of Cain was used by God-fearing people as a euphemism for the Devil. To raise Cain meant to raise the Devil, usually through loud noises or by causing lots of trouble or mischief.

TOUCH A HUNCHBACK FOR LUCK

See also TALISMAN
THEATRICAL FEARS

Hunchbacks were considered talismans against the evil eye; even the Egyptian pharaohs kept a real hunchback as a charm that they had within eyesight at all times.

It was thought that a hunchback was a particularly potent charm against the Devil because grotesques of all kinds were thought to scare evil spirits away or, to make them laugh so hard that they'd forget why they had come.

Through the centuries, as it became more impractical for people to have their own hunchbacks, the notion of touching the hump of these unfortunate people took on the same properties as the idea of having one near. Even now children will touch a hunchback for luck and then run away.

TOUCH IRON

See IRON
KNOCK ON WOOD

TOUCH OF THE BLARNEY STONE, A

See BLARNEY

TOUCHSTONE

See TALISMAN

TOUCH WOOD

See KNOCK ON WOOD

TREE OF LIFE (KNOWLEDGE)

See APPLE

TREES

See CHARTS, Trees and Their Meanings
KNOCK ON WOOD
OAK TREE

TRIANGLE

See PYRAMID

TRICK OR TREAT

See HALLOWE'EN

TRISKAIDEKAPHOBIA

See THIRTEEN

TROLLS

See SMALL FOLKS

TURNING DOWN A WEDDING PROPOSAL

When I say turning down, I mean just that. This is a custom from colonial days, when a boyfriend came calling with a courting mirror. First he'd look into the mirror, convinced that his image would stay on the glass. If the woman of his choice accepted his proposal, she smiled into the mirror; if she wanted no part of him, she'd turn the mirror face down, or more specifically, she turned him down flat.

TURQUOISE

If cold December gives you birth
The month of snow and ice and mirth
Place on your hand a turquoise blue
Success will bless whate'er you do.

(Anonymous,
Notes and Queries, 1889)

The turquoise is a stone we associate with the American Indian and his unique silver jewelry; but the stone was primarily used by the Turks (after whom it was named) and others in the Orient.

The Moslems often engraved messages from the Koran into the turquoise and then wore the stone as an amulet. The turquoise is believed to have the power to warn against poison and to draw all bad things into it, thereby protecting its wearer. It is also said to change color to indicate whether a wife has been faithful.

During the Middle Ages, when almost everything had a supernatural meaning, the turquoise was thought to be able to cure headaches and to make friends of enemies. It was a symbol of generosity, sincerity, and affection. It was also thought to preserve friendship.

Ever since the thirteenth century, the turquoise has been a talisman for horseback riders. It is believed that the stone can make a horse surefooted and prevent the rider from falling.

There is a custom that for good luck, a turquoise ought to be given, not bought.

TWINS

Ancient man was thoroughly baffled by the phenomenon of two children being born at the same time, with the same appearance, to the same woman. They made up all sorts of explanations. They said it was bad luck and drowned one of the babies. (If the twins consisted of a girl and a boy, the girl was the one to be drowned.) The birth of twins was sometimes considered good luck and a sign of the father's virility; or it could mean trouble and be a sign of the wife's infidelity.

The Romans believed it was a good sign, and so did the Egyptians. They both invented twin gods. The Romans had Romulus and Remus, who founded Rome; the Egyptians worshipped Osiris and Set. The expression by jiminy is from a Roman oath referring to the twin constellation Gemini. (*See* GEMINI)

Superstitions about twins abound:

If a husband with a pregnant wife should spill pepper, he must throw some over his right shoulder or she'll have twins.

A set of twins who marry on the same day should use different churches.

The Romans originated the idea that the first twin born is the child of love, and the second is the child of lightning. They also said that the second child is the more favored by the Fates.

Twins are never as strong as one child.

After having twins a woman is sterile.

A red streak down the middle of a pregnant woman's stomach means she'll have twins.

A set of twins only has one soul.

TWO-DOLLAR BILL

The two-dollar bill, a purely American item, has never been very popular with the public. It has always been considered bad luck and each time the Treasury department tries to reintroduce it to the public, it bombs. Fear of the two-dollar bill may have been started by gamblers, who have always equated the deuce with the Devil.

It is said that by tearing off a triangular piece from the corner of the bill the magical number of three (*See* THREE) is created as a countercharm to the bad luck of the deuce. When all four corners have been torn off, the next person who receives the bill is supposed to tear it up.

Some people are less drastic with the currency. They merely kiss it, since saliva (*See* SALIVA) is a very strong countercharm and usually keeps the Devil away.

TYING THE KNOT

See KNOTS

UMBRELLAS

See also THEATRICAL FEARS

It is unlucky to open an umbrella in the house.

If you leave your umbrella home, it will rain.

The first of these superstitions, not opening an umbrella indoors, comes from the very early (eleventh-century) idea that it was an insult to the sun to open an umbrella in the shade (especially indoors). The umbrella had originally been designed as a protection against the rays of the sun and was used this way for many centuries. It wasn't until the eighteenth century, when the umbrella came from Greece to the Continent, that it was used to protect against rain.

The pharaohs of ancient Egypt used umbrellas and considered them status symbols. They were particularly important in the protocol of the Orient, where umbrellas were a sign of royalty. One emperor of Japan had twenty-four umbrella bearers in a procession to show his absolute power.

By the sixteenth century the pope was using them, and it was said that unless the pope had conferred rank upon someone that person couldn't use an umbrella. An umbrella is still part of the pope's formal ensemble.

In the eighteenth century, umbrellas were often called Robinsons, after the one fashioned by Robinson Crusoe in the Daniel Defoe novel. Umbrellas were manufactured in earnest in England in the late 1700s and were often decorated with acorns (*See* ACORN) as protection against lightning.

UNCLE REMUS

See BRER RABBIT

UNCLE SAM

See also MASCOTS

Uncle Sam, the tall, skinny man in top hat and red-white-and-blue satin dress suit, and usually sporting a white goatee, has become the mascot of America. He started life, however, as Samuel Wilson, a meat inspector during the War of 1812. Legend has it that a group of visitors to Elbert Anderson's meat-packing plant noticed that the supplies were stamped EA-US and asked the inspector what that stood for. Wilson is supposed to have answered that the EA was for Elbert Anderson, the contractor, and the US was for Uncle Sam. The conversation was repeated and the nickname caught on.

The name first appeared in print on Sept. 7, 1813 in the *Troy* (New York) *Post*.

UNDER LADDERS

See WALKING UNDER LADDERS

UNICORNS

The unicorn is probably a mythical creation; but who's to say? There was talk of the unicorn during the Middle Ages in places like France and China. Biblical legend implies that the unicorn is now extinct because it was thrown out of the Ark.

We know what it looks like: kind of like a horse, but with a single horn growing out of the middle of its head. We think of it as a symbol of chastity, fierceness, virginity, and meekness (because it is said that the meek and the unicorn will inherit the earth).

The horn of the unicorn was said to be white at the base and black in the center, with a red tip. The horn could be used to purify water and detect poison. When ground it became a powerful ingredient in medicines. Ground unicorn horn was also reputed to be a very good aphrodisiac.

UP TO SNUFF

The snuff in up to snuff has nothing to do with the powdered tobacco people used to inhale for stimulation or enjoyment. Up to snuff used to mean a person had a highly developed ability to follow a scent or a clue because of his or her acute awareness. Today we generally mean a person who is well informed and not easily deceived (nothing very mysterious in this).

USHERS

See WEDDING CUSTOMS

VALENTINE

To-morrow is Saint Valentine's day.
All in the morning betime,
And I a maid at your window,
To be your Valentine.
(Shakespeare,
Hamlet, Prince of Denmark)

By tradition February 14 was the day that birds chose their mates for the year to come. If you should see any of these birds on Saint Valentine's Day it will indicate something about your future love life, as this little chart indicates:

Blackbird: clergy
Redbreast: sailor
Goldfinch: millionaire
Yellowbird: reasonable riches
Sparrow: love in a cottage
Bluebird: poverty
Crossbill: quarrelsome
Wryneck: no marriage
Flock of doves: good luck in every way

There is an old legend from about 240 A.D. that tells of two Christians named Valentine who were martyred on the same day,

February 14. One was a bishop, the other a priest. Both were later canonized.

One popular superstition about Saint Valentine's Day is that the first person of the opposite sex that you meet that day and kiss will be your valentine all year long.

VAMPIRES

If you believe in vampires, then you believe that:

A vampire can never rest and is doomed to roam the world in search of blood, which keeps him in his semistate of being undead.

A vampire must be home before daylight and at rest in his coffin of earth by the time the first rays of light are in the sky.

Anyone who is bitten by a vampire will become a vampire upon death.

The only effective defenses against vampires are the cross, garlic, light, iron, and bells.

A body buried with an open mouth will become a vampire.

If a vampire is suspected to have seized a body, the corpse must be dug up, and if it is not decomposed and has good color, with perhaps some drops of blood around the mouth, then you must drive a wooden stake through the heart to help that soul rest in peace. (The instructions are very specific about this procedure.)

Vampires are believed to exist in many countries. They are particularly popular in Slavic nations and especially in Hungary, Dracula's home. Dracula, the world's most famous vampire, actually did exist as Vlad the Great, a terrible and bloody ruler of Hungary during the Middle Ages.

We use the word vampire today to indicate that a person is a bloodsucker, a greedy chap who preys on others. Another use of the word is in vamp, meaning a femme fatale.

VEIL

See MOURNING VEILS
WEDDING CUSTOMS

VIOLETS

That which above all others yields the sweetest smell in the air is the violet.

(Bacon, *Essays*, XLVI, 1625)

The discovery of the violet is attributed to several sources. There are at least three Greek legends which give conflicting information:

Orpheus's lyre was resting on the grass, and when he picked it up, violets had grown under it.

Jupiter turned Io, whom he loved, into a white heifer to save her from Juno's jealousy, and then he created white violets for her to eat.

Venus was envious of Cupid's love for white violets, and so she turned them purple.

The violet was the symbol of the city of Athens. The Greeks used it extensively in their treatment of headaches and sleeplessness. The Persians crushed it for wine and the followers of Muhammad worshipped it. Medieval priests are supposed to have cultivated violets in their monastery gardens because they were powerful against evil spirits, and they used them to relieve swelling, hoarseness, and thirst.

Napoleon was known as Corporal Violette because he used the flower as his badge of honor during his exile. Consequently, the violet was banned in France for many years after Waterloo.

The blue violet is said to represent faithfulness and love; the purple variety, to mean "you occupy my thoughts"; and the wild violet, to be love in idleness.

The violet generally indicates modesty. It is the flower of those born in the month of February. It is currently used as an ingredient in treatments against inflammation and some cancers.

VIRGO

See also ZODIAC

Virgo, the virgin

Virgo the Virgin, or the Maiden, is usually pictured in a flowing gown carrying a palm leaf in her right hand and a staff of wheat in her left. Virgos are those born between August 23 and September 22. They tend to be very sure of themselves; in fact, they appear aloof or snobbish to many. They are dependable people, perfectionists, and hard workers who try to improve already established ideas rather than invent new ones. They are loyal, practical, and good with pets.

Virgo is the sign of purity, truth, and honor. Because they are under the influence of Mercury, Virgos are clear thinkers. Their birthstone is the sapphire, their best day is Wednesday, and their lucky numbers are four and eight. They should wear the color gray.

One legend is that the sphinx at Gaza was modeled with the head of Virgo and the body of Leo. In the Western world Virgo has been nicknamed the Frigid Maiden because of a seemingly untouchable quality.

Some famous Virgos are Leonard Bernstein, Raquel Welch, Ingrid Bergman, Ray Charles, H. G. Wells, Grandma Moses, and Nietzsche.

V SIGN

> *The V-sign is the symbol of the*
> *unconquerable will of the occupied*
> *territories, and a portent of the*
> *fate awaiting the Nazi tyranny.*
> (Winston Churchill, July 20, 1941)

As much as he might have liked to take credit for it, Winston Churchill did not invent the V sign. The Egyptians used it in amulets for centuries. It represents a god of ladders and staircases who, in gratitude for being saved from darkness, allowed the use of his powerful fingers as a support for faith and courage and to use for escapes. The Egyptians also believed the V (and the U shape) was a symbol of the continuity of life.

The difference between Churchill's V and some earlier ones was the direction. In the world of witchcraft, the inverted V symbolizes the Devil and his horns. Making the sign with the fingers pointing down means the Devil must stay down below where he belongs. Some people believe if you do this every morning when you get up the Devil will stay away from you all day long.

Some say Churchill didn't bring the V sign to England during World War II—the Belgians did. It is said that the sign was used as an identifying symbol among Belgian freedom fighters and that eventually a Belgian refugee introduced it on the BBC during a broadcast. In any case, Churchill could claim to have popularized it.

VULNERABLE SPOT

See ACHILLES' HEEL

WAKES

See also DEATH CUSTOMS

Sitting up with the dead was an ancient custom. It was also a time when heavy drinking was definitely encouraged. Time hasn't changed things much.

The ancients consumed vast amounts of wine because they believed that the spirits in the wine helped the soul of the deceased on its way to the spiritual world. They also believed that the wine helped to cleanse the sins of the deceased.

Watching over the corpse, one of the major reasons for a wake, comes from the notion that evil spirits from the underworld might sneak in and snatch the body before the soul has a chance to try its luck at Heaven's gates. Those who gather at a wake often sing, toast each other loudly, tell jokes, and laugh in keeping with the old beliefs that these boisterous ways will keep the evil spirits away.

This behavior, which some might consider unseemly, is believed by the Irish to make a proper and fitting send-off.

WALKING UNDER LADDERS

After you've spilled the salt and allowed a black cat to cross your path; after you've sneezed and crossed your fingers for luck; and after you've been foolish enough to sign a contract on Friday the Thirteenth, the only thing left to do is walk under a ladder. Then you've touched on most of the major superstitions.

Why do you suppose walking under a ladder would bring bad luck? There is a primitive belief that a leaning ladder formed a triangle or a pyramid, which was the universal symbol of life. Anyone walking through that sacred triangle would be punished.

In Asian countries criminals are actually hung from a leaning ladder instead of an opened ladder, because it is believed that death is contagious and that people walking under the ladder would meet the ghost of the hanged person and catch death there.

The Egyptians considered the ladder a good luck sign. Their legends tell of the sun god Osiris using a ladder to escape when captured by the spirit of darkness. Today many Egyptians carry miniature ladders as amulets for luck.

If you should walk under a ladder there are four things you can do to stop the inevitable bad luck:

Quickly make a wish while still under the ladder.
Cross your fingers until you see a dog.
Make the fig sign. (*See* FIG SIGN)
Walk backward through the ladder to where you started your ill-fated walk.

WALNUT

A very confusing history surrounds the walnut. In ancient Rome and Greece it was believed that stewed walnuts encouraged fertility. In Rumania, however,a bride who wants to remain childless places a roasted walnut into her bodice, one for each year she doesn't want to bear children. After the wedding ceremony, she buries the walnuts.

Legend has it that during the great flood God was eating walnuts. Those who were meant to be saved climbed into the walnut shells that God had discarded and sailed safely away.

There are tales of witches and evil spirits gathering under walnut trees. There's also a superstition about placing a walnut under a witch's chair so that she will become rooted to the spot.

A walnut branch is believed to be protection against lightning.

In medicine, the powers of the walnut are associated with diseases of the head. Walnuts are believed to cure anything from sore throats to thinning hair.

The black walnut has a batch of legends all its own. It was believed, particularly in England, that black walnuts contaminated apple trees. The leaves are thought to keep away ants and houseflies, and some American Indians used the bark as a very strong laxative.

WARTS

See also FROGS

Warts are mysterious things. They seem to come and go without much reason. Because of their mysterious behavior, the ancients had problems understanding them. It was said that toads and frogs caused warts. People came to this conclusion because toads have wartlike skin, and people firmly believed that like makes like.

The cures for warts are as old as the causes:

Pliny the Elder, in his *Natural History* suggested,

> *Lie on your back along a boundary line*
> *on the twentieth day of the moon, extend*
> *the hands over the head, and with*
> *whatever thing you grasp when so*
> *doing rub the warts, and they disappear.*

Steal a piece of meat, rub the wart with the meat, then bury the meat.

While a funeral procession passes, rub the wart and the wart will pass after it has rained nine times.

Tie a red string (*See* RED RIBBONS) around the wart, wrapping it three times.

WEATHER

Throughout this book there are references to weather forecasting. Every time a bird chirps or an onion grows a thick skin, someone says the weather will be either good or bad. When a cat walks this way or that, when a baby cries or a bat flies, there will either be a storm or fair weather. We all want to know what the weather will be, and the ancients, who were at the mercy of the elements in a much more real sense than we are, were especially anxious to read signs of approaching weather conditions.

Virgil in *The Aeneid,* claimed the sun was the best forecaster:

The sun who never lies,
Foretells the change of weather
In the skies.

It will rain:

if a frog sings loudly.
if a bat flies into your house.
if the cows refuse to go to the pastures.
when the cat licks its fur against the grain.
when the donkeys bray.
when a pig carries straw in its mouth.

Swallows fly high: clear blue sky.
Swallows fly high: rain we shall know.

A rainbow in the morn put your hook in the corn;
A rainbow at eve, put your hand in the sheave.
(Old nursery rhyme)

The cricket has been known for centuries as "the poor man's thermometer." (*See* CRICKETS) If you count the number of chirps within fifteen seconds, then add the number thirty-seven, you'll get the temperature in Fahrenheit degrees.

WEDDING CAKE

See WEDDING CUSTOMS

WEDDING CUSTOMS

The state of holy wedlock—how many times have you heard that expression? Wed is an old Anglo-Saxon word meaning to assign property to the bride's father as payment; lock simply means a pledge. Wedlock then is the promise to pay for the bride. What's so holy about that?

The first marriage was the kind between a caveman and the woman he chose. Most likely he'd sneak into her encampment, steal her, and hide her away in a cave until things calmed down again.

Many of our wedding customs are grounded in those early times. Some have real symbolism, others are based on a tradition whose meaning has died long ago. Here is a brief look at some of the traditions we keep alive today.

The *engagement ring*: A ring of any kind represents a pledge, and so when early Teutonic couples became engaged or betrothed, the man placed a ring on the right hand of his intended. The ring was a sign that the woman was off limits to other knights passing through the area.

June weddings: "Prosperity to the man and happiness to the maid when married in June," was an ancient Roman proverb. The Romans named June for the goddess Juno, the deity of women, and they believed that she blessed marriages that took place in her month. June is the favored month for weddings in the Western world more than in the East.

Brides: In old English the word bride came from a name for cook; that should make everything clear. *Grooms*: This was derived from male child; and *bridegroom* should mean a male cook, but it doesn't. It comes from German and means what it appears to mean—the person marrying the bride.

Bridesmaids and ushers: These come from the old Roman law that required ten witnesses to outwit the evil spirits who usually attended happy functions. The bridesmaids and ushers were dressed like the bride and groom so that the Devil wouldn't know who was getting married. Later, ushers, or groomsmen, were very useful to the bridegroom. They stood watch during the honeymoon period.

During the Middle Ages, man often reverted to the more primitive days of his caveman ancestors and simply kidnapped a woman and married her. The irate members of her family would come riding after her, but once the marriage had been consummated, nobody wanted her back. The ushers stood guard during the honeymoon. (*See* HONEYMOONS)

The accoutrements of the wedding ceremony, such as the *veil*, the *ring*, and the *white gown*, are all rooted in solid superstition. The wedding veil, for instance, probably comes from the Greeks and Romans. It is a vestige of the bridal canopy, which was constructed to keep the evil eye away from the wedding festivities. It is also an early Eastern custom that the bride should never be seen by the groom before the marriage as a sign of submission to her family. The raising of the veil is a symbol of freedom from parental control.

The *wedding ring* seals the bargain. A bride was considered property and the ring, in some societies, was a token of purchase. "With this ring" takes on a different meaning when viewed that way, doesn't it? The wedding ring is placed on the third finger of the left hand because it was believed that a love vein ran directly from the heart through that finger. If the ring was removed the love would escape from the heart.

Married when the year is new,
He'll be loving, kind and true.
When February birds do mate
You wed nor dread your fate.
If you wed when March winds blow
Joy and sorrow both you'll know.
Marry in April when you can
Joy for maiden and the man.
Marry in the month of May
And you'll surely rue the day.
Marry when June roses grow
Over land and sea you'll go.
Those who in July do wed

Must labor for their daily bread.
Whoever wed in August be,
Many a change is sure to see.
Marry in September's shine,
Your living will be rich and fine.
If in October you do marry
Love will come, but riches tarry.
If you wed in bleak November
Only joys will come, remember
When December's snows fall fast,
Marry and true love will last.
(Old English rhyme)

The *bridal bouquet* always has knots at the end of its ribbons. They represent good wishes and are also known as lover's knots, symbolic of oneness and unity. Throwing the bouquet is a fairly modern tradition: the one who catches it is sure to be the next to marry.

Orange blossoms are frequently found in the bridal bouquet. The custom goes back to the Saracens. The orange blossoms stand for chastity and purity. The orange blossom tree is an evergreen, which represents everlasting love, and is therefore the ideal symbol for a wedding.

The white gown of the bride was initially a symbol of joy. It dates at least to the Greeks, who sometimes even painted their bodies white before a marriage ceremony. The meaning of white has changed somewhat and now is purity and chastity more than joy.

"Something old, something new, something borrowed, something blue" is a rhyme that comes to us from merry old England. The something old of tradition was an old garter from a happily married woman. The notion of sympathetic magic is at work here. If the garter came from a happily married woman, it could pass along her happiness to the new couple. Something borrowed was, in olden times, a piece of gold representing the sun, the source of all life. Something blue was a compliment to the moon, the traditional protector of all women. Something new was anything at all, usually the wedding gown.

The *bridal kiss* is the groom sealing his sacred pledge. There is a tradition that it is lucky if the bride cries at about this time during the ceremony. If she doesn't cry, there will be tears throughout the marriage.

The tradition of *breaking a glass* is found in Jewish wedding ceremonies and is also important to the Hindus. The sound of the breaking glass was believed to scare the evil spirits away. It also symbolized the consummation of the marriage: the man's virility, and the woman's hyman being broken. A third interpretation is that it symbolized the destruction of the temple in Jerusalem reminding the participants how fleeting happiness can be.

When the ceremony is over it's time for the *wedding feast*, an ancient custom which is probably as old as weddings themselves. The ancients observed what they called "eating together."

At the feast there is always a *wedding cake*, which first became popular with the Romans, who broke a cake made of meal over the bride's head for luck. Then all the guests would sprinkle the crumbs from the cake over their own heads for luck. This cake sprinkling was meant to represent a wish that the newlyweds would have good things throughout their married lives. When today's bride cuts the first slice of the cake she is aided by her new husband so that he can share in the good luck. There is also an old custom of the bride putting a piece of the wedding cake under her bridal bed as a symbol of faithfulness.

The feasting is now over, and the wedding couple is ready to leave. Someone begins *throwing rice*, a symbol of fertility, prosperity, and health, and a way of chasing the evil spirits away from the couple. The Orientals began this custom, and the Romans changed it to suit their own culture. They threw nuts (a little more dangerous sometimes, but the meaning was the same) and later European cultures threw confetti (much safer and cheaper).

Then someone ties the *old shoes* to the fender of the bridal car. This custom comes from a time when the father of the bride gave the woman's old shoes to the husband as a symbol that he was now responsible for her.

The newlyweds are on their way to their *honeymoon*, a word, probably invented by the Teutons, referring to the mead mixed with honey that couples drank during the first month of marriage. (*See* HONEYMOONS)

Finally, the wedding is over, the honeymoon has been a sheer delight, and the *new husband carries his bride over the threshold*, which comes from the times when a man stole his bride away and also prevents the bride from tripping or using the wrong foot (the left foot would be the wrong foot) when stepping over the threshold of her new home. It is believed to be a sign of very bad luck for her to do either. (*See* RIGHT SIDE OF BED).

WEDDING FEAST

See WEDDING CUSTOMS

WEDDING RING

See WEDDING CUSTOMS

WEDDING VEIL

See WEDDING CUSTOMS

WEREWOLF

This was a very popular medieval superstition. The werewolf was a person turned into a wolf through an enchantment or a charm; or a person who could turn himself (or herself) into a wolf. The ways of the werewolf were frightening to those who knew about them. Werewolves roamed the countryside at night, eating babies and other human's flesh; or if they were very hungry, digging up corpses. Their skin was virtually weaponproof, unless the weapon had been specially blessed.

In the Middle Ages, the belief in werewolves was so common that a convention of theologians was called in the fifteenth century; and after careful discussion, they solemnly decreed that werewolves really existed.

The notion originated in early Greco-Roman times, and Ovid wrote of a werewolf at the beginning of the first century.

The most famous modern werewolf is probably in Robert Louis Stevenson's *Strange Case of Dr. Jekyll and Mr. Hyde,* in which the respectable Dr. Jekyll became the evil, werewolflike Mr. Hyde.

WHISTLING

Whistling in the house invites the Devil.
If little girls whistle, they'll grow beards.
Whistling aboard ship will raise a storm
Never whistle in the dressing room of a theater. (*See* THEATRICAL FEARS)
Reporters who whistle in editorial rooms or miners who whistle underground are asking for trouble.

It has been said that a woman whistled while the nails were forged for the cross, and so every time a woman whistles, the Virgin Mary's heart bleeds.

Ancient man was afraid to whistle. He thought that if you whistled, the Devil would answer. The logic behind this belief was that the Devil or evil spirits were usually thought to be responsible for all sounds that man couldn't account for, such as whistling wind through the trees and the sounds of the wind before a storm.

WHITE

See also CHARTS, Colors
WEDDING CUSTOMS
WHITE ELEPHANT
WHITE HORSE

White is the symbol of purity, simplicity, candor, innocence, truth, and hope. It was the color of the ancient clergy and of the druids. The sacrifices made to the great god Jupiter were offered while dressed in white. At Caesar's death, white was proclaimed the national mourning color. The Magi were thought to have worn white when delivering their gifts to Christ.

The Chinese still wear white for mourning. (*See* DEATH CUSTOMS)

WHITE ELEPHANT

In today's usage, a white elephant is something that's very expensive to maintain and from which no one profits. It all started in Siam (today's Thailand), where white elephants were so rare they were immediately given to the king and placed under his protection. They were revered and so could not be made to work for their upkeep. There is a story that when a nobleman did something to displease the king, he was given a white elephant by the ruler, who then waited for the expense of keeping this hungry creature to ruin the nobleman. It didn't take very long.

When P. T. Barnum bought a white elephant to appear as an attraction in his circus, it cost him 200,000 dollars just to get it to New York—now there's a white elephant for you.

WHITE GOWN

See WEDDING CUSTOMS

WHITE HORSE

Everybody knows the good guy rides a white horse; or do they?

A white horse was the symbol for purity and was thought to be able to warn of danger.

When a white horse passes, spit over your little finger for good luck and lick one thumb and stamp it into the palm of the other hand (stamping out the Devil) to avoid bad luck. At the same time, say, "Criss cross, white hoss, money for the week's done."

In some regions of England, particularly around the Midlands and in the South, the white horse is associated with murder. It seems the Saxons's banner showed a white horse, and after a bloody war in which the Saxons conquered the area, the white horse banner could be seen everywhere. When the natives saw a Saxon white horse go by they would spit over their left shoulders to banish the evil spirits created by the Saxons's presence. Many people still spit over their shoulders at the sight of a white horse.

There's an old wives' tale that white horses live longer than dark ones. Consequently, many people consider the white horse a living amulet against an early death.

WIDOW'S PEAK

Some people are born with the lovely V-shaped hairline that characterizes the widow's peak. Others try to cultivate one. Some think it's a sign of intelligence and long life; others think it's merely pretty.

There is an important superstition about the widow's peak you women ought to know before you create your own versions. It says that you will lose your first husband young and marry again soon. This goes back to 1498, when Anne of Brittany mourned the death of her husband Charles VIII. Anne married the French king Louis XII very soon after, so don't worry about her. Hat designers of the time created a new black V-shaped bonnet for Anne (who was the first for centuries to wear black for mourning) that suggested the V-shaped hairline and became known as the widow's peak.

WINE

Even back in the first century A.D., "in wine there is truth" was a common proverb. The ancients, who loved wine, believed that it contained spirits. When wine was drunk the spirits of the wine became active, and as we all know, spirits never lie. They therefore believed that spilling wine was a warning by the spirits that bad luck was on the way. The countercharm to that was to place some of the wine behind the ears with the middle finger of the right hand.

There is a superstition that spilling champagne is a bad omen. Sparkling wine was actually invented by accident when some monks put a cork in their wine bottles instead of the usual stopper. The wine mixed with the cork, creating the carbonic acid needed for carbonation. The Church believed that the carbonation was an evil spirit and champagne was outlawed for quite some time, although it never really disappeared from the market.

WISHBONES

If you believe your wish will come true when you win the break in a wishbone contest, then you're following in the footsteps of civilizations dating back to the Etruscans, 322 B.C.

Before digital clocks and morning disc jockeys, man waited for the cock to crow; and when he wanted an egg he waited for the hen to announce the coming of her product. This made the two animals mystical in that they could tell the future. They were the origins of the hen oracles.

To receive an answer to an important question from these oracles a man would draw a circle on the ground and divide it into the twenty-four letters of the alphabet. Grains of corn were placed in each section, and the cock or hen was led into the circle and then set free. It was believed that the fowl would spell out words or symbols by picking up kernels of corn from the different sections. For example, the first letter of a future husband's name would be the first kernel of corn picked. After writing the message, the fowl was sacrificed to a special deity and its collarbone was hung out to dry.

The person seeking answers made a wish on the bone. Then two other people got a chance to make a wish by snapping the dried bone in the same fashion we do, each one pulling one end. The person with the larger end of the bone was the one who got the wish. This was known as a lucky break.

The Romans picked up the wishbone habit and brought it with them when they conquered England. That's how we got it.

WISHFUL THINKING

If a man could have half his wishes
he would double his troubles.
(Benjamin Franklin,
Poor Richard's Almanack)

As we have seen, man has always believed in the concept of like makes like and it is at the heart of many superstitions. Wishful thinking is an extension of that belief. In early times man thought that he would have all his wishes answered if he made a wish while looking at or touching something that resembled the riches he prayed for; if only that were true.

WISHING ON A COIN

See WISHING WELLS

WISHING WELLS

You probably wouldn't have to wish for wealth if you still had all the coins you've tossed into wishing wells. Still you continue to throw coins in, make a wish, and wait for the results.

In the old days, the sea gods demanded tributes. The Greeks, for instance, tossed coins into their wells in the hope that they wouldn't run dry; a little bribe, you might say.

The tradition is that if you throw a coin into a well or a fountain, wait for the water to clear enough to see your own reflection, and then make your wish, the sea gods will make your wish come true. Some think it's the water spirits who accomplish this magic.

Niagara Falls, it is said, yields up thousands of dollars each year.

WISH ON A STAR

Star light, star bright
Here's the wish I wish tonight.
I wish I may, I wish I might
Have the wish I wish tonight.
(Anonymous)

People make wishes on anything new: the first star, a new moon, the first load of hay harvested, the first robin redbreast, and other firsts. There is a special magic in firsts.

Most people believe that wishes can come true if all the circumstances are right; most notably, that you keep the wish to yourself, never telling it to a soul. Of course we all know if you want something badly enough you'll probably get it, because you'll do everything you have to do to accomplish it. Then you'll go and pay tribute to some supernatural source for your success

One very popular superstition about wishing involves the moment when two people say the same word at the same time. Their wishes will come true if they link right pinkies and take turns saying:

Needles
Pins
Triplets
Twins.
When we marry
Our trouble begins.
What goes up a chimney?
Smoke
What comes down a chimney?
Santa Claus.

Then say together:

May your wish and my wish never be broken.

WITCHES AND WARLOCKS

Thou shalt not suffer a witch to live.
(The Book of Exodus)

How now, you secret, black and midnight hags!
(Shakespeare, *Macbeth*)

Those midnight hags that Shakespeare wrote of—the most famous witches in history—were created many hundreds of years before the famous playwright.

Witches were devised by pagan leaders to answer the questions evil happenings pose. The witches were originally envisioned as women who had made a pact with the Devil. Men were often sorcerers and wizards, and later in history, warlocks. Both witches and sorcerers practiced what was called black magic, which was defined as a supernatural power.

In some civilizations the medicine man and the wise old women performed the same function, that of explaining evil. It wasn't until the churches of organized religion defined good and evil that the witches of older times became truly evil, malignant creatures. It wasn't until the Middle Ages that witches and sorcerers were rooted out and burned at the stake.

The first witch in history may have been Homer's Circe, who was depicted as having the ability to cast spells on men and turn them into animals. Circe was believed to be very beautiful, and once a man heard her song he fell under her spell. Her image became corrupted through the centuries. A witch was then believed to be an ugly hag who rode broomsticks and cast spells by the light of the moon.

As the stories grew, so did the powers of the witch. She could change her form and become a cat or a rabbit. She was responsible for dead babies, sick cows, bad crops—you name it, she did it. Witches could cast spells with poisons, potions, chants and charms, or she could stick dolls with pins, as in voodoo. Witches remain an important part of many religions and there are many people today who claim to have the powers of a witch, although most of these say they are good witches who practice white magic.

If you run into a witch you will recognize her by:

eyebrows that grow together
birthmarks (particularly under the arm)
red hair
the evil eye

You'll also notice that a witch can't stand the presence of iron; has no reflection in a mirror (because she has no soul); is able to recite the Lord's prayer backward; can turn herself into birds except a robin redbreast (*See* ROBIN REDBREAST); and can't cry.

One infallible test for a witch, popular in the Middle Ages, was called the trial by water. If a suspected witch was thrown into the water and surfaced, she was guilty and was allowed to drown. If the person drowned without surfacing she was presumed innocent, but of course, by then it didn't matter for her.

WOMAN SCORNED

This is purely a literary reference, and I've included it in this book of superstitions because many think it has a base in the world of quasi fact.

Heaven has no rage like love to hatred turned,
Nor hell a fury like a woman scorned.
(Congreve, *The Mourning Bride*)

We shall find no fiend in hell
Can match the fury of a
Disappointed woman,
Scorned, slighted, dismissed
Without a parting pang.
(Colley Cibber,
Love's Last Shift, IV:1)

It's a bum rap!

WORTH HIS WEIGHT IN GOLD

See BIRTHDAYS

YANKEE DOODLE

As a British Army doctor sat watching the American troops, he doodled with words and came up with a ditty that made fun of the "homely clad colonials." Ironically "Yankee Doodle" became America's first patriotic song.

The words of Dr. Richard Shuckburg's poem were set to a traditional English folksong, or, says the folklorist Tristram Coffin, a Dutch farm song. The song was picked up by the rebels and sung with great gusto by the troops at the surrender of General Cornwallis at Yorktown.

The only confusing part of the song is "stuck a feather in his cap, and called it macaroni." Americans had little idea what it meant, but they sang it anyway. Actually, it referred to the Macaroni Club, popular in London at the time, which was a club of fops and overdressed young men who wanted to bring the lushness of Continental fashions to England.

Yankee Doodle went to London
riding on a pony,
He stuck a feather in his cap
and called it Macaroni.

Yankee Doodle keep it up,
Yankee Doodle Dandy;
Mind the music and the step
and with the girls be handy.

Father and I went down to camp,
Along with Captain Goodin'
And there we saw the men and boys
As thick as hasty puddin'.
(Edward Bangs, c. 1775)

YAWNING

There is only one yawn in the world,
and it goes from person to person.
(Anonymous)

The people of the Middle Ages believed the Devil entered your mouth through a yawn; and so it was very important either to make the sign of the cross over your mouth when you yawned or to cover your mouth.

Some early civilizations thought they'd lose their breath, never get it back, and die, so they covered their yawns, too. Some ancients believed that yawning was a sign of danger and anyone who caught the yawn was in that danger.

The Hindus are afraid of a yawn, and they snap their fingers three times to dispel the danger. In Finland, it is said that yawning horses mean rain. Hippocrates believed that intense yawning indicated the onset of a fever.

Yawning does seem to be contagious. It's actually an oxygen deficiency and if you're in a room with little oxygen it will probably not be long before several people in the room start yawning.

My particular favorite about yawning is that it is supposed to be a silent shout for help.

YELLOW

See also CHARTS, Colors
THEATRICAL FEARS

all looks yellow to the jaundic'd eye.
(Pope, *An Essay on Criticism II*)

Yellow is the color of jealousy, inconstancy, and even treachery. In France, the doors of traitors were streaked with yellow paint.

Judas is often pictured in yellow. In some countries, during ancient times, Jews were made to wear yellow to indicate their treachery in betraying Christ. The infamous star Jews were forced to wear by Nazis was yellow.

Yellow is also the color of cowards. "He has a yellow streak" means he's scared.

In Spain, executioners sometimes wore yellow.

YIN AND YANG

The dots of opposite color on each side represent the knowledge that nothing is all black or all white

Yin and yang together form the meaning of the universe in ancient Chinese philosophy. This presents two forces that are in constant opposition but form a perfect whole.

Yin, the black side of the circle, represents the female, earthly power in the universe. This is the negative, dark side, containing shadows and water. Yang, the white side of the circle, is the male, active, positive, and powerful side, representing the heavenly powers in the universe, the light, the sun, and the warmth. Together, yin, coolness of shade, and yang, warmth of the sun, make up life on earth.

The concept of yin and yang was first developed during the Chou dynasty (c. 1122–221 B.C) and later became part of Confucianism. It is, to some degree, still part of Chinese philosophy. The symbol is often seen in oriental countries.

ZODIAC

The word zodiac comes from a Latin word for animal or living being and the Greek word for life. The zodiac is an imaginary belt drawn through the sky dividing it into twelve constellations, which corresponded with the Greek animal constellations. The idea of the zodiac and its believers go back about four thousand years.

Everyone has a zodiacal sign, determined by when he or she was born. Many believe that the influence of the stars upon birth is so strong that a person's character and personality are formed by the sign under which he was born.

The twelve signs of the zodiac have been extremely popular

amulets for centuries. (*See* AMULETS) To determine which sign you should carry, consult the following chart:

Sign	*Birthday*
ARIES, *The Ram* ♈	March 21–April 20
TAURUS, *The Bull* ♉	April 21–May 20
GEMINI, *The Twins* ♋	May 21–June 20
CANCER, *The Crab* ♊	June 21–July 22
LEO, *The Lion* ♌	July 23–August 22
VIRGO, *The Virgin* ♍	August 23–September 22
LIBRA, *The Balance* ♎	September 23–October 22
SCORPIO, *The Scorpion* ♏	October 23–November 22
SAGGITTARIUS, *The Archer* ♐	November 23–December 21
CAPRICORN, *The Goat* ♑	December 22–January 19
AQUARIUS, *The Water Bearer* ♒	January 20–February 18
PISCES, *The Fish* ♓	February 19–March 20

APPENDIX

BODY PARTS GOVERNED BY ZODIAC SIGNS

CAPRICORN	(December 22–January 19)	Knees and circulation
AQUARIUS	(January 20–February 18)	Back and legs
PISCES	(February 19–March 20)	Feet
ARIES	(March 21–April 20)	Head
TAURUS	(April 21–May 20)	Throat
GEMINI	(May 21–June 20)	Arms, chest, and lungs
CANCER	(June 21–July 22)	Stomach and breast
LEO	(July 23–August 22)	Heart and the small of the back
VIRGO	(August 23–September 22)	Intestines
LIBRA	(September 23–October 22)	Liver and kidneys
SCORPIO	(October 23–November 22)	Backside and reproductive organs
SAGITTARIUS	(November 23–December 21)	Legs, hips, and thighs

BIRTHSTONES

JANUARY *Capricorn*	Garnet for Fidelity. Believed to repel flying insects, and to help the wearer be firm and steadfast.
FEBRUARY *Aquarius*	Amethyst for Sincerity. Believed to protect against drunkenness and falling in love foolishly, and to encourage calmness in the wearer.
MARCH *Pisces*	Aquamarine for Courage and Truth. Understood to bring wisdom, success, and popularity to those who wear it.
APRIL *Aries*	Diamond for Innocence. This stone will bring victory to its wearer.
MAY *Taurus*	Emerald for Happiness. It will insure a life of love and success, and protect against all eye diseases.

JUNE *Gemini*	Pearl for Long Life. This stone brings health, wealth, and good luck to the wearer.
JULY *Cancer*	Ruby for Peace of Mind. Believed to heal wounds, prevent stomachaches and bring love to its wearer.
AUGUST *Leo*	Sardonyx (a form of onyx) for Happiness. The wearer will have a happy marriage, contentment, and personal satisfaction.
SEPTEMBER *Virgo*	Sapphire for Wisdom. As a charm, this stone is believed to relieve headaches, protect the wearer from the evil eye, and clear the head of the wearer to allow wise thinking.
OCTOBER *Libra*	Opal for Hope. The wearer of this stone will receive good luck.
NOVEMBER *Scorpio*	Topaz for Loyalty. Believed to guard the wearer against calamity; and to insure faithful friendships.
DECEMBER *Sagittarius*	Turquoise for Success. This stone is supposed to protect wearers against accidents, and to make them prosperous.

COLORS

Colors speak all languages.
(Addison, in *The Spector*, 1712)

BLACK — Black, the color of pitch, was used to cover mummies. For the ancients, it was a symbol of resurrection and rebirth. In more modern times, it has come to mean death and mourning. (*See* BLACK)

BLUE — Wear blue as protection against witches and the malice of the evil eye. Blue stands for truth and creative power. Blue is the color of the heavens and is therefore one of the luckiest colors.

GREEN	Green is an unlucky color because it is the color of the small folks (*See* SMALL FOLKS), and you wouldn't want to anger the gnomes and leprechauns. It's bad luck to wear green onstage or at a Christening. It is especially unlucky for lovers. Green is the symbol of the resurrection because plants are green.
GOLD	Gold was the color of the gods in the ancient world, and largely because of the value of gold ore, it has remained a sacred color.
ORANGE	The word for orange comes from Sanskrit. It was the color of the early Church, symbol of the fruits of the earth.
RED	Red has always been considered an evil color, ever since its association with the Egyptian god Set, an evil and unlucky god. Men with red beards were considered evil. Red objects became popular amulets and talismans because it is believed that witches and the Devil are afraid of red. (*See* AMULETS; RED RIBBONS)
WHITE	White stands for joy, purity, and innocence. (*See* WHITE)
YELLOW	Yellow is an unlucky color. Actors who played the Devil in medieval plays wore yellow. In many areas Jews were forced to wear yellow as a symbol of the betrayal of Christ. Yellow is for jealousy. (*See* YELLOW)

Green's forsaken
Yellow's foresworn
Blue's the color
That shall be worn.

Blue is true
Yellow's jealous
Green's forsaken
White is love
And black is death.

EDIBLE PLANTS AND THEIR MEANINGS

ALMOND	Stupid, Indiscreet, and Thoughtless
CABBAGE	For Profit
CHICORY	Frugality
CORN	Riches
CRANBERRY	Cure for Headaches
CUCUMBER	Criticism
CURRANTS	Pleases Everyone
ENDIVE	Frugality
FIGS	Arguments and Longevity
GOOSEBERRY	Anticipation
LETTUCE	Coldhearted
MUSHROOM	Suspicion
OLIVE	Peace
PEACH	You're Terrific
PEAR	Affection
PEPPERMINT	Warm and Cordial
PINEAPPLE	You're Perfect
POMEGRANATE	Foolishness
POTATO	Benevolent
PUMPKIN	Bulkiness
QUINCE	Temptation
RASPBERRY	Remorse
RHUBARB	Advice
SPEARMINT	Warm Sentiments
STRAWBERRY	Perfect, Excellent
TURNIP	Charity
WALNUT	Intellectual
WATERMELON	Chunky
WHEAT	Prosperity
WILD GRAPES	Charity

FLOWERS
*and the months to which they are dedicated**

See also CHARTS, Language of Flowers

JANUARY	Carnations and Snowdrops
FEBRUARY	Violets and Primroses
MARCH	Daffodils and Jonquils
APRIL	Daisies and Sweet Peas
MAY	Lilies of the Valley and Hawthorn
JUNE	Roses and Honeysuckle
JULY	Water Lilies and Larkspur
AUGUST	Gladiolus and Poppies
SEPTEMBER	Morning Glories and Asters
OCTOBER	Calendula and Cosmos
NOVEMBER	Chrysanthemums
DECEMBER	Narcissus and Holly

*This is to help give the appropriate flowers to people as dictated by the months in which they were born. The above list indicates which flower is lucky to which month.

GREEK LANGUAGE OF THE MOLE

See BIRTHMARKS

IF THE MOLE IS:	IT MEANS:
Above the right eye	Wealth and a happy marriage
Above the left eye	Healthy interest in the opposite sex, which would bring joy
Temple	Happiness in love
Nose	Success in business
Chin	Lucky with friends
Cheek	Happiness without fame or fortune
Ear	Contented nature
Arms	Happy nature

Shoulders	Able to face problems with courage
Hands	Practical nature, able to care for yourself
Legs	Strong-willed
Neck	Patience

HERBS AND SENTIMENTS

ALLSPICE	Compassion
BASIL	Hatred
BAY LEAF	I change, but only in death
CHERVIL	Sincerity
CLOVES	Dignity
CORIANDER	Hidden virtues
FENNEL	Brute force
LAUREL (GROUND)	Perseverance
MARJORAM	Blushes
MINT	Virtue
MUSTARD SEED	Indifference
MYRRH	Gladness
PARSLEY	Feasting
ROSEMARY	Remembrance
SAFFRON	Don't overdo
SAGE	Domestic virtues
SORREL	Parental affection
SWEET BASIL	Good wishes
THYME	Activity

HOLIDAYS

NEW YEAR'S DAY	January 1 (*See* NEW YEAR'S DAY)
MARTIN LUTHER KING DAY	January 15
CHINESE NEW YEAR	Sometime between January 21 and February 19 depending on Chinese calendar
GROUNDHOG DAY (Candlemas Day)	February 2 (*See* GROUNDHOG DAY)
MARDI GRAS	Tuesday before Ash Wednesday
SAINT VALENTINE'S DAY	February 14 (*See* VALENTINE)
SAINT PATRICK'S DAY	March 17 (*See* SHAMROCKS)
GOOD FRIDAY	Friday before Easter
APRIL FOOLS' DAY	April 1 (*See* APRIL FOOLS' DAY)
EASTER	Sometime between March 22 and April 30; Sunday after the first full moon after the vernal equinox (*See* EASTER; EASTER EGGS)
MOTHER'S DAY	May 10 (*See* MOTHER'S DAY)
ARMED FORCES DAY	Third Saturday in May
FATHER'S DAY	Third Saturday in June
INDEPENDENCE DAY	July 4
ROSH HASHANAH AND YOM KIPPUR	Sometime between September and October determined by the Jewish calendar
COLUMBUS DAY	October 12
HALLOWE'EN	October 31 (*See* HALLOWE'EN)
VETERAN'S DAY	November 11
THANKSGIVING	Fourth Thursday in November (*See* THANKSGIVING)
HANUKKAH	Sometime between November and December determined by the Jewish calendar
CHRISTMAS DAY	December 25 (*See* CHRISTMAS)
NEW YEAR'S EVE	December 31 (*See* NEW YEAR'S DAY)

MONTHS

JANUARY

The old Welsh proverb that says, "a warm January; a cold May," has haunted farmers for centuries. Another proverb, popular among farmers, is "A mild January means bad luck for both man and beast." January was named after the Roman god Janus, keeper of the gates; a two-headed creature, Janus had the ability to see both the past and the future.

FEBRUARY

The Italians say, "February, the shortest month in the year, is also the worst." February was the Roman month of purification. It was the shortest month because Augustus Caesar took a day from February and added it to August to make that month (which had been named after him) longer than July (which had been named for Julius Caesar.)

MARCH

"Beware the ides of March" was Shakespeare's idea (in *Julius Caesar*). The month of March had been named after Mars, the God of War. It has been said that to marry in March when the winds blow means you'll experience both joy and sorrow in your wedded life. March used to hold New Year's Day, which was celebrated on March 25.

APRIL

"April's in her eyes; it is love's Spring," wrote Shakespeare (in *Antony and Cleopatra*, III:2). The month of April was probably named for the goddess Aphrodite, the Greek goddess of love. It marks the arrival of Spring and is a favorite among poets.

MAY

"The vulgar say that it is unlucky to marry in May," said Ovid (*Fasti* 5) during the opening days of the first century. This idea probably got started because the Romans honored the dead in the month of May; yet, it is a month dedicated to love. The remaining days of the month are a time for the revival of crops, flowers, and of life in general. The month may have been named for Maia, daughter of Atlas, god of growth.

JUNE

And what is so rare as a day in June?
Then, if ever, come perfect days.
(Lowell, *The Vision of Sir Launfal*, I, 1848)

As everyone knows, June is the month for marriages, probably because the month was named for Juno, wife of Jupiter and goddess of women, marriages, and childbirth. On June 1, Romans honored Juno, and ever since the month has been the luckiest to be married in.

JULY

July was named in honor of Julius Caesar in 44 B.C. It may not have been such an honor after all, since the dog days happen during July. (*See* DOG DAYS)

Hot July brings cooling showers,
Apricots and gillyflowers.
(Sara Colridge, *Pretty Lessons in Verse*, 1834)

AUGUST

Since Julius took July, his successor, Augustus Caesar, took August (seems only fair, doesn't it?). The month marked the beginning of harvest.

If the twenty-fourth of August be fair and clear,
Then hope for a prosperous Autumn that year.
(John Ray, *English Proverbs*, 1670)

SEPTEMBER

September was originally the seventh month of the Roman year and was named after the number seven. It was a time to finish harvesting and to celebrate the results.

The golden rod is yellow
The corn is turning brown
The trees in apple orchards
With fruit are bending down . . .
By all these lovely tokens
September days are here,
With summer's best of weather
And autumn's best of cheer.
(Helen Hunt Jackson, *September*)

OCTOBER

The month of October was named for the word eight, because it was the Romans' eighth month on the calendar. October means (though not literally) autumn. It is the last gasp before the winter winds set in and the snows start to fall.

October is nature's funeral month . . . The month of departure is more beautiful than the month of coming . . . Every green thing loves to die in bright colors.
(Beecher, *Proverbs from Plymouth Pulpit*, 1870)

NOVEMBER

November is Latin for ninth . . . and has been called "the month of blue devil and suicides." In the British Isles, it was once believed that November cast an evil spell over men's minds because of its monotonous, dreary days. A French novelist is supposed to have written, in 1712, "The gloomy month of November, when the people of England hang and drown themselves."

DECEMBER December, from the Latin for tenth, was dedicated to Saturn, god of seed sowing. The Saxons honored Thor, god of thunder on December 21, which they called Giul. That became yule. When the Saxons converted to Christianity, they called the month Heiligh-monath, meaning holy month. Shakespeare saw December from a different perspective:

> *When we shall hear*
> *The rain and wind beat dark December.*
> (*Cymbeline*, III:3)

MOST POPULAR SUPERSTITIONS

- Wishing on a wishbone
- Believing in the power of the horseshoe
- Carrying a rabbit's foot
- Looking for a four-leaf clover
- Seven years' bad luck from a broken mirror
- Getting out of bed on the right side
- Bad luck to walk under a ladder
- Belief in the power of Friday the thirteenth
- Looking at the new moon over your right shoulder for luck
- Consider seven a lucky number
- Won't light three cigarettes on one match
- Never invite thirteen to dinner
- Believe in the implications of spilled salt
- Never open an umbrella indoors
- Make a wish on a falling star or the first star of the night
- The wearing of amulets or talismans for luck
- Find a pin, pick it up, all day you'll have good luck
- Saying "God Bless You" to someone who sneezes

NAMES

Female

ALICE	Truth
AMY	Beloved
ANN	Full of grace, mercy, and prayer
APRIL	To open
ARLENE	A pledge
AUDREY	Strong and noble
BARBARA	Mysterious stranger
BEATRICE	She brings joy
BEVERLY	Ambitious
BRENDA	Fiery
CANDICE	Pure
CAROL	Joyous song
CAROLINE	One who is strong
COLETTE	Victorious
CYNTHIA	Moon goddess
DAISY	The day's eve
DARLENE	Dearly beloved
DOROTHY	The bee
DIANA	Pure goddess of the moon
ELAINE (Eleanor)	Light
ELIZABETH	Consecrated to God
EMILY	Industrious
ENID	Purity of soul
EVE	Life or living
FLORENCE	To flower or bloom
FRANCES	Free
GABRIELLE	Woman of God
GERALDINE	Ruler with a spear
GLORIA	The glorious
GRACE	The graceful
GWEN	White-browed
HARRIET	Mistress of the home
HELEN	Light
HILARY	Cheerful friend

HOLLY	Good luck
IDA	Happy
IRENE	Peace
ISABEL	Consecrated to God
JACQUELINE	The supplanter
JANE	God's gracious gift
JESSICA	Rice; grace of God
JOSEPHINE	She shall add
JUDITH	Admired, praised
JULIA	Youthful
KATHERINE	Pure
KIM	Noble or glorious leader
LAURA	The laurel
LEE	Meadow
LESLEY	From the gray fort
LILLIAN	A lily
LINDA	Beautiful
LOUISE	Battle maiden
LUCY	Light
LYNN	Life
MADELINE	Tower of strength
MARCIA	Of Mars
MARGARET	A pearl
MARTHA	The lady
MARY	Bitter
MELANIE	Darkness
MELISSA	Honeybee
NANCY	Full of grace
NAOMI	Sweet
NATALIE	Child of Christmas
NICOLE	Victory of the people
ORIANA	Girl of the white skin
PATRICIA	Wellborn
PAULA	Little
PHYLLIS	A green bough
PIA	Devout
POLLY	Bitter
PRISCILLA	Of ancient lineage

RACHEL	Trembling child
RENEE	Reborn
RHODA	A garland of roses
RITA	A pearl
ROSALIND	Fair rose
RUTH	Beautiful friend
SAMANTHA	Lovely flower
SARAH	Princess
SHARON	From the fertile plain
SHEILA	Musical
SOPHIA	Wisdom
STEPHANIE	Crown or garland
SUSAN	A lily
SYBIL	The prophetess
TALLULAH	Vivacious
TERESA	The harvester
TIFFANY	Manifestation of God
ULLA	Dearest of all God's burdens
VALENTINA	Vigorous and strong
VANESSA	The butterfly
VERONICA	True image
VICTORIA	Victorious
VIRGINIA	Maidenly, pure
WENDY	White-browed
YVONNE	The archer
ZOE	Life

Male

AARON	Light, high mountain
ADAM	Man of earth
ALAN	Harmony
ALEXANDER	Protector of me
ANDREW	Manly
ANTHONY	Of inestimable worth
ARTHUR	Strong as a rock
BARNABY	Son of consolation
BENJAMIN	Son of my right hand
BRIAN	Strong, powerful

CALVIN	Bald
CASEY	Valorous
CHARLES	Man
CHRISTOPHER	Christ bearer
CLARK	Scholarly
DANIEL	The Lord is judge
DAVID	Beloved
DENNIS	Lover of fine wines
DONALD	Ruler of the world
DOUGLAS	From the black stream
EDWARD	Prosperous guardian
ERIC	Kingly
FRANCIS	Free
FRANKLIN	A free man
GARY	Mighty spear
GEORGE	Farmer
GERALD	Mighty spearman
GILBERT	Bright pledge
GORDON	From the cornered hill
GREGORY	Vigilant
HENRY	Home ruler
HOWARD	Chief guardian
HUGH	Intelligence
IRA	Watcher
IRWIN	Sea friend
ISRAEL	The Lord's soldier
JAMES	The supplanter
JAY	Crow; lively
JEFFREY	Exalted by God
JOEL	Jehovah is God
JOHN	God's gracious gift
JONATHAN	Gift of the Lord
JOSEPH	He shall add
JOSIAH	He is healed by the Lord
JUSTIN	The just
KENNETH	Handsome
KEVIN	Kind, gentle
LANCE	Spear

LAWRENCE	Laurel
LEE	Meadow
LEWIS	Renowned for battle
LLOYD	Gray or dark
MARK	Warrior
MARTIN	Warrior
MATTHEW	God's gift
MICHAEL	God-like
MORGAN	From the sea
MURRAY	Sailor
NATHANIEL	Gift of God
NICHOLAS	Victory of the people
NORMAN	Man of the north
OLIVER	Peace
OSCAR	Divine spear
OTTO	Wealthy
PATRICK	Noble, patrician
PAUL	Little
PERRY	Pear tree
PETER	Rock
PHILIP	Lover of horses
RANDOLPH	Protected; advised by wolves
RAOUL	Helpful commander
RICHARD	Wealthy and powerful
ROBERT	Of bright, shining fame
RYAN	Laughing
SALVADOR	Of the Savior
SAMUEL	Name of God
SANCHO	Gallant companion
SCOTT	A Scotsman
SEAN	God's gracious gift
SHELLEY	From the ledge
SIDNEY	Follower of Saint Denis
SPENCER	Storekeeper
STANLEY	Pride of the camp
STEPHEN	Crown, garland

STEWART	Keeper of the estate
TERENCE	Tender
THEODORE	Gift of God
THOMAS	The twin
TIMOTHY	Honoring God
TODD	The fox
VAN	From
VICTOR	The conqueror
WALTER	Powerful warrior
WARNER	Protecting warrior
WAYNE	Wagon maker
WILLIAM	Determined protector
YALE	The yielder

THE SEVEN DAYS OF THE WEEK

MONDAY
(wear pearls on Monday)

"As Monday does, so goes all the week" is an old proverb which many believe. Don't move on a Monday, although Monday is a lucky day, a day of happiness and peace.

TUESDAY
(wear rubies on Tuesday)

Tuesdays are generally unlucky days. They often bring quarrels and lawsuits. Tuesdays are good days for marriage and to begin a new business deal.

WEDNESDAY
(wear sapphires on Wednesday)

The sun shines on Wednesday; but if it doesn't, there will be a terrible storm. Wednesdays are very lucky days; in fact, they are considered the best day of the week to be born.

THURSDAY
(wear garnets on Thursday)

Some believe that Thursdays are very unlucky days and that there is only one good hour (the one before dawn) on any given Thursday. But it's a good day to perform difficult tasks.

FRIDAY
(wear emeralds on Friday)

Considered the most unlucky day of the week; nothing good can happen on a Friday. Don't start anything on a Friday. So many bad things have happened on Fridays that there's no use tempting fate. (*See* FRIDAY)

SATURDAY
(wear diamonds on Saturday)

An unlucky day; nothing new should be begun on a Saturday. A new moon on a Saturday is unlucky and brings trouble. The sun will shine on a Saturday, even if only for a moment.

SUNDAY
(wear yellow stones)

Sunday is a lucky day, a favorite for weddings. It is considered a good day to be born. If you sneeze on a Sunday before breakfast you'll be in love forever.

SPORTING SUPERSTITIONS

AUTO RACING

If a (male) driver eats peanuts or talks to a woman before a race, he can expect trouble during the race.

A woman in the garage area was taboo until 1971, when two female reporters got a court order to allow them in the pit area during the Indianapolis 500.

BASEBALL

Spit into your hands before picking up a baseball bat for good luck.

A player who changes bats after two strikes will be struck out.

A piece of chewing gum stuck to the top of a baseball cap brings good luck.

Crossing bats on a baseball field will bring bad luck to the batter.

Seventh-inning stretch is an attempt by fans to bring good luck to their team by invoking the magic of the number seven. (*See* SEVEN)

BASKETBALL

The player who makes the last basket during warm-ups will do well during the game. Some believe this superstition works the other way: The player who makes the last basket at warm-up will do very badly during the game.

BOWLING

Don't fill in the score of a person who is having a string of strikes until that string has broken.

If a split occurs (that doesn't turn into a spare), the superstitious bowler will mark a heavy vertical line—called a fence—after the frame, to prevent that sort of unlucky split from happening again during the game.

GOLF

When you are teeing off, you should place a ball with the trade name or number facing up or the hole is lost.

The word socket used on the golf course, is considered bad luck by players.

All numbered balls, over the number four, are considered bad luck.

HORSE RACING

Gray horses and horses with four white hooves are unlucky.

If a photo is taken of a jockey and a horse together before they've ridden the race, they'll lose.

It is believed that a jockey's boot must never touch the floor before it is on the jockey's foot or it will slow the jockey down.

It is unlucky for the jockey to drop the whip before a race.

HORSES (NOT RACING)

Spotted horses or ones with patches of color are believed to have magical gifts.

Bells and brass disks are attached to harnesses to ward off the evil eye and other evil spirits. Another method used to protect against witches and the evil eye is placing a piece of red cloth on the forehead of the horse.

MOUNTAINEERING

Mountaintops are where the gods live, and if you attempt to climb up to their homes, evil spirits will attack you. This was a widespread belief during early times. In 1387, six clergymen climbed Mount Pilatus (in the Swiss Alps) to attempt to destroy the superstition that Pontius Pilate's spirit lived on top and that it took revenge on anyone who disturbed it. Their climb was completely successful—except that they were arrested as soon as they arrived at the bottom of the mountain!

PRIZEFIGHTING (BOXING)

If a fighter carries a pickle during a fight, the fighter will get knocked out!

TENNIS

It's bad luck to hold three tennis balls in your hand while serving.

TRACK RACING

Never let anyone step across your legs before a race.

Never let a pole come between you and a fellow competitor—both people must pass the pole on the same side or there will be unhappy results for both.

TREES AND THEIR MEANINGS

APPLE	Temptation
ASH	Grandeur
ASPEN	Grieving
BAY	Glory
BARBERRY	Sharpness
BIRCH	Meek and graceful
BOX	Stoic
CEDAR OF LEBANON	Strong and constant
CHERRY	Education
(White)	Deception
CHESTNUT	Luxury
DOGWOOD	Durable
EBONY	Blackness
ELDER	Zealousness
ELM	Dignity
FIR (Scotch)	Elevate
HAWTHORN	Hope
HAZEL	Reconciliation
LEMON	Zest
LINDEN (Lime)	Conjugal love
LOCUST	Elegance
MAPLE	Reserve
MULBERRY	
(White)	Wisdom
(Black)	I will not survive you

OAK	Hospitality
ORANGE	Generosity
PEACH BLOSSOM	I am your captive
PEAR	Comfort
POPLAR (Black)	Courage
(White)	Time
WEEPING WILLOW	Mourning

LANGUAGE OF FLOWERS
(as devised about 200 years ago)

See also CHARTS, Flowers

ACACIA	Friendship
(Pink)	Elegance
(Rose)	Platonic love
(Yellow)	Secret love
ACANTHUS	The fine arts
ALOE	Grief, bitterness, associated with religious superstitions
AMARYLLIS	Pride, haughtiness
ANEMONE (Garden)	Forsaken
(Meadow)	Sickness
(Wood)	Forlornness
ASTER	Variety
(Double)	I share your sentiments
(Single)	I will think of it
AZALEA	Temperance
BACHELOR'S BUTTON	Love and marriage
BLUEBELL	Constant kindness
BOUQUET OF FLOWERS	Gallantry
BUTTERCUP	Cheerful but ungrateful
CARNATION	Fascination; woman's love
(Striped)	Refusal
(Yellow)	Disdain
CHAMOMILE	Energy even when troubled

CHRYSANTHEMUMS	Cheerful even when troubled
(Red)	I love
(White)	Truth
(Yellow)	Slighted love
COLUMBINE	Folly
(Purple)	Resolute
(Red)	Anxious
CORNFLOWER	Delicate
CROCUS	Do not abuse
(Spring)	Youthful hope
DAFFODIL	Regard
(Great Yellow)	Chivalry
DAHLIA	Instability
DAISY	Innocence
(Double)	Participation
(Garden)	I share your sentiments
DANDELION	Love's oracle
DEAD LEAVES	Sadness, melancholy
FERN	Sincerity
FLEUR-DE-LIS	Flame
FORGET-ME-NOT	Remember; true love
FRANKINCENSE	The incense of a faithful heart
GATHERED FLOWERS	We will die together
GERANIUM	Gentility
(Dark)	Melancholy
(Rose or Pink)	Preference
(Scarlet)	Comforting; stupidity and folly
(Wild)	Steadfast piety
GOLDENROD	Careful encouragement
GRASS	Useful
HIBISCUS	Delicate beauty
HOLLY	Foresight
HOLLYHOCK	Fruitfulness
(White)	Female ambition
HONEYSUCKLE (Coral)	Color of my fate
(French)	Rustic beauty
(Monthly; Woodbine)	Bond of love; I'll answer with care

(Wild)	Inconstancy in love
HYACINTH	Sport
(Blue)	Constant
IRIS	My compliments; I have a message for you.
(German)	Ardor
(Yellow)	Flame
IVY	Friendship, fidelity
JASMINE (Cape)	Transport of joy
(Carolina)	Separation
(Indian)	I attach myself to you
(Spanish)	Sensuality
(White)	Amiability
(Yellow)	Grace and elegance
JONQUIL	I want you to love me
LAUREL (Bay)	Glory
(Common)	Perfidy
(Mountain)	Ambition, glory
LAVENDER	Distrust
LILAC (Field)	Humility
(Purple)	The first signs of love
(White)	Purity, modesty
LILY (Day)	Coquetry
(White)	Purity and sweetness
(Yellow)	Lies and gaiety
LILY OF THE VALLEY	Return of happiness
LOTUS	Eloquence
LOVE-IN-IDLENESS	Love at first sight
MAGNOLIA	Love of nature
MANDRAKE ROOT	Horror
MARIGOLD	Grief, pain, and anger
MISTLETOE	Surmount all obstacles
MOSS	Maternal love
MYRTLE	Love
NARCISSUS	Egotism, self-esteem
ORANGE BLOSSOM	You're as pure as you are lovely; chastity

PEA (Sweet)	Departure
PEONY	Shame
POPPY	Fading pleasures
(Red)	Consolation
(Scarlet)	Fantastic extravagance
(White)	Sleep helps everything
PRIMROSE	Early youth
RHODODENDRON	Danger
ROSE	Love and beauty
(Bridal)	Happy love
(Burgundy)	Unconscious beauty
(Cabbage)	Ambassador of love
(Carolina)	Love is dangerous
(Damask)	Brilliant complexion
(Deep red)	Bashful
(Full red)	Beauty
(Full white)	I am worthy of you
(Musk)	Capricious beauty
(White)	Silence
(Yellow)	Jealousy, unfaithfulness
(Red and white together)	Unity
SHAMROCK	Lightheartedness, Ireland
SNAPDRAGON	Presumption
SUNFLOWER	Adoration
(Tall)	Haughtiness
TUBEROSE	Dangerous pleasures
TULIP	Fame
(Red)	Declaration of love
(Yellow)	Hopeless love
(Variegated)	Beautiful eyes
VIOLET (Blue)	Faithfulness in love
(Dame)	You are the queen of coquettes
(Purple)	You occupy my thoughts
(Sweet)	Modesty
(Wild)	Love in idleness
WILDFLOWER	Fidelity in misfortune
ZINNIA	Thoughts of absent friends

WEDDING ANNIVERSARY GIFTS

FIRST	Paper
SECOND	Cotton
THIRD	Leather
FOURTH	Fruits and flowers (or linen)
FIFTH	Wood
SIXTH	Sugar and candy (or iron)
SEVENTH	Woolens and copper
EIGHTH	Rubber and bronze
NINTH	Pottery and willow
TENTH	Tin
ELEVENTH	Steel
TWELFTH	Silk and fine linen
THIRTEENTH	Lace
FOURTEENTH	Ivory
FIFTEENTH	Crystal
TWENTIETH	China
TWENTY-FIFTH	Silver
THIRTIETH	Pearl
THIRTY-FIFTH	Coral
FORTIETH	Ruby
FORTY-FIFTH	Sapphire
FIFTIETH	Gold
FIFTY-FIFTH	Emerald
SIXTIETH and SEVENTY-FIFTH	Diamond

BIBLIOGRAPHY

Ballou, Maturin M. *Treasury of Thought, A.*, Cambridge: Houghton Mifflin Co. (Boston)/Riverside Press, 1896.

Bartlett, John. *Bartlett's Familiar Quotations*. New York: Little, Brown & Co., 1955.

Batchelor, Julie Forsyth and Claudia DeLys. *Superstitious? Here's Why!* New York: Harcourt, Brace and World, Inc., 1954.

Benet, William Rose. *Reader's Encyclopedia, The*. 2nd ed. New York: Thomas Y. Crowell Co., 1965.

Bowser, James W., ed. *6,000 Names for Baby*. New York: Dell/Ivy Books, 1978.

Brooks, Daniel Fitzgerald. *Numerology*. New York: Franklin Watts, 1978.

Burnam, Tom. *Dictionary of Mis-Information, The*. New York: Thomas Y. Crowell Co., 1975.

Coffin, Margaret M. *Death in Early America*, New York: Thomas Nelson, Inc., 1976.

Complete Book of Astrology, Horoscope and Dreams. New York: Modern Promotions, 1979.

Cowan, Lore. *Are You Superstitious?* New York: Apex, 1969.

DeLys, Claudia. *Treasury of American Superstitions, A*. New York: Philosophical Library, Inc., 1958.

DeLys, Claudia. *Treasury of Superstitions, A*. New York: Philosophical Library, Inc., 1957.

Espy, Willard R. *Another Almanac of Words at Play*. New York: Clarkson N. Potter, 1980.

Evans, Bergen. *Comfortable Words*. New York: Random House, 1959.

Evans, Bergen. *Dictionary of Quotations*. New York: Avenel Books, 1978.

Farmer, Penelope. *Beginnings*. New York: Atheneum, 1979.

Funk & Wagnall's Standard Dictionary of Folklore, Myth and Legend. New York: Funk & *Wagnall, 1972*.

Funk, Charles Earle. *Heavens To Betsy!* New York: Harper & Brothers, 1955.

Garden, Nancy. *Witches*. New York: J.B. Lippincott Co., 1975.

Garrison, Webb. *How it Started*. New York: Abingdon Press, 1972.

Gray, Eden. *Tarot Revealed, The.* New York: Bell Publishing Co., 1960.

Greenway, Kate and Jean Marsh. *Illuminated Language of Flowers.* New York: Holt, Rhinehart & Winston, 1978.

Heaps, Willard A. *Superstition!* New York: Thomas Nelson, Inc., 1972.

Helfman, Elizabeth S. *Signs and Symbols Around the World.* New York: Lothrop, Lee & Shepard, Co., 1967.

Hollingsworth, Buckner. *Flower Chronicles.* New Jersey: Rutgers University Press, 1958.

Huggett, Richard. *Supernatural on Stage.* New York: Taplinger Publishing Co., 1975.

Krythe, Maymie R. *All About the Months.* New York: Harper & Row Publishers, Inc., 1966.

Lawson, J. Gilchrist. *World's Best Conundrums and Riddles of All Ages, The.* New York: Harper & Brothers, 1924.

Lewis, Linda Rannells. *Birthdays.* New York: Little, Brown & Co. (An Atlantic Monthly Press Book), 1976.

Louis, David. *More Fascinating Facts.* New York: Ridge Press/ Crown Publishing Co., 1976.

Mann, Peggy. *Telltale Line, The.* New York: Macmillan Publishing Co., 1976.

Maple, Eric. *Superstition and the Superstitious.* New York: A.S. Barnes & Co., 1972.

Mercatante, Anthony S. *Who's Who in Egyptian Mythology.* New York: Clarkson N. Potter, Inc., 1978.

Morris, William and Mary. *Dictionary of Word and Phrase Origins.* New York: Harper & Row Publishers, 1962.

Morrison, Lillian, (compiled by). *Touch Blue.* New York: Thomas Y. Crowell Co., 1958.

Myers, Robert. *Celebrations.* New York: Doubleday & Co., Inc., 1972.

Powell, Claire. *Meaning of Flowers, The.* London: Jupiter Books, 1977.

Rachleff, Owen S. *Secrets of Superstition, The.* New York: Doubleday & Co., Inc., 1976.

Radford, Edwin. *Encyclopedia of Superstition.* London: Hutchinson, 1961.

Reader's Digest's American Folklore and Legend. Pleasantville, New York: Reader's Digest Association, 1978.

Reader's Digest's Stories Behind Everyday Things. Pleasantville, New York: Reader's Digest Association, 1980.

Rosten, Leo. *Joys of Yiddish, The.* New York: McGraw-Hill Book Co., 1968.

Ruoff, Henry W. *Century Book of Facts, The.* King-Richardson Co., 1902.

Sarnoff, Jane and Reynold Ruffins. *Take Warning!* New York: Charles Scribner's and Sons, 1978.

Schwartz, Alvin. *Cross Your Fingers, Spit in Your Hat.* New York: J.B. Lippincott Co., 1974.

Seymour, John. *Gardner's Delight.* New York: Harmony Books, 1979.

Stevenson, Burton. *Home Book of Proverbs, Maxims and Familiar Phrases, The.* New York: Macmillan Co., 1948.

Stimpson, George. *Book About a Thousand Things, A.* New York: Harper and Brothers, 1946.

Sullivan, George. *Sports Superstitions.* New York: Coward, McCann & Geoghegan, Inc., 1978.

Van Druten, John. *Bell, Book and Candle.* New York: Random House, Inc., 1951.

Vogel, Malvina (compiled and edited by). *Big Book of Amazing Facts, The.* New York: Playmore, Inc./Waldman Publishing Corp., 1980.

Wagner, Edward A. *Sun-Sign Handbook, The.* New York: Dell/Ivy Book, 1979.

Wallechinsky, David and Irving Wallace. *People's Almanac #2, The.* New York: Bantam Books, Inc., 1978.

Zolar's Horoscope and Lucky Number Dream Book. New York: Prestige Books, Inc., 1980.